Technology for a Sustainable Environment

Edited by

Bhupinder Dhir

School of Sciences
Indira Gandhi National Open University,
New Delhi, India

Technology for a Sustainable Environment

Editor: Bhupinder Dhir

ISBN (Online): 978-981-5124-03-3

ISBN (Print): 978-981-5124-04-0

ISBN (Paperback): 978-981-5124-05-7

Published by Bentham Science Publishers Pte. Ltd. Singapore. All Rights Reserved.

First published in 2023.

need for a court order if at any point you breach any terms of this License Agreement. In no event will any delay or failure by Bentham Science Publishers in enforcing your compliance with this License Agreement constitute a waiver of any of its rights.

3. You acknowledge that you have read this License Agreement, and agree to be bound by its terms and conditions. To the extent that any other terms and conditions presented on any website of Bentham Science Publishers conflict with, or are inconsistent with, the terms and conditions set out in this License Agreement, you acknowledge that the terms and conditions set out in this License Agreement shall prevail.

Bentham Science Publishers Pte. Ltd.
80 Robinson Road #02-00
Singapore 068898
Singapore
Email: subscriptions@benthamscience.net

BENTHAM SCIENCE

CONTENTS

FOREWORD

Environmental conditions play an important role in the survival of living beings. Clean and safe environmental conditions ensure good health and well-being of living organisms. Rapid industrialization, urbanization and massive population growth have led to the deterioration of the environment. Various components of the environment, *i.e.*, air, water and soil, have shown deterioration in their quality due to increased anthropogenic activities. Therefore, it becomes necessary that appropriate steps are taken to remediate the environment and achieve the goal of creating a sustainable environment. Various technologies have been developed globally in this direction and offer huge potential to accomplish environmental sustainability.

In this book, Dr. Bhupinder Dhir has drawn great attention to an important issue of environmental degradation. She used her expertise to develop a book that can provide readers with updated knowledge about the current environmental scenario. Furthermore, emerging technologies promoting environmental sustainability have been discussed. The chapters in the book deal with some of the modern-day technologies, such as nanotechnology, renewable sources of energy and alternate sources of energy and fuels that contribute to the sustainability of the environment. Recent topics such as the development of green technologies, biodegradable polymers, and plastics have been dealt with in this book in detail. She has also tried to highlight the role of biotechnology in tackling problems related to environmental degradation. These modern-day technologies help in the remediation of environmental pollutants, save non-renewable sources of energy and prevent environmental degradation. This book by Dr. Bhupinder Dhir will undoubtedly contribute significantly to inspiring researchers, students, policymakers and environmentalists working in the related area.

Pooja Ghosh
Centre for Rural Development and Technology IIT,
Delhi, India

PREFACE

Rapid population growth and increasing urbanization have posed a threat to natural resources. Soil, air and water are facing degradation in quality due to overexploitation and getting damaged at a higher rate due to an increase in pollution. Researchers and scientists worldwide are engaged in developing strategies and techniques that help us achieve a sustainable environment. Focus has been shifted to new techniques such as bioremediation, nanotechnology and biotechnology. Besides this development of alternate fuels, eco-friendly materials, or the use of non-conventional sources of energy has also been emphasized.

This book presents an overview of various methods and techniques that can be adapted to get a sustainable environment. It provides a detailed study about the role of biotechnology, nanotechnology and other techniques that help in achieving minimum degradation of environment and utilization of natural resources to achieve sustainability in terms of energy, food and water security. The author uncovers the various environmental problems caused by anthropogenic activities. The book suggests various ideas for getting solutions to various environmental problems, whether related to the monitoring of environmental pollutants, their removal from various sectors of the environment, or remediation *via* various techniques.

The book has been divided into three sections. The first section focuses on the use of biotechnological techniques in improving the quality of air, water and soil. The second section discusses the use of nanotechnology in achieving environmental sustainability. Various alternate sources of energy, fuels and other technologies that are eco-friendly and do not harm the environment have been elaborated on in section three of the book. The book provides a comprehensive overview of the key concepts of sustainability and ideas for students. The book is suitable for students having introductory interdisciplinary courses in the fields of environmental science, engineering, biotechnology sociology. The book is of use to researchers, faculty members, policyholders and planners who play a major role in encouraging the development of methods to ensure the long-term sustainability of the planet. The text includes material on environmental sustainability of water, food, and energy that are related to social and economic stability.

Bhupinder Dhir
School of Sciences
Indira Gandhi National Open University,
New Delhi,
India

List of Contributors

Avinash Tomer	Division of Vegetable Science, ICAR- Indian Agricultural Research Institute, New Delhi-110012, India
Akanksha Bhardwaj	Department of Environmental Science and Technology, Central University of Punjab, VPO-Ghudda, Punjab 151401, India
Bhupinder Dhir	School of Sciences, Indira Gandhi National Open University, New Delhi-110078, India
Bhoirob Gogoi	Department of Microbiology, Assam University, Silchar, Assam, India
Chanderkant Chaudhary	Department of Plant Molecular Biology, University of Delhi, South Campus, New Delhi, 110021, India
Chandrika Ghoshal	Division of Genetics, ICAR- Indian Agricultural Research Institute, New Delhi-110012, India
Hemen Sarma	Department of Botany, N NSaikia College, Titabar, Assam, India
Joginder Singh	Department of Chemistry, Maharishi Markandeshwar (Deemed to be University), Mullana, Ambala 133207, India
Jayaraman Nagendra Babu	Department of Chemistry, Central University of Punjab, VPO- Ghudda, Punjab 151401, India
Monu Sharma	Department of Biotechnology, Maharishi Markandeshwar (Deemed to be University), Mullana, Ambala 133207, India
Neehasri Kumar Chowdhury	Department of Zoology, Gauhati University, Guwahati, Assam, India
Puneeta Pandey	Department of Environmental Science and Technology, Central University of Punjab, VPO-Ghudda, Punjab 151401, India
Raman Kumar	Department of Biotechnology, Maharishi Markandeshwar (Deemed to be University), Mullana, Ambala 133207, India
Ruby Tiwari	Department of Genetics, University of Delhi South Campus, New Delhi-110021, India
Reshma Choudhury	Department of Biotechnology, Royal Global University, Guwahati, Assam, India
Rashmi Verma	Department of Genetics, University of Delhi South Campus, New Delhi, India
Shashi Pandey	Division of Genetics, ICAR- Indian Agricultural Research Institute, New Delhi-110012, India
Sonu Sharma	Department of Biotechnology, Maharishi Markandeshwar (Deemed to be University), Mullana, Ambala 133207, India
Sekar Hamsa	Department of Genetics, University of Delhi, South Campus, New Delhi, 110021, India
Suprity Shyam	Department of Life Sciences, Dibrugarh University, Dibrugarh, Assam, India

CHAPTER 1

Non-renewable Resources and Environmental Sustainability

Sonu Sharma[1], Monu Sharma[1], Joginder Singh[2], Bhupinder Dhir[3] and Raman Kumar[1,*]

[1] *Department of Biotechnology, Maharishi Markandeshwar (Deemed to be University), Mullana, Ambala 133207, India*

[2] *Department of Chemistry, Maharishi Markandeshwar (Deemed to be University), Mullana, Ambala 133207, India*

[3] *School of Sciences, Indira Gandhi National Open University, New Delhi, India*

Abstract: Growing need for energy for sustaining increasing population has resulted in overexploitation of natural resources and over use of fossil fuel-based energy sources (coal, oil and gas). The consumption of non–renewable resources such as coal, petroleum and natural gas has increased tremendously resulting in environmental problems and climatic changes. Emission of greenhouse gases and other environmental concerns have increased. The decline in the quantity of non-renewable resources has generated the search of alternate energy sources. Switch to alternate sources of energy and fuel can be a sustainable option to this problem. Solar, tidal, geothermal, wind are some of the renewable sources of energy that are being focused to curtail the energy crisis and ensure sustainability for environment. A framework based on fulfilling the SDGs need to be developed which can contribute for more profitable, responsible path of economic growth and development.

Keywords: Carbon emission, Economic growth, Environmental quality, Non-renewable resources.

INTRODUCTION

Pollution, soil degradation, climatic changes, depletion of natural resources are some of the major environmental problems the world is facing today. The main cause of all these alterations is rapid increase in population, industrialization, burning of fossil fuels and many other anthropogenic activities. Ecosystems irrespective terrestrial, aquatic (freshwater, marine) are get affecting by the rapid

* **Corresponding author Raman Kumar:** Department of Biotechnology, Maharishi Markandeshwar (Deemed to be University), Mullana, Ambala 133207, India; E-mail: ramankumar4@gmail.com

change in the environmental conditions and the rate of degradation is anticipated to speed up in the coming decades.

CO_2 emissions coming from industrial processes, fossil fuel combustion, power stations, cement producing units and refineries has increased at an alarming rate. Increase in emissions of greenhouse gases has contributed to increase in global mean temperature by about 1.09°C in the last few decades. Wildfires across many regions of the world have also contributed to increase in global temperature. Loss in the forest covers at a rapid rate to meet the demand for more food, shelter and cloth for the growing population and other anthropogenic activities has resulted in increase in levels of carbon dioxide. Events such as La Niña altered rainfall seasons creating drought like conditions in many areas and floods in the other ones world. Drought over large parts of Africa, Asia, and Latin America due to heat waves, severe storms, cyclones and hurricanes in other parts of the world are the some of the major changes noted during the last decade. This has affected the agricultural sector to a great extent. The change in climatic conditions exerts diverse impact on the fauna and flora throughout the globe. The change in temperature has resulted in melting of glaciers and associated increase in the water level in seas and oceans. Ocean warming *via* thermal expansion of sea water has increased threat to aquatic life. Global warming has also lead to decrease in pH of ocean water.

Natural resources are support to any civilization. They are beneficial to humans and were naturally classified as agricultural land, fisheries, mineral resources, fuels. Their classification into renewable and non-renewable resources came into existence quiet late and it was based on their existence. Natural resources have noted depletion at an unprecedented rate. The quality of soil, water, air has shown a significant change over the past few decades. Pollution has declined the quality of natural resources such as water, air or land. Erosion, overgrazing, pollution, monoculture planting, soil compaction, land-use conversion has affected the quality of soil and lead to its deterioration. Depletion in soil quality has affected its productivity thereby threatening the food security at global level. According to an estimate, about 12 million hectares of farmland are getting degraded each year. All these changes have severely affected the survival of the living beings.

Non renewable resources are being utilized by humans in very high quantities resulting in their significant decrease. Environmental degradation is posing as a threat to survival of living beings and leading to extinction of species. A significant loss in biodiversity has been noted every corner of the globe. This has lead to the generation of the concept of sustainability.

Non Renewable Resources

Non-renewable resources such as fossil fuels mainly coal, petroleum, natural gas and oil are limited in amount. They contribute to bout 85% of the energy consumed all over the world. These resources are formed from organic material from plant and animal remains that existed millions of years ago. Since the materials took millions of years to form, they also require millions of years to replenish. They are highly combustible, hence a rich source of energy. Non-renewable resources are affordable as they are cost effective. These resources can be used to form various products. They provide a major source of energy and medium to carry out various anthropogenic activities. These are consumed faster sources and will eventually get depleted. Their amount has become limited to their misuse and over-utilization, therefore it becomes important that they are sustained for use of future generation. Changes in the status of non-renewable material affect the size of the economy [1, 2].

Non-renewable resources are used as major source of energy. Many countries of the world use them for industrial, urban and anthropogenic activities which result to increase in pollution and global warming in the environment [3, 4]. The depletion of natural resources is supposed to increase energy crisis. Over use of fossil fuel results in the emission of greenhouse gases, this is primarily responsible for global warming and climate change. Use of non- renewable energy resources leads to environmental degradation and affect the economic growth of the world. A10% rise in consumption of non- renewable energy resources result in 2.11% rise in GDP [5]. The main reason of de*via*tion from non- renewable resources to renewable energy resources is their environmental effects [6 - 10]. Utilization of energy has both governmental as well as environmental consequences [11 - 14]. With increase in non- renewable energy resources GDP rises.

Non-renewable resources, especially metals, can be expanded by reprocessing. This process involves gathering and processing unused industrial and household products to recover renewable materials including metals and plastics. In oil producing countries, natural gas and petroleum and are the main drivers of economic growth [15, 16]. Non- renewable energy resources such as coal, petroleum and natural gas cannot be replicated once exhausted [17, 18]. Industrial energy supplies are mostly based on the use of non- renewable resources. Use of fossil fuel releases certain amount of residue in the form of solid substances and gases. This residue causes environmental pollution.

Around 300-360 fossil fuels were formed during the time of carboniferous period. About 10 feet of solid vegetation got flattened, heated and created foot of coal.

Round about 50% electricity is produced from coal. According to the U.S Energy Information Administration, the global transportation of crude oil is enough to meet human demand through the year 2050.

Categories of Non-renewable Resources

Metals

Pure elemental substances or mixture of numerous metals and non- metals have many uses Fig. (**1**). They can be used in industries are zinc, aluminum, chromium, iron, lead, manganese, uranium, mercury, nickel. Ore is obtained by mining and metal is extracted from ore containing oxygen and sulphur. If the metals have sulphide minerals, they are treated in presence of oxygen at high temperature. This releases gaseous sulphur leaving the metal behind.

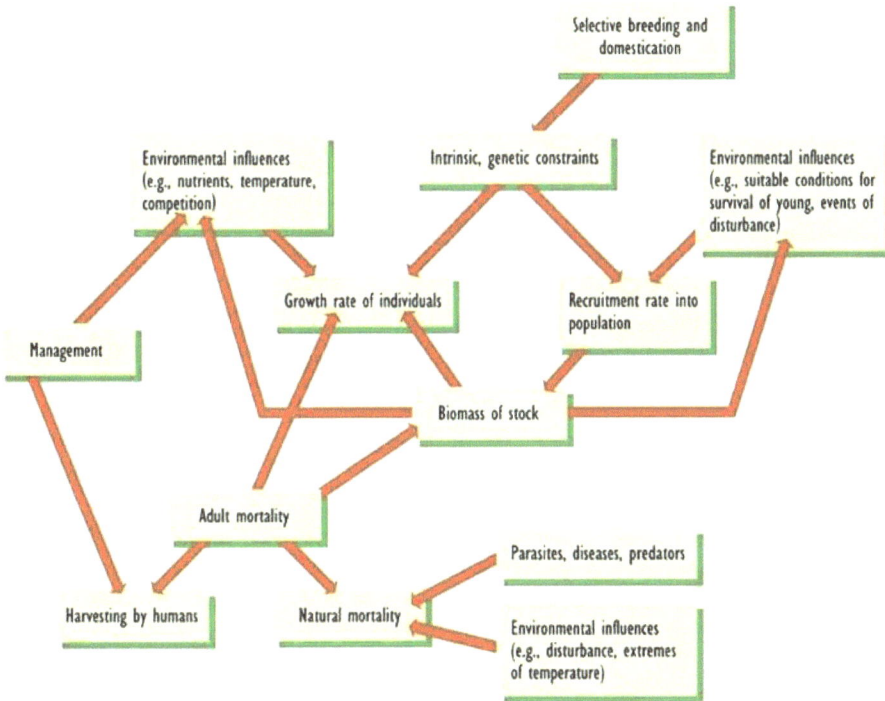

Fig. (1). Metal Mining and Use Source: Modified from Freedman [19].

This diagram depicts major stages of the mining, manufacturing, use, re-use of metals, and associated emissions of waste gases, particulates to the environment. The diagram represents a flow-through system, with recycling to extent.

Fossil Fuels

Fossil fuel includes coal, natural gas, oil send and petroleum. These materials are obtained from the biomass of dead plants and microorganisms buried under ground hundreds of millions of years ago Fig. (**2**). Nowadays fossil fuels are manufactured by subjecting dead biomass to high temperature and pressure. Their reserves will get diminished when they are extracted from the environment.

Fig. (2). An oil pump in southeastern Saskatchewan. Source: B. Freedman [20].

Coal

Coal contains carbon and hydrocarbon matter. Coal is extracted by digging up the ground and processed for energy. Coal is one of the one of the inexpensive sources of fuel used for generating electric energy and many other uses. Coal occurs in various forms such as anthracite, lignite, bituminous, and sub-bituminous. Bituminous contains 45% to 86% of carbon. It has high heat content and is used in generating energy and in making steel and iron. It is the energy resource and used in electric power plants to generate electricity. Bituminous coal reserves are found in good amount in Illinois, U.S.A. Total amount of coal reserve in Illinois base is 104.2 billion tons [21]. Generally it contains high sulphur content as it releases gases that harm the environment. The curable reserve *i.e.* amount of coal that can be recovered from existing coal reserves is approximately 1.24 billion tons [22]. According to an estimate that the reserves of coal are likely to last till 2170.

Petroleum

Petroleum is a liquid mixture of hydrocarbons with some contamination such as organic compound that contain nitrogen, sulphur and vanadium. Petroleum must be heated to form an extremely light liquid that immediately volatizes into the atmosphere. 49.29 million barrels of distillate fuel oils, 29.75 million barrel of jet fuels, 21.1 million barrels of liquefied petroleum gas and 1033,000 barrels of residual fuel oils and other (includes asphalt and road oil, aviation, kerosene) [23]. The fragment may be used as a fluid fuels, or they can be assembled into many useful materials such as plastics and other pigments. It occurs in various forms with different uses. A light hydrocarbon combination known as Gasoline is used to fuel automobile. Moderate weight fraction, includes diesel which is used as a fuel by trucks, trains and homes. Kerosene, another form of petroleum is used for cooking and as a fuel for airplanes. Thick residual oil, is used a fuel that used in oil- fired power plants and large ships. Semi-solid form is used to pave roads and manufacture roofing products.

Reserves of petroleum had been decreasing over the last 20 year period. Its cost has been increasing making it non-viable economic resources [24]. Fossil fuels are used at a large scale mainly in advanced and rapidly developing economies.

Oil

Oil is the third amongst the fuel used for power generation in whole world. Around 50% of oil *i.e.* crude oil in the world is turned into gasoline. Oil and natural gas also have a more limited availability and are estimated to last until 2041 and 2071.

Use of non- renewable resources exerts bad effect on environment by the way they processed, used and disposed off. Burning of fossil fuel produces high amount of carbon dioxide which is a major cause of global warming [25, 26]. Most of the countries fulfil their energy needs from fossil fuels such as coal and natural gas and consumption of energy increases, emission of carbon is increasing as well [27]. About 66% of energy in the world is obtained from fossil fuels, while 8% is generated from nuclear energy [28 - 32]. Boontome *et al.* [33] used ECM Granger causality technique to indicate that expenditure of non- renewable energy expand carbon emission but had trivial collision on economic growth. According to GMM and pooled mean group estimates, fossil fuel utilization enhanced carbon dioxide emission but delayed economic growth in advanced economies.

A possible approach to decrease emission from coal- fired power plant is through enhanced energy efficiency and conservation program [34]. One procedure to restraint emission from coal-fired power plants is through a request response program, which needs consumers to use minimum amount of energy and consequently construction of fewer coal- fired power plants.

Nuclear Fuels

Nuclear fuels are obtained through mining and refining of uranium ore. Nuclear fuels mostly include unstable isotopes of heavy elements such as Uranium and Plutonium. In nuclear reactions atoms combine into dissimilar compounds without changing their internal structure. Nuclear fission liberates energy in the form of heat. Numerous environmental problems are associated with nuclear power generation. An unrestrained chain reaction can result in a devastating nuclear explosion [35]. It generates power through a process known as nuclear fusion. It is used for running turbines and generates nuclear power.

Apart from fossil fuel other classes of non- renewable resources include metals, minerals such as potash and gypsum. Limestone, glass sand, peat, copper, gold are other non renewable resources. The stocks of these resources can be maintained if managed properly.

Effect of Non- renewable Resources on Environment

Fossil fuels burn to produce carbon dioxide which results in greenhouse effect [36]. Apart from greenhouse effect, coal-fired plants are the highest source of mercury emission which lead to neurological and neurobehavioral effect in embryo and young children. Increase in release of carbon impels the central authorities make effort towards execution and unification of energy strategies to stimulate the utilization of cleaner and renewable energy to minimize the consequence of carbon [37].

Burning of fossil fuels release pollutants that enter the atmosphere and sulphur and other chemicals get back earth in the form of acid rain. This rain proves harmful to fish and aquatic animals and also causes damage to tree as well as forest ecosystem.

Mining waste dumped into soil and other location results in poor soil quality and enters into food chain. Oil producing companies involved in natural gas and petroleum production are not able to regulate the carbon dioxide emissions. The carbon balance needs to be maintained otherwise it can result in decline in economic growth [38, 39].

Energy Use

In the earlier times, wood was collected and plant biomasses collected were used as the source of fuel or as a source of energy. Biomass fuels are the renewable resource of energy because the rate at which they were being harvested is much smaller than that rate at which new biomass was being produced by vegetation.

The energy is used to run industrial operations, machines, to keep cozy in winter, cool in summer and to produce food.

High amount of energy is mainly used in advanced countries such as Canada and USA. Energy consumption is linked with gross domestic product GDP [40].

Protection of Non- renewable Resources

Most of the people in the world totally depend on non-renewable resources. It is very essential to encourage alternative energy resources such as renewable resources like solar and wind power. Decreasing our dependency on non-renewable resources and increasing usage of renewable energy is one of the solutions to a sustainable future [41]. Efforts are made to shift to renewable sources of energy like solar, wind, biogas and geothermal energy. Use of renewable energy may implicit a dilemma between growth and environmental quality.

Action like driving electric and hybrid vehicle, installing solar panels and appositely insulating home and business and using energy efficient appliances are all smaller- scale changes to reduce usage of non-renewable resources. Enhanced response means encouraging consumers to reduce usage of electricity at their place, as consequences of comprehensive prices of electricity. Electricity is very important and useful type of energy used in industrial societies, extensively administered to industries and homes. Solar energy can also produce electricity directly, *via* photovoltaic technology.

Renewables are projected to be the fastest-growing energy source and nuclear power is expected to be the world's second fastest-growing source of energy. Renewable fuels are expected to grow faster than fossil fuels by the year 2040 [42]. Natural gas is expected to be the fastest-growing fossil fuel in the future because it is abundant natural gas resources and increased production includes supplies of tight gas, shale gas, and coal bed methane. The rise in oil price is expected to make consumers opt for more energy-efficient technologies and move away from liquid fuels.

ENVIRONMENTAL SUSTAINABILITY

World commission on Environment and development of the United Nation defined sustainability development as "development that meets the needs of the present without compromising the ability of future generous to meet their own needs". Sustainability is defined as the development relates to three interconnecting goals: environmental, economic and social. Sustainability requires protection of environment, economic growth and responsibility towards social life. The abstraction of sustainability has been more commonly associated with planet earth. The concept of sustainability growth development was non-segregated into various disciplines.

The consumption of nonrenewable resources may only be sustainable if the existing supply of the resource does not decline in the future. This can only be possible if technological advancement allows for a meaningful increase in the efficiency of the consumption of these resources in the future and the increased development of renewable sources.

Environmental sustainability is very biggest matter faced by the human being at present. Growing population together with enormous growth in anthropogenic activities has increased the matter of sustainability of natural resources on our planet. It has become important to cover the ever growing energy need in deposition to achieve sustainable development and reduce the pollution of resources used [43]. The implementation of environmental sustainability assists to make sure that the needs of today's population are encountered without threatening the ability of future generations to meet their needs Fig. (3). Many countries are implementing strategies for development of superior environmental quality by decreasing further worsening of situation. Such strategies are established both by the government's direct and indirect monitoring interferences. The consciousness for more aware environmental living can be spread among the society by means such as environmental education, behavior and moral preaching [44]. Production using energy inputs that increase pollution can be administered by enforcing taxes and fines [45-48]. Our environmental strategy focus to reduce the construction of waste, exclude practices that involve the use of dangerous chemicals, with an aim to clean environmental set up.

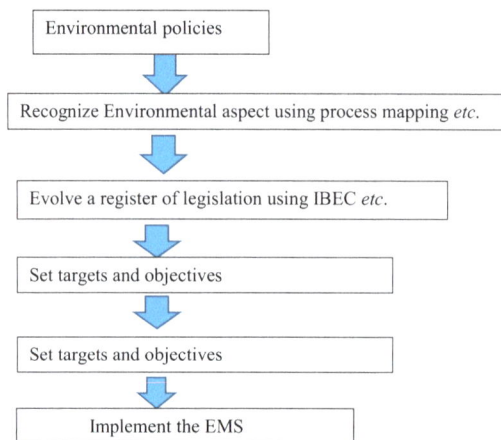

Fig. (3). Flow chart showing various steps leading to environmental sustainability.

Around 17 Sustainable Development Goals (SDGs) which included 169 targets were proposed by The United Nations General Assembly at the UN in New York by the Open Working Group. The 17 (SDGs) can transform the world as they aim at reducing poverty, hunger, promoting good health and well-being, providing quality education, maintain gender equality, provision of clean water and sanitation, affordable and clean energy and economic growth. The SDGs add great value and simultaneously demand to the scientific community. The set of global SDGs were set with the aim that climate change and its impacts will be combated, and a sustainable future is ensured for future generations. In addition to climate change, focuss has also been given on renewable energy, food, health and water provision through a coordinated global monitoring and modelling of socially, economically and environmentally oriented factors.

The replacement of fossil fuel-based energy sources with renewable energy sources such as solar energy, geothermal energy, hydropower, wind and ocean energy (tide and wave) will help in achieving sustainability. The role of governments, intergovernmental agencies, parties and individuals need to be ensured for achieving a sustainable future. It is noted that implementation of renewable energy technologies will generate about 2.3 million jobs all over the world. These will be in the sectors such health, education, energy and environmental safety. Seventh SDG ensures affordable and clean energy to all which can be achieved by using renewable energy source since they are generally distributed across the globe.

Improvement in education, awareness, and human institutional capacity environmental issues is required. Countries need to incorporate decarbonization policies and strategies into various sectors such as energy, forest, health, agriculture, water resource, and many other sectors. Partnership between developed, developing and least developed countries will help in promoting the distribution and transfer of environmentally friendly technologies, innovations and will help in achieving sustainability in a shorter possible time.

Five key steps to environmental sustainability:

- Environmental compliance
    - Establishing legal environmental compliance.
    - Environmental strategy and commitment register.
    - Environmental Management
    - Recognition and Execution as "Best Practice".
    - Sectoral Benchmark (Working towards ISO 14001)
- Environmental management system (ISO14001)
    - Aligance and Planning

   ◦ Execution and Management
   ◦ Quantitative and Qualitative indicators of performance
   ◦ Ongoing review and upgrade improvement
- Environmental Performance
   ◦ Environmental Performance Evaluation
   ◦ Basic life cycle assessment
   ◦ Environmental reporting
- Environmental Sustainability
   ◦ Back- casting and Ecological foot printing
   ◦ Emission and Impact Reduction strategies
   ◦ Change of reasonable and Innovation

The term "Sustainable Development" was first popularized in report given by the World Commission on Development and environment, an agency of United Nation. According to an estimate, about 5-to-10 fold expands in the human economy could be sustained. The solution of sustainable development is possible only by means conventional and economic policies so that sustainable human economy is ascertained.

The concept of Sustainable development modified total economy [49 - 52]. Consequently, the modified net economy (savings) variable is used as the sustainable development variable Fig. (4). Environmental economies help in detecting the costs of following kinds of impairment.

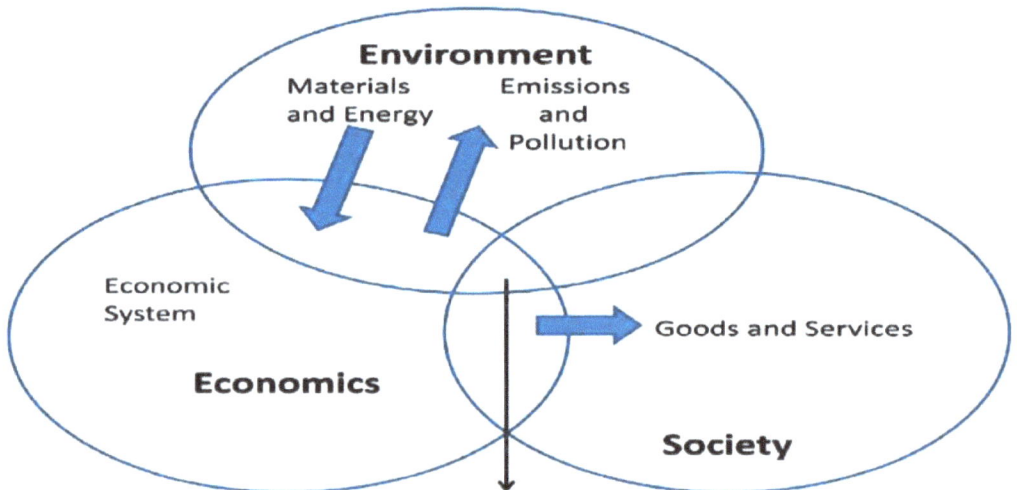

Fig. (4). Diagram showing the integration of Environment and Economics in the society.

- The expenditure of natural resources, involve its longer term inference for the continuation of future generation.
- Disruption that cause harm to natural ecosystems.
- Impairment and destruction of species.
- Ecological and pollution and human health effect.

Environmental destruction is though recognized as essential but its advantages are only partly captured by conventional economies.

CONCLUSION

Non-renewable resources decline at an alarming rate because of their tremendous use as source of energy and fuel. Non-renewable resources can be used at very large extant to achieve the economic growth but are not the basis of sustainable economy. Non- renewable resources cause harm to environment to a greater extent. The life of fossil fuels is only few hundred years. Hence use of non-renewable resources need to be restricted to save environment and secure future of coming generations. Besides this alternate sources of energy and fuel need to be explored. Renewable sources of energy provide a viable option to this problem. Use of renewable energy sources and other strategies for minimal use of non-renewable resource can help in achieving the SDGs and secure a better environment top live in.

AUTHOR'S CONTRIBUTION

All authors have equally contributed to this chapter.

REFERENCES

[1]   Adelman MA. Mineral Depletion, with Special Reference to Petroleum. Rev Econ Stat 1990; 72(1): 1-10.
[http://dx.doi.org/10.2307/2109733]

[2]   Australasian Institute of Mining and Metallurgy. Australasian Code for Reporting of Mineral Resources and Ore Reserves (The JORC code). Prepared by the Joint Ore Reserves Committee of The Australasian Institute of Mining and Metallurgy, Australian Institute of Geoscientist and Minerals Council of Australia (JORG); 1999.

[3]   Mahalik MK, Mallick H, Padhan H. Do educational levels influence the environmental quality? The role of renewable and non-renewable energy demand in selected BRICS countries with a new policy perspective. Renew Energy 2021; 164: 419-32.
[http://dx.doi.org/10.1016/j.renene.2020.09.090]

[4]   Mohsin M, Rasheed AK, Sun H, *et al.* Developing low carbon economies: An aggregated composite index based on carbon emissions. Sustain Energy Technol Assess 2019; 35: 365-74.
[http://dx.doi.org/10.1016/j.seta.2019.08.003]

[5]   Adams S, Klobodu EKM, Apio A. Renewable and non-renewable energy, regime type and economic growth. Renew Energy 2018; 125: 755-67.

[http://dx.doi.org/10.1016/j.renene.2018.02.135]

[6]   Dogan E. Analyzing the linkage between renewable and non-renewable energy consumption and economic growth by considering structural break in time-series data. Renew Energy 2016; 99: 1126-36.
[http://dx.doi.org/10.1016/j.renene.2016.07.078]

[7]   Destek MA. Renewable energy consumption and economic growth in newly industrialized countries: Evidence from asymmetric causality test. Renew Energy 2016; 95: 478-84.
[http://dx.doi.org/10.1016/j.renene.2016.04.049]

[8]   Destek MA, Aslan A. Renewable and non-renewable energy consumption and economic growth in emerging economies: Evidence from bootstrap panel causality. Renew Energy 2017; 111: 757-63.
[http://dx.doi.org/10.1016/j.renene.2017.05.008]

[9]   Ozcan B, Ozturk I. Renewable energy consumption-economic growth nexus in emerging countries: A bootstrap panel causality test. Renew Sustain Energy Rev 2019; 104: 30-7.
[http://dx.doi.org/10.1016/j.rser.2019.01.020]

[10]   da Silva PP, Cerqueira PA, Ogbe W. Determinants of renewable energy growth in Sub-Saharan Africa: Evidence from panel ARDL. Energy 2018; 156: 45-54.
[http://dx.doi.org/10.1016/j.energy.2018.05.068]

[11]   Adedoyin FF, Alola AA, Bekun FV. An assessment of environmental sustainability corridor: The role of economic expansion and research and development in EU countries. Sci Total Environ 2020; 713: 136726. a
[http://dx.doi.org/10.1016/j.scitotenv.2020.136726] [PMID: 32019050]

[12]   Adedoyin FF, Gumede MI, Bekun FV, Etokakpan MU, Balsalobre-lorente D. Modelling coal rent, economic growth and CO_2 emissions: Does regulatory quality matter in BRICS economies? Sci Total Environ 2020; 710: 136284. b
[http://dx.doi.org/10.1016/j.scitotenv.2019.136284] [PMID: 31923665]

[13]   Etokakpan MU, Adedoyin FF, Vedat Y, Bekun FV. Does globalization in Turkey induce increased energy consumption: insights into its environmental pros and cons. Environ Sci Pollut Res Int 2020; 27(21): 26125-40.
[http://dx.doi.org/10.1007/s11356-020-08714-3] [PMID: 32358749]

[14]   Udi J, Bekun FV, Adedoyin FF. Modeling the nexus between coal consumption, FDI inflow and economic expansion: does industrialization matter in South Africa? Environ Sci Pollut Res Int 2020; 27(10): 10553-64.
[http://dx.doi.org/10.1007/s11356-020-07691-x] [PMID: 31939028]

[15]   Shahbaz M, Hye QMA, Tiwari AK, Leitão NC. Economic growth, energy consumption, financial development, international trade and CO_2 emissions in Indonesia. Renew Sustain Energy Rev 2013; 25: 109-21.
[http://dx.doi.org/10.1016/j.rser.2013.04.009]

[16]   Awodumi OB, Adeleke AM. Non-Renewable Energy and Macroeconomic Efficiency of Seven Major Oil Producing Economies in Africa. Zagreb International Review of Economics and Business 2016; 19(1): 59-74.
[http://dx.doi.org/10.1515/zireb-2016-0004]

[17]   Suwono A, Mansoori GA. Proceedings of Fluid and Thermal Energy Conversion (FTEC) Conferences www.uic.edu/labs/trl/ftec.home/ 2003, 2006.

[18]   Araujo PLBD, Mansoori GA, Araujo ESD. Diamondoids: occurrence in fossil fuels, applications in petroleum exploration and fouling in petroleum production. A review paper. Int J Oil Gas Coal Technol 2012; 5(4): 316-67.
[http://dx.doi.org/10.1504/IJOGCT.2012.048981]

[19]   Freedman B. Environmental Ecology. 2nd ed., San Diego, CA: Academic Press 1995.

[20]   Freedman B, Hutchings J, Gwynne D, *et al.* Ecology: A Canadian Context. 2nd ed., Nelson Canada, Toronto, ON 2014.

[21]   IDCEO Illinois Department of Commerce and Economic Opportunity (2010) 'The Illinois coal Industry', Report of the office of Coal Development, 2010 Report, p.1-31.

[22]   IDCEO Illinois Department of Commerce and Economic Opportunity (2008) 'The Illinois Coal Industry', Report of the Office of Coal Development, 2008 Report, p.1-75.

[23]   US-EIA U.S. Energy Information Administration, (2011) 'State Energy Data System: Illinois', Independent Statistics and Analysis, Technical Notes and Documentation. Data Release Date: 2011-06-20.

[24]   Harris JM, Roach B. Environmental and Natural Resource Economics: A Contemporary Approach. 3rd ed., New York, NY: Routledge 2014.

[25]   Dogan E, Seker F. Determinants of CO_2 emissions in the European Union: The role of renewable and non-renewable energy. Renew Energy 2016; 94: 429-39.
[http://dx.doi.org/10.1016/j.renene.2016.03.078]

[26]   Apergis N, Ben Jebli M, Ben Youssef S. Does renewable energy consumption and health expenditures decrease carbon dioxide emissions? Evidence for sub-Saharan Africa countries. Renew Energy 2018; 127: 1011-6.
[http://dx.doi.org/10.1016/j.renene.2018.05.043]

[27]   Elum ZA, Momodu AS. Climate change mitigation and renewable energy for sustainable development in Nigeria: A discourse approach. Renew Sustain Energy Rev 2017; 76: 72-80.
[http://dx.doi.org/10.1016/j.rser.2017.03.040]

[28]   Bolt K, Matete M, Clemens M. Manual for calculating adjusted net savings. Unpublished Working Paper Environmental Department. Washington (DC): The World Bank, 2002.

[29]   Arrow K, Dasgupta P, Goulder L, *et al.* Are we consuming too much? J Econ Perspect 2004; 18(3): 147-72.
[http://dx.doi.org/10.1257/0895330042162377]

[30]   Hamilton K. Sustaining economic welfare: estimating changes in per capita wealth. World Bank Policy Research Working Paper No: 2498. Washington DC: The World Bank, 2005.

[31]   Ahmad M, Akram W, Ikram M, *et al.* Estimating dynamic interactive linkages among urban agglomeration, economic performance, carbon emissions, and health expenditures across developmental disparities. Sustainable Production and Consumption 2021; 26: 239-55. a
[http://dx.doi.org/10.1016/j.spc.2020.10.006]

[32]   Alvarado R, Deng Q, Tillaguango B, *et al.* Do economic development and human capital decrease non-renewable energy consumption? Evidence for OECD countries. Energy 2021; 215: 119147.
[http://dx.doi.org/10.1016/j.energy.2020.119147]

[33]   Boontome P, Therdyothin A, Chontanawat J. Investigating the causal relationship between non-renewable and renewable energy consumption, CO 2 emissions and economic growth in Thailand 1 1This is a preliminary work. Please do not quote or cite without permission of the authors. Energy Procedia 2017; 138: 925-30.
[http://dx.doi.org/10.1016/j.egypro.2017.10.141]

[34]   Taib S, Al-Mofleh A. Tools and Solution for Energy Management, Energy Efficiency - The Innovative Ways for Smart Energy, the Future Towards Modern Utilities. Moustafa Eissa, IntechOpen 2012.
[http://dx.doi.org/10.5772/48401]

[35]   Owusu PA, Asumadu-Sarkodie S, Dubey S. A review of renewable energy sources, sustainability issues and climate change mitigation. Cogent Eng 2016; 3(1): 1167990.
[http://dx.doi.org/10.1080/23311916.2016.1167990]

[36]   Halder PK, Paul N, Joardder MUH, Sarker M. Energy scarcity and potential of renewable energy in

Bangladesh. Renew Sustain Energy Rev 2015; 51: 1636-49.
[http://dx.doi.org/10.1016/j.rser.2015.07.069]

[37]   Sharif A, Mishra S, Sinha A, Jiao Z, Shahbaz M, Afshan S. The renewable energy consumption-environmental degradation nexus in Top-10 polluted countries: Fresh insights from quantile-o--quantile regression approach. Renew Energy 2020; 150: 670-90.
[http://dx.doi.org/10.1016/j.renene.2019.12.149]

[38]   Shahbaz M, Kumar Tiwari A, Nasir M. The effects of financial development, economic growth, coal consumption and trade openness on CO_2 emissions in South Africa. Energy Policy 2013; 61: 1452-9.
[http://dx.doi.org/10.1016/j.enpol.2013.07.006]

[39]   Wolde-Rufael Y. Coal consumption and economic growth revisited Zagreb. Int Rev Econ Bus 2016; 19(1): 59-74.

[40]   Caraiani C, Lungu CI, Dascălu C. CO_2 emissions, renewable and non-renewable energy consumption, and economic growth: evidence from panel data for developing countries. Inter Econ 2015; 36(1): 553-9.

[41]   Chofreh AG, Goni FA, Shaharoun AM, Ismail S, Klemeš JJ. Sustainable enterprise resource planning: imperatives and research directions. J Clean Prod 2014; 71: 139-47.
[http://dx.doi.org/10.1016/j.jclepro.2014.01.010]

[42]   Rungtusanatham MJ, Choi TY, Hollingworth DG, Wu Z, Forza C. Survey research in operations management: historical analyses. J Oper Manage 2003; 21(4): 475-88.
[http://dx.doi.org/10.1016/S0272-6963(03)00020-2]

[43]   Ozturk I, Acaravci A. Electricity consumption and real GDP causality nexus: Evidence from ARDL bounds testing approach for 11 MENA countries. Appl Energy 2011; 88(8): 2885-92.
[http://dx.doi.org/10.1016/j.apenergy.2011.01.065]

[44]   Xia Z, Abbas Q, Mohsin M, Song G. Trilemma among energy, economic and environmental efficiency: Can dilemma of EEE address simultaneously in era of COP 21? J Environ Manage 2020; 276: 111322.
[http://dx.doi.org/10.1016/j.jenvman.2020.111322] [PMID: 32891035]

[45]   Sun H, Pofoura AK, Adjei Mensah I, Li L, Mohsin M. The role of environmental entrepreneurship for sustainable development: evidence from 35 countries in Sub-Saharan Africa. Sci Total Environ 2020 741: 140132.

[46]   Al Asbahi AAMH, Gang FZ, Iqbal W, Abass Q, Mohsin M, Iram R. Novel approach of principal component analysis method to assess the national energy performance *via* energy trilemma index. Energy Rep 2019; 5: 704-713.

[47]   Mohsin M, Zhang J, Saidur R, Sun H, Sait SM. Economic assessment and ranking of wind power potential using fuzzy-TOPSIS approach. Environ Sci Pollut Res Int 2019; 26(22): 22494-511.
[http://dx.doi.org/10.1007/s11356-019-05564-6] [PMID: 31161545]

[48]   Sun H, Tariq G, Haris M, Mohsin M. Evaluating the environmental effects of economic openness: evidence from SAARC countries. Environ Sci Pollut Res Int 2019; 26(24): 24542-51.
[http://dx.doi.org/10.1007/s11356-019-05750-6] [PMID: 31236865]

[49]   Aidt TS. Corruption, institutions, and economic development. Oxf Rev Econ Policy 2009; 25(2): 271-91.
[http://dx.doi.org/10.1093/oxrep/grp012]

[50]   Aidt TS. Corruption and sustainable development. Cambridge, MA: Faculty of Economics and Jesus College, University of Cambridge. CWPE 2010; p. 1061.

[51]   Spaiser V, Scott K, Owen A, Holland R. Consumption-based accounting of CO_2 emissions in the sustainable development Goals Agenda. Int J Sustain Dev World Ecol 2019; 26(4): 282-9.
[http://dx.doi.org/10.1080/13504509.2018.1559252]

[52]   Güney T. Governance and sustainable development: How effective is governance? J Int Trade Econ
Dev 2017; 26(3): 316-35.
[http://dx.doi.org/10.1080/09638199.2016.1249391]

CHAPTER 2

Role of Biotechnology in Treatment of Solid Waste

Bhupinder Dhir[1,*]

[1] *School of Sciences, Indira Gandhi National Open University, New Delhi, India*

Abstract: Waste management has become a major global concern. The rapid rise in the rate of population has increased the generation of waste at a tremendous pace. Improper disposal of agricultural, household, municipal and industrial wastes can pose a threat to the health of living beings and the environment. Industrial waste, in particular, is highly hazardous as it contains toxic chemicals and metals. Many methods of waste disposal have been adopted, but most of them produce various kinds of after-effects, therefore, biological methods have been adopted because of their eco-friendly and sustainable nature. Sustainable waste management aims to minimize the amount of waste generation. Waste is treated in a proper way, involving the steps such as segregation, recycling and reuse. Biotechnological methods such as composting, biodegradation of xenobiotic compounds and bioremediation have been tried. These methods have proved useful in treating waste in an eco-friendly way. More research studies need to be carried out to standardize the method for the proper treatment of waste so that environmental sustainability can be achieved.

Keywords: Anaerobic degradation, Biogas, Bioremediation, Combustion, Composting, Fermentation, Incineration, Recycle, Reduce, Reuse, Xenobiotic.

INTRODUCTION

Waste released from municipal, domestic and industrial processes is becoming a problem worldwide. Solid waste includes wastes obtained from household, commercial and demolition activities. In the present era, e-waste (electronic) has also become an issue of concern. Besides these, medical waste, radioactive waste and many other forms of waste are generated. Globally, municipal, domestic, industrial, plastic and electronic waste has been discarded at an alarming rate by people worldwide. It is becoming difficult to manage waste generated in huge quantities. Increasing waste is proving a threat to the environment. The main reason for this is rapid industrialization and urbanization-change in the lifestyle of individuals. Recent reports indicated that the rate of waste generation at present is estimated to be about 0.25 to 0.66 kg/person/day [1]. Studies suggest that the rate

* **Corresponding author Bhupinder Dhir:** School of Sciences, Indira Gandhi National Open University, New Delhi, India; E-mail: bhupdhir@gmail.com

of disposal of municipal solid waste (MSW) all over the world is around one billion metric tons. The amount of waste disposal is expected to reach 2.2 billion by the year 2025. Plastic is one of the major components of MSW. The wastes are both biodegradable (organic waste) and non-biodegradable (Table **1**).

Table 1. Different types of wastes.

Type of waste	Examples
Non-biodegradable waste	
E-waste	Disposed/discarded computers, TV, music cassettes, disks, printer cartridges, electronic items
Liquid waste	Waste from industries, tanneries, distilleries, thermal power plants
Plastic waste	Bags, bottles, cans, packaging *etc*
Metal waste	Metal sheets, metal scraps
Nuclear waste	Waste released from nuclear power plants
Miscellaneous	Foil, wrappings, pouches, sachets, tetra packs, discarded clothing, old/broken furniture, and discarded equipments
Biodegradable or wet waste	
Household waste	Vegetable waste, Fruit peels kitchen waste (cooked and uncooked food items)
Garden waste	Green/dry leaves
Sanitary wastes	Tissue papers, napkins, toilet papers

The wastes are generally disposed of in open dumps, landfills, or subjected to incineration or composting. The practices of disposal of waste in open dumps or incineration affect the various components of the environment, such as soil, land and affect public health *via* the outbreak of diseases. Waste treatment is important before it is released into the environment to prevent ecological imbalance. There is a need to develop technologies so that waste is treated in a manner that does not bring any harm to the environment. Generally direct combustion or crushing of the waste (such as agricultural, household, municipal and certain biomass materials) after sorting has been followed, but these activities release gases such as CO and CO_2 into the atmosphere. If wastes are not treated properly, it emits gases like methane (CH_4) and carbon dioxide (CO_2), which cause air and water pollution [2]. The problem can be mitigated if wastes are treated and processed by adopting environment-friendly technologies before disposal.

Many waste management policies and technologies have been developed, and efforts have been made to tackle waste in an effective manner by public and private agencies. Waste management is an activity that helps in organizing waste from its production to the process of final treatment. Waste management includes processes such as collection, sorting, treatment and recycling of waste materials. Each and every kind of waste needs to be managed in an efficient manner (Fig. **1**). Sustainable waste management strategies aim to reduce the production of waste or follow activities that engross the reuse of waste materials in some or the other way. Less amount of waste generation will prevent contamination and the spread of diseases. Waste management is considered a key element for a sustainable environment [3, 4]. Sustainable waste management reduces the number of products such as plastic along with an increase in means of their recycling at the same time.

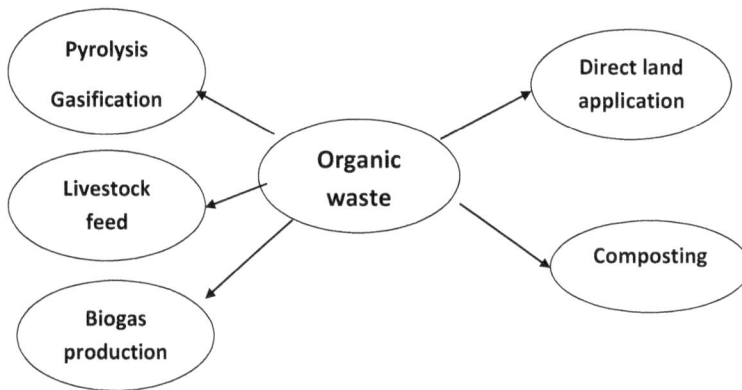

Fig. (1). Various ways in which organic waste can be treated.

Besides the physcio chemical methods, biowaste treatment technologies, including direct land application, animal feed, combustion, physical and chemical treatment processes, temperature treatment (such as pyrolysis, liquefaction, gasification) and biological treatment (composting, vermicomposting, anaerobic digestion, fermentation) have also shown promising results with an aim for achieving waste management [5, 6]. Biowaste treatment reduces threats to human health and environmental deterioration [7. 8]. Conversion of waste-to-energy (WTE) is suggested as another way of handling waste effectively. This technology is supposed to produce a good amount of energy *via* the utilization of waste, thus reducing its quantity and lessening the chances for soil and land pollution. It also suggests a way of safe disposal of waste. The waste-to-energy technique has been adopted by many developed countries. Biotechnology deals with waste management by degrading harmful elements and toxic chemicals with the help of microbes [9]. Biodegradable waste can be effectively treated and converted into useful products using microorganisms. Microbes that possess the potential to metabolize any kind of waste can be engineered *via* molecular techniques.

APPROACHES FOR SOLID WASTE MANAGEMENT (SWM)

Solid waste management has received recognition as a technology by government agencies and local authorities in most countries of the world as a viable strategy for the treatment of waste. Public-private, community-public and private-private partnerships have been working to manage the waste. The private sector includes registered enterprises. These agencies help in the transport, treatment, disposal and recycling of waste. Wastepickers, waste buyers, and traders form an important component of small-scale private sector units that play very important in management of waste. Public sector undertakings involve local departments and at the city level. Activities related to solid waste management are carried out by various NGOs.

Steps involved in waste management include

1. Segregation of waste at the source which involves the separation of biodegradable and non-biodegradable solid waste.

2. Treatment of different types of solid wastes by using suitable techniques.

Biowaste treatment is done in different types, which include direct use, physico-chemical treatment, thermochemical treatment and biological treatment.

Combustion

Combustion or open burning is the burning of waste in open or dump sites. The volume of waste gets reduced after combustion but results in the emission of harmful gases into the environment [10]. This practice is followed as one of the primary practices by developing or undeveloped countries all over the world because it is very easy, requires no prior technical knowledge, and is cost-effective [11, 12]. The emission of gases resulting from complete and/or incomplete combustion of waste poses a threat to human and environmental health. Stockholm Convention declared combustion as an "environmentally unacceptable process" [12].

Thermal Treatment

Heat is also used for the treatment of waste materials. This method helps in recovering energy for electricity or heating. Advantages such as quick reduction in the volume of waste, less cost of transportation and low greenhouse gas emissions make this approach more acceptable. The treatment process has proved very useful in the treatment of non-biodegradable waste with low moisture content. Incineration and pyrolysis/gasification are the two main techniques included in this category. This process results in the de-composition of organic matter present in the waste to produce heat energy or fuel oil or gas. The method is practiced by many local authorities. It acts as an inexpensive solution for the management of solid waste.

Thermal waste treatment includes processes such as Incineration and Gasification and Pyrolysis. In the process of incineration, the combustion of waste material occurs in the presence of oxygen. The waste gets reduced in volume after heating. Combustion of waste produces heat and electricity. Incineration works well, ideally in a situation where waste does not want to get stored. One of the drawbacks of the method is that heat produced during the process pollutes the air by the release of gases. In Singapore, 8,200 tons of garbage/ waste is incinerated daily. This results in a reduction of its volume by 90%. Incineration plants produce over 2,500 MWh of energy daily. Reusable metals can also be recovered *via* the process that can be sold for profit. The process of pyrolysis is mainly used in industrial processes where the combustion occurs in the absence of oxygen.

Organic waste materials get decomposed after exposure to low amounts of oxygen and high temperature. The production of gas from waste involves the process of gasification. Waste is combined with oxygen and/or steam to produce 'syngas', a gas that can then be used to make numerous useful products such as transport fuels, fertilisers or electricity. The gasification process requires a lot of energy, and the reactors need regular maintenance. A very low amount of oxygen is used in gasification [13, 14]. In the process of gasification, burning recovers energy but does not cause air pollution.

Bio-chemical Conversion

This process works well for wastes with a high percentage of bio-degradable (organic) matter and high moisture/water content. These conditions stimulate microbial activity. The main methods that come in this category are anaerobic digestion (also known as Bio-methanation). The enzymatic decomposition of organic matter by microbial action produces methane gas or alcohol [15, 16].

Biological Treatment

Biological treatment processes involve the use of living organisms for waste degradation. The process takes place in moist conditions. Biochemical processes are comparatively slow and require less energy in comparison to thermochemical conversion [17]. Other methods include the treatment and disposal of solid waste by applying biotechnology in controlled conditions. Xenobiotic compounds can be removed *via* the process of biodegradation. Waste can also be treated following a bioremediation process that involves the use of living organisms, particularly microorganisms, for the treatment of chemical compounds and other pollutants.

Some studies have also shown that larvae of insects show a capacity to convert waste into protein products. For example - house flies and blow flies help in managing slaughterhouse waste [18].

Composting

Solid organic waste can be treated by the process of composting. The process involves using microorganisms to decompose waste under controlled conditions [19]. Organic matter is broken down by microorganisms to produce carbon dioxide, water and heat. The decomposition of organic matter in composting is carried out under controlled aerobic conditions. Humus is a stable organic product formed during the process. Composting can be done for different types of organic solid wastes. Leaves, grass, food and agricultural waste, newspapers, vegetable, cereals and organic waste are some of the materials that can easily be composted [20]. Composting involves various steps such as separation of the waste, compaction in the presence of moisture and optimum temperature to get humus-like substance. The three phases in composting are the mesophilic phase which lasts for a couple of days, the thermophilic phase, which lasts for a few weeks to several months, and the cooling and maturation phase, which lasts for several months [20]. The metabolic reactions carried out by microorganisms help in the degradation of the material during the thermophilic phase.

Moist conditions support the microbial activity required for the process of composting [21, 22]. Compost formed by the process is a material rich in nutrients such as nitrogen, potassium, phosphorus, a range of beneficial minerals and humus. All these things benefit the growth of plants [22]. Compost is used in the amendment of damaged soils, restoration of landfill cover and remediation or restoration of land [23].

Rapid composting involves the addition of chemicals such as bauxite and phosphogypsum to solid waste to accelerate the rate of waste biodegradation. The addition of chemical compounds such as bauxite increases the temperature and pH of the composting mixture. It also acts as a catalyst that provides the proper aeration required for carrying out composting. The rate of decomposition of the composting mixture can also be increased by adding glucose, a carbon source.

Vermicomposting

Vermicomposting method involves the use of microorganisms and earthworms for the breakdown of solid waste. In this process, organic waste gets degraded and stabilized under controlled aerobic conditions. Manure (vermicompost) rich in nutrients is formed [24, 25]. A pre-composting phase occurs before vermicomposting. Microorganisms are present in the gut and intestine of the worms. The earthworms feed on waste and promote microbial activity. The fecal material produced by microbes proves beneficial in improving the quality of the vermicompost and helps in the quick degradation of organic waste [28]. Household waste, organic municipal waste, sewage sludge and organic waste residues can be converted into manure by the process of vermicomposting [26, 27]. The method of vermicomposting is less energy-consuming, cost-effective and economically feasible as compared to conventional technologies.

Anaerobic Degradation

Anaerobic degradation/biomethanization or biomethanation process results in the decomposition of organic matter under an anaerobic environment [29, 30]. Anaerobic conditions support the decomposition of waste. Anaerobic digestion degrades high levels of sugar, starch, proteins, or fats present in the substrates. The substrates vary according to organic content present, carbon: nitrogen ratio, percentage of dry matter and volatile solids present in the dry matter and biogas. The process is used mainly in the treatment of municipal solid waste and sewage sludge, industrial wastewater and agricultural residues [31, 32]. Methane generated during the process can be used in cooking, generating heat and producing electricity. This is a process similar to composting.

The microbial communities carry out anaerobic digestion in a bioreactor. They degrade organic matter present in waste compounds into products, such as methane and carbon dioxide, collectively called biogas [33, 34]. Biogas production occurs through a batch or continuous process or can be a single or multi-step process. The biogas is mainly composed of methane (CH_4), carbon dioxide (CO_2), and sulfuric elements (H_2S) (Table **2**). The reduction of waste protects the environment. Assessment of the activity of microorganisms and factors that affect biogas production can help to enhance the fermentation process and reduce the cost of processing. Animal waste, agriculture residues, municipal waste, sewage sludge and organic wastes from food are the main residues used in biogas production.

Table 2. Components of biogas.

Components of Biogas	Percentage
Methane (CH_4)	55-65
Carbon dioxide (CO_2)	35-45
Nitrogen (N_2)	0-3
Hydrogen (H_2)	0-1
Hydrogen sulphide (H_2S)	0-1

(Table 2) cont.....

Components of Biogas	Percentage
Ammonia (NH$_3$)	0-1
Oxygen (O$_2$)	0-2

Biodegradation

Microorganisms possess a high capacity to degrade xenobiotic compounds that generally show resistance to biodegradation. Some of the xenobiotic compounds that can be degraded by microbes are listed in (Table **3**).

Table 3. List of Common Microorganisms for Xenobiotic Compounds.

Xenobiotic Compounds	Microorganisms Degrading Xenobiotic Compound
Petroleum hydrocarbons	*Bacillus sp.* S6 and S35
Pesticides Glyphosate	*Pseudomonas putida, P. aeruginosa* and *Acinetobacter faecalis*
Organochlorine – DDT	*Morganella morganii, Stenotrophomonas maltophilia, Actinomycetes*
Tetrachlorvinphos	*Proteusvulgaris, Stenotrophomonas maltophilia*
Chlorpyrifos	Bacterial strains
Atrazine	*Providencia spp, Enterobacter spp*

Bioremediation

The process in which living organisms, *viz.* microbes and plants, remove waste or pollutants from various components of the environment, such as water and soil, is called bioremediation. It is an environment-friendly method. The process of bioremediation can occur in natural conditions (*in situ)* or outside the habitat or nature (*ex situ).* In the process of *in situ* remediation, treatment in the contaminated soil or water takes place at the site without removing containments. The treatment/ degradation of compounds is carried out by bacteria. In the other type, *i.e.*, *ex-situ* bioremediation, the contaminated soil is removed and transferred to other places for treatment. It includes composting, bio piling and land forming. In composting, the degradation of waste is carried out by controlled conditions, the temperature (55-65°C). In the process of land forming, the contaminated soil is excavated and tilled at regular time intervals to initiate the process of degradation. Bio piling is a process in between land forming and composting. The process is effectively used in the treatment of petroleum hydrocarbons. The time duration of the whole process is about 2 to 3 months.

Biosparging is the process in which waste, such as petroleum products like diesel, gasoline, and lubricating oil, is treated. The concentration of oxygen is increased by injecting the air below groundwater under pressure. The air pressure needs to be controlled properly to avoid the liberation of volatile particles to the atmosphere that causes air pollution. Solid waste compounds get degraded aerobically *via* the process of bioventing. The rate of treatment of waste varies for each site as the composition of hydrocarbons varies, and soil texture shows differences. Oxygen and nutrients like phosphorus and nitrogen are injected into contaminated sites, which increases the removal process. Microorganisms are introduced into the contaminated site to accelerate the degradation process in bioaugmentation. The rate of degradation of contaminants present in the groundwater and soil is supported by the specific metabolic activity of microbes. The microorganisms degrade contaminants to non-toxic compounds like chloride and ethylene.

Fermentation

In the process of fermentation, microorganisms produce energy in the form of adenosine triphosphate (ATP) by the degradation of organic nutrients anaerobically. The process results in the degradation of waste along with the production of bio-ethanol (ethyl alcohol) [35]. Bio-ethanol (gasoline) is an environmental-friendly fuel [36]. It is produced from various sources such as sugars, starch and lignocellulose-based biomass using different conversion technologies [37 - 40]. It is mainly produced from corn-derived (starch-based) and sugarcane-derived (saccharose-based) feedstocks [41 - 43]. Bioethanol was produced using vegetable waste with the help of the yeast *Saccharomyces cerevisiae* [44]. The residues such as rice straw, potato waste, sawdust, corn stalks, sugarcane bagasse and sugar beet from agro-waste industries are used for the production of biofuels [45 - 47]. Agro-industrial residues act as cheap resources that contain a high amount of constituents and possess the unlimited potential to be used as an alternative substrate for fermentation. Bioethanol can be produced from edible (1st generation) and non-edible lignocellulosic (2nd generation) feedstock [48, 49]. Research studies are being conducted to get bio-ethanol from low-grade lignocellulosic biomass [50].

SUSTAINABLE APPROACHES FOR TACKINLG SOLID WASTE

Refuse, reduce, reuse and recycle are some of the sustainable strategies suggested for tackling solid waste. Refuse means denying buying anything which we don't require, while reduce is related to minimizing the amount of waste generated either by reducing its quantity, decreasing its production, changing the form of waste, or extracting useful materials (such as packaging materials) from it. Reuse involves using waste materials to get other useful things, *i.e.*, using waste things to their maximum capacity [15]. For example, instead of paper plates, solid metallic/plastic plates can be used that can be cleaned and used again. The process of recycling involves the use of waste components again in another manufacturing process [16]. Recycling it requires energy, money, and resources to turn waste back into usable materials. These four approaches prove useful in minimizing the quantity of waste generated. Garbage can be recycled to get useful products and provides raw materials for new products. Plastics and steel can be treated using these methods. Lead from used batteries can be recycled and used in new batteries. These approaches to solid waste management are comprehensive, cost-effective, environment-friendly and sustainable.

BIOTECHNOLOGICAL METHODS FOR THE TREATMENT OF SOLID WASTE

Solid-state Fermentation (SSF)

Solid waste from food, beer and wine, agriculture, paper, textiles, detergent, and animal feed industries mainly serves as a substrate for solid-state fermentation (SSF) [51, 52]. Cereal grains (rice, wheat, barley, and corn), legume seeds, wheat bran, lignocellulose materials, such as straws, sawdust or wood shavings, and a wide range of plant and animal materials, act as common substrates for SSF [53]. These substrates are polymeric in nature and are insoluble or sparingly soluble in water. Less amount of water or its absence offers several advantages such as easy product recovery, low cost of production, less time for fermentation, reduced downstream processing, and decrease in requirement of energy for stirring and sterilization [54]. Some SSF processes require the growth of selective microbes, such as molds. Single pure cultures, mixed cultures, or consortiums form a major part of the SSF process. The process of SSF is regulated by processes such as microorganisms, type of solid support used, water, temperature, aeration and type of fermenter used. Extracellular enzymes secreted by microorganisms carry out at low moisture levels. Microorganisms like fungi, yeasts, and bacteria are used in the process of SSF. They require high moisture content for carrying out efficient fermentation [55]. Molds are frequently used in SSF as they grow naturally on solid substrates, such as pieces of wood, seeds, stems, and roots and help in maximizing the production of value-added products.

The process of SSF involves many steps starting with the selection of substrate, and treatment of substrate by mechanical, chemical or biochemical means. This improves the availability of the nutrients present and also reduces the size of the components. Thereafter hydrolysis of polymeric substrates such as polysaccharides and proteins occurs followed by fermentation. Finally, the products are purified and quantified.

The technique of SSF is gaining importance because of the use of different types of organic wastes and the large-scale production of value-added products [56]. The process of SSF has emerged as a sustainable and green process that converts organic wastes into valuable products. The process of SSF is relatively simple and involves the use of abundant low-cost biomaterials with minimum or no prior treatment for conversion, and less wastewater generation [57]. Biologically active metabolites are produced during the conversion of organic solid wastes by SSF. Enzymes, organic acids, biofertilizers, biopesticides, biosurfactants, bioethanol, aroma compounds, animal feed, pigments, vitamins, and antibiotics are some of the products of the SSF process [58]. Agricultural wastes, in particular, are used to produce large value-added products using this process [59]. Different substrates produce different valuable products based on the differences in their composition.

Enzyme Production

Wastes, particularly agro-industrial wastes, are used as a raw material as they support the growth of microorganisms [58, 60 - 65]. Fermentation results in the production of many valuable enzymes (Table **4**). Amylase is one of the important enzymes that is used in the processing of starch obtained from industries. It helps in the degradation of polysaccharides into sugar components. Different cellulolytic enzymes, such as endoglucanase and β-glucosidase, are produced from agricultural wastes by solid-state cultivation [66]. Amylase and glucoamylase are produced from waste such as peel, seed, oil cakes, and field residues (such as rice and wheat bran) by *A. awamori* in solid-state fermentation [67 - 69]. *Aspergillus niger* MTCC 104 has shown the production of α-amylase *via* solid-state fermentation [45, 70, 71]. When oil cakes are used as the substrate for agro-industrial waste, the production of lipase enzymes is noted [72]. *Aspergillus ibericus* was used for the production of lipase.

Table 4. Enzymes produced by microorganisms using agro-industrial wastes.

Substrates	Enzymes	Microorganisms
Papaya waste	α-Amylase	*A. niger*
Groundnut oil cake	Lipase	*C. rugosa*
Wheat bran and orange peel	Pectin methyl esterase	*P. notatum*
Linseed oil cake	Lipase	*P. aeruginosa*
Orange peel	α-Amylase	*A. niger*
Coconut oil cake	α- Amylase	*A. oryzae*
Rice bran	α-Amylase	*Bacillus* sp.
Corn bran	α-Amylase	*Bacillus* sp.

(Table 4) cont.....

Substrates	Enzymes	Microorganisms
Rice bran, wheat bran, black gram bran, and soybean	α-Amylase	*A. niger*
Fruits peel waste	Invertase	*A. niger*

Single-cell Protein Production

The production of single-cell protein (SCP) from fruit wastes has been reported [73]. Cucumber and orange peels were used as the substrate for the production of SCP along with *Saccharomyces cerevisiae* by the method of submerged fermentation. Microbes show the capacity to convert fruit wastes into SCP. Orange peel waste has also been used for the production of Poly (3-HB) [74]. Biosurfactants and xanthan (food additive, basically an exopolysaccharide produced from *Xanthomonas* species) can be produced along with enzymes from agro-waste such as castor oil, sunflower oil, barley bran, peanut cake, and rice bran [75].

Enzymatic Degradation

Enzymes possess tremendous potential for degrading pollutants. Enzymes degrade a wide range of recalcitrant materials *via* oxidation. They carry out degradation at a wide range of temperatures, pH and salinity.

ENERGY GENERATION THROUGH WASTE

The process of conversion of waste into energy, either fuel or in the form of heat or electricity, is called as Waste-to-Energy or WTE. The process results in the production of clean and renewable energy (US EPA). Biowaste-derived energy has been gaining importance due to the increasing demand for energy [76 - 80]. Sweden has been converting 100% of their garbage into clean energy. WTE provides benefits such as a reduction in waste going to landfills by 60% to 90% depending upon the composition of waste, and a reduction in environmental pollution. The utilisation of garbage for energy can reduce waste by almost 90% and recover fuel gas for cooking and electricity at the same time. If waste is converted to energy, the production of methane is reduced, and at the same time, the consumption of natural resources such as coal, oil, and gas is reduced. Waste to energy is thus a very good technology aimed at better sustenance of the environment, with minimum damage to the environment. WTE sector can thus support the global economy by creating approximately 14,000 jobs.

CONCLUSION

Solid waste generation has noted a tremendous rise in the last few decades. Since the dumping or disposal of waste exerts a tremendous effect on natural resources such as soil, land and water, various technologies have been explored to treat waste and prevent pollution. Composting has been found to be the most suitable method for the simple degradation of organic waste. Bioremediation is another method that helps in the degradation of various types of solid waste. Its potential for further research in achieving sustainable solid waste treatment needs to be explored. Biotechnology is among one the approaches recommended for the efficient treatment and disposal of solid waste. The technique has proved useful in maintaining the sustainability of soil and land resources. Microorganisms also prove useful in the removal or treating waste *via* the process of fermentation. Reuse, recycle, reduce and refuse are four ways to help create wealth from waste and reduce waste input into landfills. Agro-industrial wastes, in particular, can be used as solid support in SSF processes, which can lead to the production of a range of beneficial products. New approaches need to be adopted to get cost-effective treatment of solid waste and maintain the sustainability of natural resources such as soil and land resource.

REFERENCES

[1]   UNEP UNEP Year Book : emerging issues in our global environment United Nations Environment Programme. 2011.

[2]   Sánchez A, Artola A, Font X, *et al.* Greenhouse gas emissions from organic waste composting. Environ Chem Lett 2015; 13(3): 223-38.
        [http://dx.doi.org/10.1007/s10311-015-0507-5]

[3]   Scheinberg A, Wilson DC, Rodic L. Solid waste management in the World's cities.UN-Habitat's third global report on the state of water and sanitation in the World's cities. London: Earthscan for UN Habitat 2010.

[4]   Wilson DC. Global Waste Management Outlook International Solid Waste Association (ISWA) and United Nations Environmental Programme (UNEP). Nairobi: United Nations Environment Programme 2015.

[5]   Sood G, Chitre N, Dabhadkar KC. Review of Biological Methods for Hazardous Waste Treatment'. IOSR J Biotechnol Biochem 2016; 2: 50-5.

[6]   Dwivedi V. Eco-Friendly Environment by Managing Waste Disposal. Int J Trend Sci Res Dev 2019; 3: 123-5.

[7]   Ahmad R, Jilani G, Arshad M, Zahir ZA, Khalid A. Bio-conversion of organic wastes for their recycling in agriculture: an overview of perspectives and prospects. Ann Microbiol 2007; 57(4): 471-9.
        [http://dx.doi.org/10.1007/BF03175343]

[8]   Friedrich E, Trois C. Quantification of greenhouse gas emissions from waste management processes for municipalities – A comparative review focusing on Africa. Waste Manag 2011; 31(7): 1585-96.
        [http://dx.doi.org/10.1016/j.wasman.2011.02.028] [PMID: 21450453]

[9]   Dille D, Archana K. Biotechnology for solid waste management – a critical review. Open Access Int J Sci Eng 2017; 2: 2456-3293.

[10]   Estrellan CR, Iino F. Toxic emissions from open burning. Chemosphere 2010; 80(3): 193-207.
         [http://dx.doi.org/10.1016/j.chemosphere.2010.03.057] [PMID: 20471058]

[11]   Smith JU, Fischer A, Hallett PD, *et al.* Sustainable use of organic resources for bioenergy, food and
         water provision in rural Sub-Saharan Africa. Renew Sustain Energy Rev 2015; 50: 903-17.
         [http://dx.doi.org/10.1016/j.rser.2015.04.071]

[12]   UNEP Guidelines on best available techniques and provisional guidance on best environmental
         practices— open burning of waste, including burning of landfill sites Relevant to Article 5 and Annex
         C of the Stockholm Convention on Persistent Organic Pollutants United Nations Environmental
         Programme, Geneva. 29. 2007; p.

[13]   Babu BV. Biomass pyrolysis: a state-of-the-art review. Biofuels Bioprod Biorefin 2008; 2(5): 393-
         414.
         [http://dx.doi.org/10.1002/bbb.92]

[14]   Arena U. Process and technological aspects of municipal solid waste gasification. A review. Waste
         Manag 2012; 32(4): 625-39.
         [http://dx.doi.org/10.1016/j.wasman.2011.09.025] [PMID: 22035903]

[15]   Onyelowe KC, Bui Van D, Ubachukwu O, *et al.* Recycling and reuse of solid wastes; a hub for
         ecofriendly, ecoefficient and sustainable soil, concrete, wastewater and pavement reengineering. Int J
         Low Carbon Technol 2019; 14(3): 440-51.
         [http://dx.doi.org/10.1093/ijlct/ctz028]

[16]   Troschinetz AM, Mihelcic JR. Sustainable recycling of municipal solid waste in developing countries.
         Waste Manag 2009; 29(2): 915-23.
         [http://dx.doi.org/10.1016/j.wasman.2008.04.016] [PMID: 18657963]

[17]   Basu P. Biomass gasification, pyrolysis and torrefaction. Boston: Academic Press 2013.

[18]   Diener S, Studt Solano NM, Roa Gutiérrez F, Zurbrügg C, Tockner K. Biological treatment of
         municipal organic waste using black soldier fly larvae. Waste Biomass Valoriz 2011; 2(4): 357-63.
         [http://dx.doi.org/10.1007/s12649-011-9079-1]

[19]   Insam H, de Bertoldi M. Microbiology of the composting process.Waste management series.
         Amsterdam: Elsevier 2007; pp. 25-48.

[20]   Epstein E. The science of composting. Boca Raton: CRC Press 1997.

[21]   Cooperband L. The art and science of composting – a resource for farmers and compost producers.
         Madison: Center for Integrated Agricultural Systems, University of Wisconsin 2002.

[22]   Polprasert C. Organic waste recycling—technology and management. London: IWA Publishing 2007.

[23]   Farrell M, Jones DL. Critical evaluation of municipal solid waste composting and potential compost
         markets. Bioresour Technol 2009; 100(19): 4301-10.
         [http://dx.doi.org/10.1016/j.biortech.2009.04.029] [PMID: 19443214]

[24]   Ndegwa PM, Thompson SA, Das KC. Effects of stocking density and feeding rate on
         vermicomposting of biosolids. Bioresour Technol 2000; 71(1): 5-12.
         [http://dx.doi.org/10.1016/S0960-8524(99)00055-3]

[25]   Ali U, Sajid N, Khalid A, *et al.* A review on vermicomposting of organic wastes. Environ Prog Sustain
         Energy 2015; 34(4): 1050-62.
         [http://dx.doi.org/10.1002/ep.12100]

[26]   Aalok A, Tripathi AK, Soni P. Vermicomposting: a better option for organic solid waste management.
         J Hum Ecol 2008; 24(1): 59-64.
         [http://dx.doi.org/10.1080/09709274.2008.11906100]

[27]   Garg VK, Suthar S, Yadav A. Management of food industry waste employing vermicomposting
         technology. Bioresour Technol 2012; 126: 437-43.

[http://dx.doi.org/10.1016/j.biortech.2011.11.116] [PMID: 22197330]

[28] Singh RP, Singh P, Araujo ASF, Hakimi Ibrahim M, Sulaiman O. Management of urban solid waste: Vermicomposting a sustainable option. Resour Conserv Recycling 2011; 55(7): 719-29.
[http://dx.doi.org/10.1016/j.resconrec.2011.02.005]

[29] Vögeli Y, Lohri CR, Gallardo A, Diener S, Zurbrügg C. Anaerobic digestion of biowaste in developing countries—practical information and case studies Swiss Federal Institute of Aquatic Science and Technology. Dübendorf: Eawag 2014.

[30] Zhang C, Su H, Baeyens J, Tan T. Reviewing the anaerobic digestion of food waste for biogas production. Renew Sustain Energy Rev 2014; 38: 383-92.
[http://dx.doi.org/10.1016/j.rser.2014.05.038]

[31] Esposito G, Frunzo L, Giordano A, Liotta F, Panico A, Pirozzi F. Anaerobic co-digestion of organic wastes. Rev Environ Sci Biotechnol 2012; 11(4): 325-41.
[http://dx.doi.org/10.1007/s11157-012-9277-8]

[32] Jimenez J, Latrille E, Harmand J, *et al.* Instrumentation and control of anaerobic digestion processes: a review and some research challenges. Rev Environ Sci Biotechnol 2015; 14(4): 615-48.
[http://dx.doi.org/10.1007/s11157-015-9382-6]

[33] Mao C, Feng Y, Wang X, Ren G. Review on research achievements of biogas from anaerobic digestion Renew Sustain Energy Rev 2015.

[34] da Silva OB, Carvalho LS, de Almeida GC, de Oliveira JD, Carmo TS, Parachin NS. Biogas - Turning Waste into Clean Energy. Fermentation Processes, Angela Faustino Jozala, IntechOpen 2017.
[http://dx.doi.org/10.5772/64262]

[35] Mussatto SI, Dragone G, Guimarães PMR, *et al.* Technological trends, global market, and challenges of bio-ethanol production. Biotechnol Adv 2010; 28(6): 817-30.
[http://dx.doi.org/10.1016/j.biotechadv.2010.07.001] [PMID: 20630488]

[36] Balat M, Balat H. Recent trends in global production and utilization of bio-ethanol fuel. Appl Energy 2009; 86(11): 2273-82.
[http://dx.doi.org/10.1016/j.apenergy.2009.03.015]

[37] El-Tayeb TS, Abdelhafez AA, Ali SH, Ramadan EM. Effect of acid hydrolysis and fungal biotreatment on agro-industrial wastes for obtainment of free sugars for bioethanol production. Braz J Microbiol 2012; 43(4): 1523-35.
[http://dx.doi.org/10.1590/S1517-83822012000400037] [PMID: 24031984]

[38] Bhutto AW, Harijan K, Qureshi K, Bazmi AA, Bahadori A. Perspectives for the production of ethanol from lignocellulosic feedstock – A case study. J Clean Prod 2015; 95: 184-93.
[http://dx.doi.org/10.1016/j.jclepro.2015.02.091]

[39] Chen H, Fu X. Industrial technologies for bioethanol production from lignocellulosic biomass. Renew Sustain Energy Rev 2016; 57: 468-78.
[http://dx.doi.org/10.1016/j.rser.2015.12.069]

[40] Maiti S, Sarma SJ, Brar SK, *et al.* Agro-industrial wastes as feedstock for sustainable bio-production of butanol by *Clostridium beijerinckii*. Food Bioprod Process 2016; 98: 217-26.
[http://dx.doi.org/10.1016/j.fbp.2016.01.002]

[41] Najafi G, Ghobadian B, Tavakoli T, Yusaf T. Potential of bioethanol production from agricultural wastes in Iran. Renew Sustain Energy Rev 2009; 13(6-7): 1418-27.
[http://dx.doi.org/10.1016/j.rser.2008.08.010]

[42] Sarkar N, Ghosh SK, Bannerjee S, Aikat K. Bioethanol production from agricultural wastes: An overview. Renew Energy 2012; 37(1): 19-27.
[http://dx.doi.org/10.1016/j.renene.2011.06.045]

[43] Gupta A, Verma JP. Sustainable bio-ethanol production from agro-residues: a review Renew Sustain

Energy Rev 41: 550-67.2015;

[44]   Mushimiyimana I, Tallapragada P. Bioethanol production from agro wastes by acid hydrolysis and fermentation process. J Sci Ind Res 75: 383-8.2016;

[45]   Duhan JS, Kumar A, Tanwar SK. Bioethanol production from starchy part of tuberous plant (potato) using Saccharomyces cerevisiae MTCC-170 Afr J Microbiol Res 7: 5253-60.2013;

[46]   Kumar A, Duhan JS, Gahlawat SK. Production of ethanol from tuberous plant (sweet potato) using Saccharomyces cerevisiae MTCC- 170 Afr J Biotechnol 13(28): 2874-83.2014;

[47]   Kumar A, Sadh PK. Bio-ethanol production from sweet potato using co-culture of saccharolytic molds (Aspergillus spp) and Saccharomyces cerevisiae MTCC170 J Adv Biotechnol 6(1): 822-7.2016;

[48]   Saini JK, Saini R, Tewari L. Lignocellulosic agriculture wastes as biomass feedstocks for second-generation bioethanol production: concepts and recent developments. 3 Biotech 2015; 5(4): 337-53. [http://dx.doi.org/10.1007/s13205-014-0246-5] [PMID: 28324547]

[49]   Vohra M, Manwar J, Manmode R, Padgilwar S, Patil S. Bioethanol production: Feedstock and current technologies. J Environ Chem Eng 2014; 2(1): 573-84. [http://dx.doi.org/10.1016/j.jece.2013.10.013]

[50]   Hahn-Hägerdal B, Galbe M, Gorwa-Grauslund MF, Lidén G, Zacchi G. Bio-ethanol – the fuel of tomorrow from the residues of today. Trends Biotechnol 2006; 24(12): 549-56. [http://dx.doi.org/10.1016/j.tibtech.2006.10.004] [PMID: 17050014]

[51]   Bhargav S, Panda BP, Ali M, Javed S. Solid-state fermentation: an overview. Chem Biochem Eng 2008; 22(1): 49-70.

[52]   Brown D, Li Y. Solid state anaerobic co-digestion of yard waste and food waste for biogas production. Bioresour Technol 2013; 127: 275-80. [http://dx.doi.org/10.1016/j.biortech.2012.09.081] [PMID: 23131652]

[53]   Sadh PK, Duhan S, Duhan JS. Agro-industrial wastes and their utilization using solid state fermentation: a review. Bioresour Bioprocess 2018; 5(1): 1. [http://dx.doi.org/10.1186/s40643-017-0187-z]

[54]   Pandey A. Solid-state fermentation. Biochem Eng J 2003; 13(2-3): 81-4. [http://dx.doi.org/10.1016/S1369-703X(02)00121-3]

[55]   Buenrostro J, Ascacio A, Sepulveda L, De la Cruz R, Prado-Barragan A. Aguilar- Gonzalez MA, Rodriguez R, Aguilar CN. Potential use of different agro-industrial by products as supports for fungal ellagitannase production under solid state fermentation. Food Bioprod Process 2013; 92: 376-82. [http://dx.doi.org/10.1016/j.fbp.2013.08.010]

[56]   Pandey A, Soccol CR, Mitchell D. New developments in solid state fermentation: I-bioprocesses and products. Process Biochem 2000; 35(10): 1153-69. [http://dx.doi.org/10.1016/S0032-9592(00)00152-7]

[57]   Singhania RR, Patel AK, Soccol CR, Pandey A. Recent advances in solid-state fermentation. Biochem Eng J 2009; 44(1): 13-8. [http://dx.doi.org/10.1016/j.bej.2008.10.019]

[58]   Dharmendra KP. Production of lipase utilizing linseed oilcake as fermentation substrate. Int J Sci Environ Technol 2012; 1: 135-43.

[59]   Kaur PS, Kaur S, Kaur H, Sharma A, Raj P, Panwar S. Solid substrate fermentation using agro industrial waste: new approach for amylase production by *Bacillus licheniformis*. Int J Curr Microbiol Appl Sci 2015; 4: 712-7.

[60]   Sodhi HK, Sharma K, Gupta JK, Soni SK. Production of a thermostable α-amylase from Bacillus sp. PS-7 by solid state fermentation and its synergistic use in the hydrolysis of malt starch for alcohol production. Process Biochem 2005; 40(2): 525-34. [http://dx.doi.org/10.1016/j.procbio.2003.10.008]

[61]   Gayen S, Ghosh U. Pectinmethylesterase Production from mixed agro- wastes by Penicillium notatum NCIM. 923 in Solid-State fermentation. J Bioremediat Biodegrad 2011; 2(2): 119.
        [http://dx.doi.org/10.4172/2155-6199.1000119]

[62]   Sharanappa A, Wani KS, Pallavi P. Bioprocessing of food industrial waste for α-amylase production by solid state fermentation. Int J Adv Biotechnol Res 2011; 2: 473-80.

[63]   Rekha KSS, Lakshmi C, Devi SV, Kumar MS. Production and optimization of lipase from Candida rugosa using groundnut oilcake under solid state fermentation. Int J Res Eng Technol 2012; 1(4): 571-7.
        [http://dx.doi.org/10.15623/ijret.2012.0104008]

[64]   Sindiri MK, Machavarapu M, Vangalapati M. Alfa-amylase production and purifcation using fermented orange peel in solid state fermentation by *Aspergillus niger.* Int J Appl Res 2013; 3: 49-51.

[65]   Mehta K, Duhan JS. Production of invertase from *Aspergillus niger* using fruit peel waste as a substrate. Int J Pharm Biol Sci 2015; 5(2): B353-60.

[66]   Kalogeris E, Christakopoulos P, Katapodis P, *et al.* Production and characterization of cellulolytic enzymes from the thermophilic fungus Thermoascus aurantiacus under solid state cultivation of agricultural wastes. Process Biochem 2003; 38(7): 1099-104.
        [http://dx.doi.org/10.1016/S0032-9592(02)00242-X]

[67]   Ellaiah P, Adinarayana K, Bhavani Y, Padmaja P, Srinivasulu B. Optimization of process parameters for glucoamylase production under solid state fermentation by a newly isolated Aspergillus species. Process Biochem 2002; 38(4): 615-20.
        [http://dx.doi.org/10.1016/S0032-9592(02)00188-7]

[68]   Negi S, Banerjee R. Optimization of extraction and purification of glucoamylase produced by Aspergillus awamori in solid-state fermentation. Biotechnol Bioprocess Eng; BBE 2009; 14(1): 60-6.
        [http://dx.doi.org/10.1007/s12257-008-0107-3]

[69]   Suganthi R, Benazir JF, Santhi R, *et al.* Amylase production by Aspergillus niger under solid state fermentation using agro-industrial wastes. Int J Eng Sci Technol 2011; 3: 1756-63.

[70]   Kumar A, Duhan JS, Tanwar SK. Screening of Aspergillus spp. for extra cellular α-amylase activity. In: Khanna AK, Chopra G, Matta VS, Bhutiani R, Eds. Impact of global climate change on earth ecosystem. 205-14.New Delhi: Biotech Books 2013; pp. a

[71]   Kumar A, Duhan JS, Tanwar SK. Screening of Aspergillus spp. for extra cellular α-amylase activity. In: Khanna AK, Chopra G, Matta VS, Bhutiani R, Eds. Impact of global climate change on earth ecosystem. 205-14.New Delhi: Biotech Books 2013; pp. b

[72]   Oliveira F, Souza CE, Peclat VROL, *et al.* Optimization of lipase production by *Aspergillus ibericus* from oil cakes and its application in esterification reactions. Food Bioprod Process 2017; 102: 268-77.
        [http://dx.doi.org/10.1016/j.fbp.2017.01.007]

[73]   Mondal AK, Sengupta S, Bhowal J, Bhattacharya DK. Utilization of fruit wastes in producing single cell protei. Int J Sci Environ Technol 2012; 1(5): 430-8.

[74]   Sukan A, Roy I, Keshavarz T. Agro-industrial waste materials as substrates for the production of poly (3-hydroxybutyric acid). J Biomater Nanobiotechnol 2014; 5(4): 229-40.
        [http://dx.doi.org/10.4236/jbnb.2014.54027]

[75]   Vidhyalakshmi R, Vallinachiyar C, Radhika R. Production of xanthan from agro-industrial waste. Int J Adv Sci Res 2012; 3: 56.

[76]   Anis M, Siddiqui T. Waste to Energy: A Green Paradigm in Solid Waste Management. Curr World Environ 2015; 10(3): 764-71.
        [http://dx.doi.org/10.12944/CWE.10.3.06]

[77]   Pham TPT, Kaushik R, Parshetti GK, Mahmood R, Balasubramanian R. Food waste-to-energy conversion technologies: Current status and future directions. Waste Manag 2015; 38: 399-408.

[http://dx.doi.org/10.1016/j.wasman.2014.12.004] [PMID: 25555663]

[78] Lohri CR, Rodić L, Zurbrügg C. Feasibility assessment tool for urban anaerobic digestion in developing countries. J Environ Manage 2013; 126: 122-31.
[http://dx.doi.org/10.1016/j.jenvman.2013.04.028] [PMID: 23722149]

[79] Lohri CR, Diener S, Zabaleta I, Mertenat A, Zurbrügg C. Treatment technologies for urban solid biowaste to create value products: a review with focus on low- and middle-income settings. Rev Environ Sci Biotechnol 2017; 16(1): 81-130.
[http://dx.doi.org/10.1007/s11157-017-9422-5]

[80] Ahmad AA, Zawawi NA, Kasim FH, Inayat A, Khasri A. Assessing the gasification performance of biomass: A review on biomass gasification process conditions, optimization and economic evaluation. Renew Sustain Energy Rev 2016; 53: 1333-47.
[http://dx.doi.org/10.1016/j.rser.2015.09.030]

CHAPTER 3

Role of Biotechnology in Afforestation and Land Rehabilitation

Bhupinder Dhir[1,*] and **Ruby Tiwari**[2]

[1] *School of Sciences, Indira Gandhi National Open University, New Delhi, India*

[2] *Department of Genetics, University of Delhi South Campus, New Delhi-110021, India*

Abstract: Increased requirements for food and commodities have generated immense pressure on land resources. Landforms and forest areas have been converted to agricultural lands and rehabilitation areas to support the needs of a growing population. Owing to these changes, an urgent need for afforestation and land restoration has been generated. Various methodologies have been tried to restore the degraded land and increase the forest cover. Clonal propagation aiming at rapid multiplication and large-scale production of plants *via* selected clones has been successfully implemented. This approach has proved useful in raising commercial plantations. The use of biotechnological approaches such as molecular markers and advanced breeding programmes proved useful in raising clones for achieving afforestation and land rehabilitation on a large scale. The present chapter provides a detailed account of biotechnological techniques and processes that have played a significant role in afforestation and land rehabilitation.

Keywords: Genetic markers, Micropropagation, Plantation forests, Restoration, Reclamation.

INTRODUCTION

Forests are natural resources that form an important part of the ecosystem because of innumerable goods and services provided by them. Besides providing the oxygen required for the survival of living beings, they offer various materialistic things such as food, timber, paper, medicines, and fuelwood. Ecological roles played by them mainly include providing habitat to fauna, cycling nutrients, protecting watersheds, and regulating climate [1]. According to an estimate, the economic benefits of forests amount to about US\$ 130 million per year.

* **Corresponding author Bhupinder Dhir:** School of Sciences, Indira Gandhi National Open University, New Delhi, India; E-mail: bhupdhir@gmail.com

Forests all over the world are facing a threat due to conversion to other landforms such as croplands, residential areas, industrial setups, and other developmental setups [2]. Studies have shown that about 30% of the Earth's forest cover has been converted for other uses, and about 20% (about two billion hectares) has been degraded. Studies have shown that about 2.3 million km^2 of the forest area has been lost since the year 2000, and around 170 million ha will be lost by 2030 [2, 3]. According to an estimate, about 25% of forests have been cleared in the last 100 years [4 - 6]. The change in the distribution and structure of forests has been caused mainly due to anthropogenic activities. Deforestation, overexploitation, pests, diseases, and pollution have been considered the major causes of the reduction in forest cover. Disturbances such as climate change, flooding, droughts, and fires are some of the other factors responsible for bringing change in forest cover. An increase in the concentration of atmospheric CO_2 and temperature has affected the growth and production of forests across the globe [7]. Climatic changes such as global warming have brought a change in species composition, productivity, and biodiversity of forests in recent years. The change in forest structure and productivity affect biodiversity significantly and results in the loss of vulnerable species.

Land is another important natural resource covering a large part of the earth, including bare soil, vegetation, habitation, and impervious surfaces. Soil erosion, salinization and over-irrigation have resulted in the degradation of land significantly. An increase in the cropland areas in the last 50 years decreased land productivity due to the excessive application of fertilizers. Overloading of nitrogen and phosphorus has been noted in such areas [8]. Extensive mining in many areas has also brought alteration in the characteristics of land and components of soil (soil horizons, structure, soil microbe populations, nutrient cycles) [9 - 12]. The changes in land use and degradation lead to a reduction in soil organic carbon stocks which reduces the productive potential. According to an estimate, the net primary production of the terrestrial area globally has been reduced by 23% [5, 13, 14].

Prevention of land and forest degradation, restoration as well as the resilience of these natural resources is required for sustaining a healthy ecosystem. Various practices have been followed all over the globe to achieve environmental sustainability by preserving land and forest cover [15]. Afforestation programmes have been carried out in various regions of the world [16]. Besides various conventional technologies, such as raising plantations, biotechnological approaches have also been tried to restore degraded land and maintain the forest cover. In recent years, the biotechnological approaches have been proven as a milestone in the recovery of forest areas and degraded lands.

STRATEGIES FOR AFFORESTATION AND LAND REHABILITATION

Remediation and restoration techniques effectively maintain the productivity and biotic function of degraded lands and forest areas. Restoration brings the damaged or degraded area back to its natural state. The reclamation and revegetation mainly include clearing of an area, replacement of species through selection, and re-establishment of vegetation. The process of reclamation helps in restoring the ecological integrity of disturbed areas. These methods help in recreating an ecosystem and establishing a functionally effective and self-sustaining system. Brancalion *et al.* [17] proposed a strategy for the restoration of forests and landscapes. According to this, importance should be given to

- Protection of existing forest area,
- Restoration of native vegetation of an area,
- Restoration of degraded and low-productive lands.

The main prerequisite before the management of land or forest area is an assessment of the condition of an area [18-21].

Various methods have been practiced for a long time to restore degraded forests and lands [22]. Techniques for forest restoration mainly include natural regeneration and afforestation *i.e.*, planting of native tree seedlings, monoculture plantations, and mixed-species plantations. Afforestation converts non-forest land to forests through plantations using seedlings or other plant sources [23]. The technique of afforestation has been adopted by many countries around the world. This is because afforestation helps in preventing drastic climate change, protecting natural forests/vegetation, recovering the loss of trees, and meeting the requirements/products provided by forests [24]. The rate of afforestation has seen a rapid increase in the past decades.

The plantations increased by 277.9 million ha in 2015, which accounts for 6.95% of the global forest area [25]. Studies have shown that plantation forests occupy around 135 million ha of land areas all over the globe. Most (about 75%) of the plantation forests have been established in temperate regions. About 90% of plantation forests have been established on degraded and deforested areas primarily to meet the requirements of timber, fiber, fuelwood for industrial use, and environmental protection.

Plantation forests were established using monocultures and species/interspecific hybrids. Planting of short or long-term species and monocultures proves useful in restoring native forest ecosystem species [26]. Mixed-species plantations increase biodiversity, enhance watershed protection and induce more carbon storage [27,

28]. Monocultures and mixed-species plantations have been successfully established in Southeast Asia [29]. Commercial tree plantations have been established in various countries of the globe [30]. China has contributed to afforestation which accounts for 28% of the total global area and is the most in the world [25, 31]. Monoculture plantations of fast-growing exotics such as *Acacia*, *Eucalyptus,* and *Gmelina* have been raised throughout Southeast Asia, mainly for pulp, fuelwood, and low-cost lumber.

Agroforestry is another way adapted for afforestation and land rehabilitation. Agroforestry is the simultaneous cultivation of woody plants (trees or shrubs) and herbaceous crops [32]. Agroforestry practices prove useful in enhancing microbial biomass and water-holding of the soil, thereby helping in land rehabilitation [33]. The understory consists of annual or perennial plants.

Many approaches have been followed for the rehabilitation of the land. These mainly include the removal of toxins and other harmful substances, improvement in soil conditions by using various additives and adding new vegetation/flora. The success of land reclamation depends on the occurrence and distribution of soil microflora, the functioning of mycorrhizal symbiosis, and various enzymatic activities in soil. Microbial activity plays an important role in decomposition and nutrient cycling. Soil microbe populations and their metabolic activity determine the stability of a restored ecosystem. A mycorrhiza is a mutualistic association between plants and fungi.

Besides these conventional approaches, biotechnology has emerged as a technique that has played a major role in the restoration of land and forests *via* methods such as mass propagation, genetic improvement, and biomass utilization.

ROLE OF BIOTECHNOLOGY IN AFFORESTATION

Modern biotechnology techniques have proven useful in forestry and afforestation. Biotechnology aims to increase forest cover by increasing the number of taxa. Tissue culture and somatic hybridization have proved useful in enhancing the productivity of forests [34]. Biotechnological techniques have been applied to improve the functioning of tree genetic diversity and develop new tree varieties well adapted to changing environmental conditions and with the capacity to tolerate abiotic stress [35 - 37]. Genetic engineering methods have helped in raising tree species with tolerance to various abiotic and biotic stresses. Biotechnology-based forest biodiversity conservation programmes have been followed in many countries of the world Table (**1**).

Table 1. Various Programmes launched all over the world to get genetic improvement in tree species.

Name of the programme	Year	Objectives of the Programme
ArborGen	1999	Enable research on the genetic modification for herbicide resistance, growth, fibre quality for paper pulp in *Eucalyptus, Pinus radiata, Populus, Pinus taeda,* and *Liquidambar* sp.
GenForSA	1999	Inducing disease resistance of *Pinus radiate* and improving wood quality and formation.
Monfori Nusantra Indonesia	1996	Mass production of tissue cultures of *Tectona grandis, Acacia,* and *Eucalyptus* for field trial establishment and commerc ialization.
Tree Genetic Engineering Res earch Cooperative (TGERC), Oregon State University	2001	GM poplar plantations
Poplar Molecular Genetics Coop erative (PMGC)	1995	Improving knowledge of genetic and molecular mechanisms responsible for productivity and quality trait variations in hybrid poplars.
Forest Biotechnology Industrial Research Consortium (FORBIRC) mission, North Carolina State University	2003	Integrate genome technology, metabolic engineering, tree breeding and wood and paper science to create superior wood for use as raw material.

Biotechnological techniques have been useful in achieving breeding and large-scale clonal propagation of tree varieties. Large-scale vegetative propagation (micropropagation) or somatic embryogenesis has acted as a rapid and effective means to develop superior tree varieties [38]. Somatic embryogenesis has been used to achieve about 65% of micropropagation followed by cell/tissue culture (17%), *in vitro* culture of micro cuttings (13%), cryopreservation/conservation (4%), and embryo (1%) [39]. Large-scale production of uniform plant material using micropropagation technique has been restricted to a limited number of taxa such as *Eucalyptus, Pinus taeda, P. radiata, P. pinaster* and *Populus* Table (**2**).

Table 2. Micropropagation carried out in forest tree species.

Tree species	Countries	References
Populus alba *P. deltoids* *P. tremula*	Germany, India, Spain, Lithuania	[65]
Eucalyptus camaldulensis E. globulus *E. grandis* *E. nitens* *E. tereticornis* *E. urophylla*	South Africa, India, Spain, Portugal, Vietnam, Australia, South America	[66, 67, 68]

(Table 2) cont.....

Tree species	Countries	References
Acacia mangium *A. melanoxylon* *A. mangium × A. auriculiformis*	Malaysia, South Africa	[68-70]
Tectona grandis	India, Vietnam, Brazil, Indonesia, Thai land, Costa Rica, Malaysia, Australia,	[71-75]
Larix	Canada, France	[76]
Pinus radiata, P. taeda, P. pinaster, Pseudotsuga menziesii	France, New Zealand, United States, Canada	[76, 77]
Sequoia sempervirens, Sequoiadendron giganteum	France	-
Anogeissus latifolia, A. pendula	India	[78]
Betula pendula	Norway, Finland	-
Paulownia fortunei	Australia	-
Platanus acerifolia	China	[79]
Acacia mangium, A. mangium × A. auriculiformis	Malaysia, Indonesia, Vietnam	[69]

Biotechnology approaches help in developing genetic pools of tree species in very little time. Genetic modifications have proven useful in the conservation of forest tree species and gene pools. Genomics and proteomic approaches boost conventional tree breeding and improve the efficiency of existing forestry programmes. Biotechnology techniques enable breeders to access traits of interest for the improvement of varieties and the development of new varieties [40]. Tree species have been genetically modified to develop varieties with traits such as herbicide, metal, salt, or insect resistance, or cold tolerance. The *c*haracterization of genes involved in flowering, lignin, cellulose biosynthesis, and the identification of specific promoters of transgenic expression has been made in forest trees. The trees with reduced lignin content have been successfully produced to be used in the pulp and paper industry. This also helps to get economic benefits and environmental sustainability [41]. Genes regulating lignin formation have been identified in *Arabidopsis thaliana* [42]. The genetic basis of improving wood quality has been identified in forest tree species such as *Eucalyptus, Pinus,* and *Populus.*

Genetic Modification in the Forest Tree Species

Marker-assisted selection (MAS) has proved useful in tree breeding and selection [43]. The selection process based on the use of molecular markers speeds up the evaluation of genotypes. The use of molecular markers in phenotypic selection led to the development of marker-assisted selection (MAS). Molecular markers are nucleotide sequences. These markers are closely related to the target gene. The molecular markers act as signs or flags and can be investigated through the polymorphism present between the nucleotide sequences of different individuals. Insertion, deletion, point mutations, duplication, and translocation are the basis of polymorphism which might affect the activity of genes.

A gene or DNA sequence controls a particular gene or trait. A DNA marker is generally evenly distributed throughout the genome, highly reproducible, and has the ability to detect a higher level of polymorphism. Various DNA molecular markers have been developed and successfully applied in genetics and breeding activities in various tree species [44]. Markers such as RAPDs (random amplified polymorphic DNAs), RFLPs (restriction fragment length polymorphisms), AFLPs (Amplified fragment length polymorphism), and microsatellites help in assessing the degree of relation to other species and identifying the zones of the variability of quantitative traits [45, 46]. The variation in genes noted during selection is linked to the trait of interest. Isozymes and DNA markers are also used for assessing genetic diversity and quantification of genetic variation, which help in gene conservation and breeding.

Markers help in developing breeding programmes for species with industrial uses and some non-industrial species. Marker studies have been focused on tree species such as *Pinus, Eucalyptus* spp. including hybrids and *Picea* [47 - 55]. In forest trees, inherited traits such as disease resistance have been studied using MAS. However, the economically important traits of trees such as wood properties, trunk straightness, growth rate show quantitative, complex inheritance patterns, and hence, MAS was less effective for them.

MAS and genomics studies have been performed in many regions of the world. The studies have been carried out in Europe (43%), North America (34%), Asia (11%), Oceania (8%), South America (2%), and Africa (2%). The work carried out in Africa (South Africa and Congo) and South America (Brazil) relates to *Eucalyptus*.

Meuwissen *et al.* [56] proposed an alternative approach based on an analysis of all QTL effects. The genomic selection is based on the analysis of quantitative trait loci (QTLs) using a large number of molecular markers distributed throughout the genome. Each QTL is responsible for phenotypic variability. Such traits are linked with multiple QTLs. For example, in *Pinus taeda*, QTLs explained the phenotypic variance in wood traits but less genetic variability present in the entire population of loblolly pine. MAS is effective for simple traits controlled by a few QTLs. High-density markers are used in GS. In GS, preliminary information about MAS phenotype-marker linkage (MAS), localizations of QTLs in the genome, and their relative effect on the phenotype is not required. Next-generation sequencing (NGS) and the use of small nucleotide protein (SNP) markers also proved useful in the analysis of genomic selection.

Marker-assisted strategies can be implemented at pre-breeding stages in parents and offspring [57]. These also approaches aid conservation and germplasm tracing [58, 59].

GS (Genome sequencing) allows simultaneous and early selection for traits among a large number of individuals. This is generally difficult in the case of conventional tree breeding [60]. GS can significantly shorten the breeding cycle because the marker-based assessment of genotypes can be done at a very early stage. The genomic selection eliminates the need for lengthy, expensive field testing of progeny required for phenotypic evaluation. GS can significantly reduce the breeding cycle, especially in the case of slow-growing boreal coniferous species with a breeding cycle of 20 to 30 years. Stimulation in flowering or early flowering shortens the breeding cycle. The genotyping of large and poorly studied genomes of trees such as conifers is difficult. In the case of trees, GS can provide a higher genetic gain per unit time, a faster succession of generations in breeding programs, and faster creation of genotypes better adapted to environmental changes (the global climate change, the spread of diseases), and can better meet the industrial demand for wood. The use of GS reduced the field testing stage, which took several months. GS nearly halved the breeding cycle of loblolly pine. GS reduced the breeding cycle of *Eucalyptus* from 10 to 5 years. The shortening of the breeding cycle by 50% in *Eucalyptus* species increased the efficiency gains by 50–100% while shortening by 75% increased the efficiency by 200–300%. *Eucalyptus* breeding achieved flowering in 1–4 years in comparison to 4–8 years when raised under natural conditions. Although GS can potentially improve the efficiency of tree breeding programs, the outcomes would strongly depend on species and traits of interest. GS is not used frequently in forest tree breeding. GS proves much more efficient in forest breeding due to the low degree of forest tree domestication, long traditional breeding cycles, and large genetic variability of any trait. Improvement in the genetic basis of breeding programs for wild populations of woody perennials has been achieved using genomics-assisted breeding [61]. These approaches have proved useful in avoiding genetic erosion and increasing long-term adaptability to climate change and tolerance against abiotic stresses (such as drought and heat). Analysis of genomic diversity helps in capturing rare variants of germplasm [62]. Phylogenomic and species diversity may help in identifying novel alleles that support selective breeding for traits such as wood quality [63]. The phenotype was mainly assessed based on growth and wood traits, pulp yield, and species-specific traits such as essential oils and their components in *Eucalyptus polybractea, and* rubber production in *Hevea brasiliensis*. More recently, studies for resistance to stresses, such as diseases, pests, and drought, have also been done using GS.

GS has attracted interest among forest breeders around the world. More than 80% of them were performed in trees from the genera *Eucalyptus, Picea,* and *Pinus*. More of the common forest species, such as Douglas-fir, Japanese cedar, rubber tree, *etc.*, have also been included in the list in the last 3–4 years. Genomic studies carried out so far have been restricted to about 40 genera, among which mainly include *Pinus, Populus, Eucalyptus,* and *Picea*. The studies have been related to traits such as improvement in wood properties (lignin composition), abiotic resistance to cold or drought stress, genetic diversity, growth, flowering, and biotic resistance. The study has been based on gene products, the construction of genetic maps, the identification of genes, and QTLs (Quantitative trait Loci). Genetic mapping has been done for about 25 forest species. These mainly include P*inus*, *Picea*, *Eucalyptus*, *Acacia*, *Quercus*, *Populus*, *Larix,* and *Tectona grandis*.

Genetically Modified (GM) tree species have been developed by introducing desired genes into the genome of a forest tree species. Transformation methods consist of inserting a mutated gene or a gene from another organism into the genome of the host plant either by microinjection/projection (direct transformation) or through a vector such as *Agrobacterium tumefaciens* (indirect transformation). Successful genome modification has been done to raise species for herbicide, disease, and insect resistance. Genetic modification studies have been done to target gene expression, *in vitro* regeneration of plant tissue, lignin properties, herbicide, and biotic resistance. Genetic modification has been done in 29 genera of trees. Most of the work on forest tree species has been done in North America, followed by Europe, Asia, Oceania, Africa, and South America.

The gene encoding the enzymes that target the herbicide glyphosate has been inserted in *Populus alba* × *P. grandidentata, P. trichocarpa* × *P. deltoides, Eucalyptus grandis, Larix decidua,* and *Pinus radiata*. Similarly, genes encoding an enzyme for the detoxification of the herbicide chlorosulfuron have been inserted in *Populus tremula* and *Pinus radiata*. A microbial gene encoding an enzyme for the detoxification of the herbicide has been applied to *Populus alba, P. alba* × *P. tremula, P. tremula* × *P. alba, P. trichocarpa* × *P. deltoides,* and *Eucalyptus camaldulensis* and *Pinus radiata* and *Picea abies* [64].

A gene coding for a protease inhibitor that modifies insect digestion, causing the death of the pest, has also been introduced in tree species. Genes encoding an endotoxin that binds to the receptors in the intestine of *Lepidoptera, Coleoptera,* and *Diptera*, lysing the organ and killing the insect, have been inserted in tree species. This was done in *Populus alba* × *P. grandidentata, P. tremula* × *P.*

tremuloides and *Picea glauca*. GM *Populus* trees and *Pinus radiata* expressing the *Bacillus thuringiensis* endotoxin '*Bt*' have been obtained. *A. tumefaciens* mediated transformation of potato *viz.P. alba × P. grandidentata* by the introduction of gene *pin*2, a protease inhibitor, has been done. The gene of a rice protease inhibitor has also been introduced into *P. tremula × P. tremuloides*.

Genetic engineering has also been done to introduce other useful traits, such as

- Lignin reduction in pulp species
- Cold tolerance (particularly in *Eucalyptus)*
- Insect resistance in poplars

Genetically transformed lines have been regenerated for *Populus, Eucalyptus, Pinus* (including *P. taeda),* and *Picea.*

ROLE OF BIOTECHNOLOGY IN THE RESTORATION OF DEGRADED LANDS

Various biotechnological methods have been followed to restore degraded lands. Some of the methods include micropropagation, improvement in soil fertility using mycorrhizal association, development of stress-tolerant plants, preservation and multiplication of native endangered wild plant species [43, 80, 81].

Micropropagation

Mass-scale production of desired genotypes can be obtained using axillary or adventitious buds. Species developed by this technique possess the capacity to grow and establish themselves on degraded lands. Mass multiplication of species can be achieved using shoot apical meristems or buds followed by rooting *in vitro* conditions followed by transfer, acclimatization, and adaptation of field conditions. The technique has been used to achieve large-scale plantations. Micropropagation of commercial species has been used successfully done using this technique in countries such as Europe, North America, and Asia. In Asia, micropropagation accounted for 38%, followed by Europe (33%), North America (16%), South America (7%), and Africa (3%).

USE OF MYCORRHIZAE

Revegetation of severely disturbed mine lands has been achieved using mycorrhizal fungi [82, 83]. These fungi have been inoculated with tree seedlings, shrubs, and grasses. Fungal inocula help in forming mycorrhizal associations in the roots of seedlings. Mycorrhizae are symbiotic non-pathogenic associations between plant roots and fungi. Mycorrhizal associations increase the strength of the root. The establishment of mycorrhizal cultures helps in improving the growth and survival of seedlings by enhancing water and nutrient uptake and protecting against pathogens. Mycorrhizal fungi play a crucial role in plant nutrient uptake, water relations, ecosystem establishment, plant diversity, and productivity of plants [84, 85]. Ectomycorrhizal fungi associations, in particular, improve the water status of host plants, reduce root pathogen infestation and mobilize essential plant nutrients directly from the soil. Ectomycorrhiza (EM) forms extensive mycelia connected by different hyphal strands called rhizomorphs, which help in the transport of water and nutrients over long distances. Hence, mycorrhizal association can prove useful in the restoration and improvement of revegetation of disturbed mined lands.

Eucalyptus is associated with numerous species of EM [86]. *Eucalyptus* species such as *E. camaldulensis* form arbuscular mycorrihzae (AM). This type of mycorrhizae support the growth of legume trees, enhancing the uptake of phosphorus (P), nitrogen (N) and other nutrients. The mycorrhizae-mediated hydraulic redistribution increases the soil stability. Mycorrhizal species, *Pisolithus tinctorius* has been used widely in afforestation. Inoculation of *E. camaldulensis* with mycorrhizal species such as *Pisolithus reticulata* and *A. peregrina* has shown improvement in plant growth *via* more height and diameter growth.

IMPROVEMENT OF SOIL INFERTILITY

Biofertilizers, including nitrogen-fixing bacteria such as *Rhizobium,* form a symbiotic association with legumes. Actinomycetes bacteria such as *Frankia* form an association with non-leguminous species, thereby enhancing the fertility and productive capacity of the soil. About 160 species of angiosperms form nitrogen-fixing root nodules with *Frankia*. Species of *Frankia* help in improving the fertility of the soil and, therefore, can prove useful in reforestation. The use of nitrogen-fixing bacteria also improves economic feasibility as the expense of costly nitrogen fertilizers can be avoided.

The addition of various supplements such as hay, sawdust, bark, mulch, wood chips, wood residues, sewage sludge, and animal manures improves the productivity of soil by stimulating microbial activity (bacteria and mycorrhiza), providing nutrients such as nitrogen, phosphorus, and organic carbon to the soil.

REVEGETATION

Revegetation helps in the reclamation, protection of damaged/degraded soils, and reduction in erosion. It also helps in developing mineral resources. The revegetation is carried out with plants selected based on their ability to survive and regenerate in the local environment and possess the ability to stabilize the soil structure. Revegetation facilitates the development of nitrogen-fixing bacteria and mycorrhizal association [87, 88]. This approach also helps in maintaining soil quality by mediating the processes of organic matter turnover and nutrient cycling. Planting of native species, grasses, trees, and rotation with legumes helps in improving soil nutrient-deficient conditions and accelerating ecological succession.

DEVELOPMENT OF PLANTS TOLERANT TO ABIOTIC STRESS

Tissue culture and genetic engineering techniques help in developing plants resistant to various abiotic stresses such as salinity, water scarcity, acidity, heavy metal contamination, *etc.* Plant varieties showing traits of resistance to specific abiotic stress can be developed and thus, the technique can prove useful in establishing vegetation and plantation in the areas of stress. Plants with traits of tolerance to salt stress can be used for plantations on degraded lands. *Brassica spp., Citrus aurantium, Nicotiana tabacum* are some of the plant species that grow on degraded lands because of their resistance against abiotic stress. Plant species with resistance to abiotic stresses such as drought, and acidity can be successfully used in the restoration of degraded lands. Acidic soils can be reclaimed and restored by planting metal-tolerant plants. These plants possess the ability to in nutrient-deficient soil.

"Triticale", a crop developed using biotechnology technique, has shown the potential of resistance against acidity, drought, dry, sandy, alkaline, and calcareous soils. It has also been found suitable to be grown on degraded and deteriorated lands deficient in minerals. It has been successfully grown in countries like Kenya, Ethiopia, Ecuador, Mexico, and Brazil.

REMOVAL OF CONTAMINANTS

Biotechnology has proved useful in developing methods to remove heavy metal contamination from the land and restore the soil. Microbes such as bacteria, algae, and fungi showing the capacity to sequester metals have been developed *via* genetic engineering. These microbes have shown high metal biosorption potential. Fungal species such as *Rhizopus, Aspergillums, Penicillium, Neurospora, Ganoderma lucidum* are good absorbers of heavy metals and thus can be used for the removal of heavy metals from contaminated sites. The metal scavenging activity of microorganisms occurs in different ways. Microorganisms bioaccumulate metals on their cell walls or transport them to intracellular and intercellular free space and cellular organelles or sequester them by synthesizing metal-binding proteins or peptides. Metal ions get adsorbed/bound to negatively charged cell surfaces of microorganisms. Organic acids such as citric acid, oxalic acid, gluconic acid, lactic acid, malic acid assist in the complexation of metal ions. In some bacteria, metals get precipitated as hydroxides, sulfates or get immobilized on calcium oxalate crystals. *Desulfovibrio* and *Desulfotomaculum* show extracellular precipitation of metals as sulfides during the transformation of SO_4 to H_2S. Some microbes transform metals through enzymatic activity. Example- *Bacillus subtilis* has shown a capacity to reduce gold from Au^{3+} to Au^0.

CONSERVATION OF BIODIVERSITY

Many wild species are threatened or on the verge of extinction due to the destruction of habitats. Biotechnological methods have proved very useful in the conservation of plant biodiversity. Genes from wild species can be conserved using biotechnological approaches. One of the strategies is the establishment of "gene banks" for '*ex-situ* conservation'. The plant populations are preserved in resource centers, botanical gardens, national parks, collection centers, *etc.* The tissue culture method has been used successfully to increase the number of endangered/threatened plant species.

Cryopreservation is another method that effectively maintains the juvenility and genetic gains offered by clonal forestry with industrial species. The technology is thus applicable mainly in developing breeding programmes and clonal forestry.

LAND REHABILITATION SUCCESS STORIES

Many land rehabilitation efforts have been made successfully. The success of land rehabilitation varies from region to region and also according to the level of contamination or degradation. The U.S. Department has completed numerous land rehabilitation projects. A project named 'Restoration of the West Branch of the Grand Calumet River in Indiana' was taken up by Environmental Protection Agency. The project was launched in 2010 and aimed to remove contaminated sediment along a stretch of the river and restore the river shoreline with native grasses, flowers, trees, and shrubs. Another land restoration project is the 'Kubuqi Ecological Restoration Project'. This project aimed to combat desertification in China's Kubuqi desert, located south of the Gobi Desert. This project aimed to stabilize the desert and lead to afforestation. Nearly 200,000 acres of the desert area were planted with pine forests (which account for about 31% of the land) by the year 2017.

CONCLUDING REMARKS

In the current scenario of rapid deforestation, researchers are looking for ways to generate and develop forest areas in a rapid way so as to make the environment sustainable for future generations. Biotechnological techniques can prove useful in achieving this goal *via*the technique of micropropagation and selection of useful traits. Biotechnological approaches help in generating tree species having resistance to various stresses *via* the insertion of genes controlling traits for specific characters. The use of mycorrhizal systems in the rhizosphere, landscape regeneration, and bioremediation of contaminated soils are other techniques used in the restoration of degraded land. Biotechnology and conventional selection breeding programmes mainly aim for the improvement of various traits such as wood quality, growth, and induce stress resistance in tree species. The knowledge about the molecular control of these traits is limited since some traits, such as wood production, growth rate, and adaptability are under polygenic control and depend on more than one gene. Marker-assisted selection is appropriate for species for advanced breeding, creation, and population maintenance. The application of marker-assisted selection can prove advantageous in clonal forestry programmes. The use of GP Genomics programmes (GP) for germplasm characterization to date has been restricted to a few tree species such as *Populus, Pinus,* and *Eucalyptus.* More efforts are required to get restoration of land at the landscape level to maintain the sustainability of the environment.

REFERENCES

[1]   Mori AS, Lertzman KP, Gustafsson L. Biodiversity and ecosystem services in forest ecosystems: a research agenda for applied forest ecology. J Appl Ecol 2017; 54(1): 12-27.
[http://dx.doi.org/10.1111/1365-2664.12669]

[2]   Hansen MC, Potapov PV, Moore R, *et al.* High-resolution global maps of 21st-century forest cover change. Science 2013; 342(6160): 850-3.
[http://dx.doi.org/10.1126/science.1244693] [PMID: 24233722]

[3]   Potapov P, Hansen MC, Laestadius L, *et al.* The last frontiers of wilderness: Tracking loss of intact forest landscapes from 2000 to 2013. Sci Adv 2017; 3(1): e1600821.
[http://dx.doi.org/10.1126/sciadv.1600821] [PMID: 28097216]

[4]   Hurtt GC, Frolking S, Fearon MG, *et al.* The underpinnings of land-use history: three centuries of global gridded land-use transitions, wood-harvest activity, and resulting secondary lands. Glob Change Biol 2006; 12(7): 1208-29.
[http://dx.doi.org/10.1111/j.1365-2486.2006.01150.x]

[5]   Hurtt GC, Chini LP, Frolking S, *et al.* Harmonization of land-use scenarios for the period 1500–2100: 600 years of global gridded annual land-use transitions, wood harvest, and resulting secondary lands. Clim Change 2011; 109(1-2): 117-61.
[http://dx.doi.org/10.1007/s10584-011-0153-2]

[6]   Waring RH, Running SW. Forest ecosystems: analysis at multiple scales. Elsevier 2010.

[7]   Schoene DHF, Bernier PY. Adapting forestry and forests to climate change: A challenge to change the paradigm. For Policy Econ 2012; 24: 12-9.
[http://dx.doi.org/10.1016/j.forpol.2011.04.007]

[8]   Tilman D, Lehman C. Human-caused environmental change: Impacts on plant diversity and evolution. Proc Natl Acad Sci USA 2001; 98(10): 5433-40.
[http://dx.doi.org/10.1073/pnas.091093198]

[9]   Sheoran AS, Sheoran V, Poonam P. Remediation techniques for contaminated soils. Environ Eng
         Manag J 2008; 7(4): 379-87.
         [http://dx.doi.org/10.30638/eemj.2008.054]

[10]   Alvarez-Berríos NL, Mitchell Aide T. Global demand for gold is another threat for tropical forests.
         Environ Res Lett 2015; 10(1): 014006.
         [http://dx.doi.org/10.1088/1748-9326/10/1/014006]

[11]   Murguía DI, Bringezu S, Schaldach R. Global direct pressures on biodiversity by large-scale metal
         mining: Spatial distribution and implications for conservation. J Environ Manage 2016; 180: 409-20.
         [http://dx.doi.org/10.1016/j.jenvman.2016.05.040] [PMID: 27262340]

[12]   Sonter LJ, Barrett DJ, Moran CJ, Soares-Filho BS. A Land System Science meta-analysis suggests we
         underestimate intensive land uses in land use change dynamics. J Land Use Sci 2015; 10(2): 191-204.
         [http://dx.doi.org/10.1080/1747423X.2013.871356]

[13]   Ellis EC, Klein Goldewijk K, Siebert S, Lightman D, Ramankutty N. Anthropogenic transformation of
         the biomes, 1700 to 2000. Glob Ecol Biogeogr 2010; 19(5): no.
         [http://dx.doi.org/10.1111/j.1466-8238.2010.00540.x]

[14]   Van der Esch S, Ten Brink B, Stehfes E, *et al.* 2017.

[15]   Wang Q, Wang X, Wang X, Ma H, Ren N. Bioconversion of kitchen garbage to lactic acid by two
         wild strains of Lactobacillus species. Journal of Environmental Science and Health. 2005 Oct
         1;40(10):1951-62 ftp://ftp.fao.org/docrep/fao/012/ak593e/ak593e00.pdf

[16]   Hua F, Wang X, Zheng X, *et al.* Opportunities for biodiversity gains under the world's largest
         reforestation programme Nature Comm 72016; Article number

[17]   Brancalion PHS, viani RAG, Calmon M, Carrascosa H, Rodrigues RR. How to organize a large-scale
         ecological restoration program? The framework developed by the Atlantic Forest Restoration Pact in
         Brazil. J Sustain For 2013; 32(7): 728-44.
         [http://dx.doi.org/10.1080/10549811.2013.817339]

[18]   Lindenmayer DB, Noss RF. Salvage logging, ecosystem processes, and biodiversity conservation.
         Conserv Biol 2006; 20(4): 949-58.
         [http://dx.doi.org/10.1111/j.1523-1739.2006.00497.x] [PMID: 16922212]

[19]   Brockerhoff EG, Jactel H, Parrotta JA, Quine CP, Sayer J. Plantation forests and biodiversity:
         oxymoron or opportunity? Biodivers Conserv 2008; 17(5): 925-51.
         [http://dx.doi.org/10.1007/s10531-008-9380-x]

[20]   Brockerhoff EG, Jactel H, Parrotta JA, Ferraz SFB. Role of eucalypt and other planted forests in
         biodiversity conservation and the provision of biodiversity-related ecosystem services. For Ecol
         Manage 2013; 301: 43-50.
         [http://dx.doi.org/10.1016/j.foreco.2012.09.018]

[21]   Hunter M Jr, Westgate M, Barton P, *et al.* Two roles for ecological surrogacy: Indicator surrogates and
         management surrogates. Ecol Indic 2016; 63: 121-5.
         [http://dx.doi.org/10.1016/j.ecolind.2015.11.049]

[22]   Lamb D, Erskine PD, Parrotta JA. Restoration of degraded tropical forest landscapes. Science 2005;
         310(5754): 1628-32.
         [http://dx.doi.org/10.1126/science.1111773] [PMID: 16339437]

[23]   The Physical Science Basis S Solomon et al (eds), Cambridge University Press 2007.

[24]   Bastin JF, Finegold Y, Garcia C, *et al.* The global tree restoration potential. Science 2019; 365(6448):
         76-9.
         [http://dx.doi.org/10.1126/science.aax0848] [PMID: 31273120]

[25]   Payn T, Carnus JM, Freer-Smith P, *et al.* Changes in planted forests and future global implications.
         For Ecol Manage 2015; 352: 57-67.

[http://dx.doi.org/10.1016/j.foreco.2015.06.021]

[26] Elliott S, Blakesley D, Hardwick K. Restoring tropical forests - a practical guide. Royal Botanic Gardens, Kew, UK, p. 2013.

[27] Human Energy Requirement FAO. Report of a joint FAO/WHO/UNU expert consultation. Food and Nutrition Technical Report 2001. FAO, Rome

[28] Kanowski J, Catterrall C, Proctor H, Reis T, Tucker N, Wardell-Johnson G. Biodiversity values of timber plantations and restoration plantings for rainforest fauna in tropical and subtropical Australia.Reforestation in the tropics and subtropics of Australia using rainforest tree species. Rainforest CRC 2005; pp. 183-205.

[29] Jia Y, Milne RI, Zhu J, *et al.* Evolutionary legacy of a forest plantation tree species (Pinus armandii) 132646: 2646-62.2020; Implications for widespread afforestation

[30] Lamb D, Stanturf J, Madsen P. What is forest landscape restoration? Forest landscape restoration. Dordrecht: Springer 2012; pp. 3-23.

[31] Hua F, Xu J, Wilcove DS. A New Opportunity to Recover Native Forests in China. Conserv Lett 2018; 11(2): e12396.
[http://dx.doi.org/10.1111/conl.12396]

[32] Mercer DE. Adoption of agroforestry innovations in the tropics: a review 61-62: 311-28.2004;

[33] Tully K, Ryals R. Nutrient cycling in agroecosystems: Balancing food and environmental objectives. Agroecol Sustain Food Syst 2017; 41(7): 761-98.
[http://dx.doi.org/10.1080/21683565.2017.1336149]

[34] Jones N. Somatic embryogenesis as a tool to capture genetic gain from tree breeding strategies: Risks and benefits. Southern African Forestry Journal 2002; 195(1): 93-101.
[http://dx.doi.org/10.1080/20702620.2002.10434610]

[35] Ramachandran S. Applications of Biotechnology in Forestry.Applications of Biotechnology in Forestry and Horticulture. Boston, MA: Springer 1989.
[http://dx.doi.org/10.1007/978-1-4684-1321-2_1]

[36] Fenning TM, Gershenzon J. Where will the wood come from? Plantation forests and the role of biotechnology. Trends Biotechnol 2002; 20(7): 291-6.
[http://dx.doi.org/10.1016/S0167-7799(02)01983-2] [PMID: 12062973]

[37] Sonnino A. Role of biotechnology in afforestation and land rehabilitation. Current and potential application of biotechnology in forestry: A critical review. Asian Biotechnol Dev Rev 2016; 18(3): 41-85.

[38] G P, C S, L N, C A. Somatic embryogenesis and plant regeneration in *Eucalyptus globulus* Labill. Plant Cell Rep 2002; 21(3): 208-13.
[http://dx.doi.org/10.1007/s00299-002-0505-5]

[39] Park YS. Implementation of conifer somatic embryogenesis in clonal forestry: technical requirements and deployment considerations. Ann For Sci 2002; 59(5-6): 651-6.
[http://dx.doi.org/10.1051/forest:2002051]

[40] Seppälä R. Forest biotechnology and the global forest sector.Forest biotechnology in Europe: The challenge, the promise, the future. Institute of Forest Biotechnology 2003; pp. 39-43.

[41] Christensen JH, Baucher M, O'Connell A, van Montagu M, Boerjan W. Control of lignin biosynthesis.Molecular biology of woody plants. Dordrecht, Netherlands: Kluwer 2000; Vol. 1: pp. 227-67.
[http://dx.doi.org/10.1007/978-94-017-2311-4_9]

[42] Goujon T, Sibout R, Eudes A, MacKay J, Jouanin L. Genes involved in the biosynthesis of lignin precursors in Arabidopsis thaliana. Plant Physiol Biochem 2003; 41(8): 677-87.
[http://dx.doi.org/10.1016/S0981-9428(03)00095-0]

[43] Yanchuk AD. The role and implications of biotechnological tools in forestry. Unasylva 2001; 204(52): 53-61.

[44] Nadeem MA. Nawaz MA, Shahid MQ, Doğan Y, Comertpay G, Yıldı M. DNA molecular markers in plant breeding: current status and recent advancements in genomic selection and genome editing. Rev Agric Environ Biotechnol 2017; 32: 261-85.

[45] Marques CM, Araújo JA, Ferreira JG, *et al.* AFLP genetic maps of *Eucalyptus globulus* and *E. tereticornis.* Theor Appl Genet 1998; 96(6-7): 727-37.
[http://dx.doi.org/10.1007/s001220050795]

[46] Gerber S, Mariette S, Streiff R, Bodénès C, Kremer A. Comparison of microsatellites and amplified fragment length polymorphism markers for parentage analysis. Mol Ecol 2000; 9(8): 1037-48.
[http://dx.doi.org/10.1046/j.1365-294x.2000.00961.x] [PMID: 10964223]

[47] McDougall GJ. A comparison of proteins from the developing xylem of compression and non-compression wood of branches of sitka spruce (*Picea sitchensis*) reveals a differentially expressed laccase. J Exp Bot 2000; 51(349): 1395-401.
[PMID: 10944153]

[48] Thamarus KA, Groom K, Murrell J, Byrne M, Moran GF. A genetic linkage map for *Eucalyptus globulus* with candidate loci for wood, fibre, and floral traits. Theor Appl Genet 2002; 104(2): 379-87.
[http://dx.doi.org/10.1007/s001220100717] [PMID: 12582710]

[49] Brown GR, Bassoni DL, Gill GP, *et al.* Identification of quantitative trait loci influencing wood property traits in loblolly pine (*Pinus taeda* L.). III. QTL Verification and candidate gene mapping. Genetics 2003; 164(4): 1537-46.
[http://dx.doi.org/10.1093/genetics/164.4.1537] [PMID: 12930758]

[50] Garnier-Géré P, Bedon F, Pot D, Austerlitz F, Léger P, Kremer A, Plomion C. DNA sequence polymorphism, linkage desequilibrium and haplotype structure in candidate genes of wood quality traits in maritime pine (Pinus pinaster). In Tree biotechnologies, Umea, Sweden, 7–12 June 2003, s6.19.

[51] Le Dantec L, Chagné D, Pot D, Bedon F, Géré-Garnier P, De Daruvar A, Plomion C. Data mining in pine ESTs: II. Development of SNP markers. In Tree biotechnologies, Umea, Sweden, 7–12 June 2003

[52] Plomion C, Pionneau C, Baillères H. Identification of tension-wood responsive proteins in the developing xylem of *Eucalyptus.* Holzforschung 2003; 57: 353-8.
[http://dx.doi.org/10.1515/HF.2003.053]

[53] Kirst M, Myburg AA, De León JPG, Kirst ME, Scott J, Sederoff R. Coordinated genetic regulation of growth and lignin revealed by quantitative trait locus analysis of cDNA microarray data in an interspecific backcross of *eucalyptus.* Plant Physiol 2004; 135(4): 2368-78.
[http://dx.doi.org/10.1104/pp.103.037960] [PMID: 15299141]

[54] Ranik M, Creux N, Myburg AA. Transcriptome analysis of xylogenesis in Eucalyptus using cDNA-AFLP and Li-Cor automated DNA analysers.

[55] McMillan LK, Cato SA, Wilcox PL, Echt CS. Candidate gene selection in Pinus radiata finding the needles in the haystack. In Plant & Animal Genomes XII Conference San Diego, CA, 10–14: 2004

[56] Meuwissen THE, Karlsen A, Lien S, Olsaker I, Goddard ME. Fine mapping of a quantitative trait locus for twinning rate using combined linkage and linkage disequilibrium mapping. Genetics 2002; 161(1): 373-9.
[http://dx.doi.org/10.1093/genetics/161.1.373] [PMID: 12019251]

[57] De Dato G, Teani A, Mattioni C, Marchi M, Monteverdi MC, Ducci F. Delineation of seed collection zones based on environmental and genetic characteristics for *Quercus suber* L. in Sardinia, Italy. IForest (Viterbo) 2018; 11(5): 651-9.
[http://dx.doi.org/10.3832/ifor2572-011]

[58]   Chiocchini F, Mattioni C, Pollegioni P, *et al.* Mapping the genetic diversity of *Castanea Sativa*: exploiting spatial analysis for biogeography and conservation studies. J Geogr Inf Syst 2016; 8(2): 248-59.
[http://dx.doi.org/10.4236/jgis.2016.82022]

[59]   Mattioni C, Martin MA, Chiocchini F, *et al.* Landscape genetics structure of European sweet chestnut (*Castanea sativa* Mill): indications for conservation priorities. Tree Genet Genomes 2017; 13(2): 39.
[http://dx.doi.org/10.1007/s11295-017-1123-2]

[60]   Lebedev VG, Lebedeva TN, Chernodubov AI, Shestibratov KA. Genomic Selection for Forest Tree Improvement: Methods, Achievements and Perspectives. Forests 2020; 11(11): 1190.
[http://dx.doi.org/10.3390/f11111190]

[61]   Migicovsky Z, Myles S. Exploiting wild relatives for genomics-assisted breeding of perennial crops. Front Plant Sci 2017; 8: 460.
[http://dx.doi.org/10.3389/fpls.2017.00460] [PMID: 28421095]

[62]   Piot A, Prunier J, Isabel N, *et al.* Genomic Diversity Evaluation of *Populus trichocarpa* Germplasm for Rare Variant Genetic Association Studies. Front Genet 2020; 10: 1384.
[http://dx.doi.org/10.3389/fgene.2019.01384] [PMID: 32047512]

[63]   Wang M, Zhang L, Zhang Z, *et al.* Phylogenomics of the genus *Populus* reveals extensive interspecific gene flow and balancing selection. New Phytol 2020; 225(3): 1370-82.
[http://dx.doi.org/10.1111/nph.16215] [PMID: 31550399]

[64]   Bishop-Hurley SL, Zabkiewicz RJ, Grace L, Gardner RC, Wagner A, Walter C. Conifer genetic engineering: transgenic *Pinus radiata* (D. Don) and *Picea abies* (Karst) plants are resistant to the herbicide Buster. Plant Cell Rep 2001; 20(3): 235-43.
[http://dx.doi.org/10.1007/s002990100317]

[65]   Bueno A, Gomez A, Manzarena JA. Propagation and DNA markers characterization of Populus tremula L. and Populus alba L.Micropropagation of woody trees and fruits, forestry sciences. Dordrecht, Netherlands: Kluwer 2003; pp. 37-74.
[http://dx.doi.org/10.1007/978-94-010-0125-0_2]

[66]   Watt MP, Blakeway FC, Termignoni R, Jain SM. Somatic embryogenesis in Eucalyptus grandis and E. dunnii.Somatic embryogenesis in woody plants. UK: Kluwer 1999; Vol. 5: pp. 63-78.
[http://dx.doi.org/10.1007/978-94-011-4774-3_4]

[67]   Watt MP, Blakeway FC. Mokotedi Meo, Jain SM. Micropropagation of Eucalyptus.Micropropagation of woody trees and fruits. Dordrecht, Netherlands: Kluwer 2003; pp. 217-44.
[http://dx.doi.org/10.1007/978-94-010-0125-0_8]

[68]   Monteuuis O, Alloysius D, Garcia C, Goh D, Bacilieri R. Field behavior of an *in vitro*-issued *Acacia mangium* mature selected clone compared to its seed-derived progeny. Aust For 2003; 66: 87-9.
[http://dx.doi.org/10.1080/00049158.2003.10674894]

[69]   Galiana A, Goh D, Chevallier MH, *et al.* Micropropagation of *Acacia mangium* × *A. auriculiformis* hybrids in Sabah. Bois For Trop 2003; 275: 77-82.

[70]   Quoirin M. Micropropagation of Acacia species.Micropropagation of woody trees and fruits. Dordrecht, Netherlands: Kluwer 2003; pp. 245-68.
[http://dx.doi.org/10.1007/978-94-010-0125-0_9]

[71]   Nicodemus A, Nagarajan B, Mandal AK, Suberbramanian K. Genetic improvement of teak in India.

[72]   Kjaer ED, Kaosa-Ard A, Suangtho V. Domestication of teak through tree improvement: Options, potential gains and critical factors In Site, technology and productivity of teak plantations, FORSPA Publication 161-89.2000;

[73]   Schmincke KH. Teak plantations in Costa Rica: precious woods experience. Unasylva 2000; 201: 29-35.

[74] Goh D, Monteuuis O. Production of tissue-cultured teak: the plant biotechnology laboratory experience.

[75] Goh DKS, Alloysius D, Gidiman J, Chan HH, Mallet B, Monteuuis O. 2005.

[76] Lelu MA, Thompson D. L'embryogenèse somatique des conifères: une filière pour produire du matériel forestier de reboisement C. R. Colloq. Sainte Catherine 2000; pp. 167-74.

[77] Rahman MS, Messina MG, Newton RJ. Performance of loblolly pine (Pinus taeda L.) seedlings and micropropagated plantlets on an east Texas site. For Ecol Manage 2003; 178(3): 245-55.
[http://dx.doi.org/10.1016/S0378-1127(02)00482-6]

[78] Saxena S, Dhawan V. Large-scale production of *Anogeissus pendula* and *A. Latifolia* by micropropagation. In Vitro Cell Dev Biol Plant 2001; 37(5): 586-91.
[http://dx.doi.org/10.1007/s11627-001-0103-1]

[79] Liu G, Bao M. Adventitious shoot regeneration from in vitro cultured leaves of London plane tree (Platanus acerifolia Willd.). Plant Cell Rep 2003; 21(7): 640-4.
[http://dx.doi.org/10.1007/s00299-002-0569-2] [PMID: 12789413]

[80] Pilate G, Pâques M, Leplé JC, Plomion C. Les Biotechnologies chez les arbres forestiers. Rev For Fr 2002; 54(2): 161-80.
[http://dx.doi.org/10.4267/2042/4910]

[81] Tripathi V, Edrisi SA, Chen B, *et al.* Biotechnological Advances for Restoring Degraded Land for Sustainable Development. Trends Biotechnol 2017; 35(9): 847-59.
[http://dx.doi.org/10.1016/j.tibtech.2017.05.001] [PMID: 28606405]

[82] Quoreshi Ali M. The Use of Mycorrhizal Biotechnology in Restoration of Disturbed Ecosystem.Mycorrhizae: Sustainable Agriculture and Forestry. Dordrecht: Springer 2008; pp. 303-20.
[http://dx.doi.org/10.1007/978-1-4020-8770-7_13]

[83] de Carvalho AMX, de Castro Tavares R, Cardoso IM, Kuyper TW. Mycorrhizal associations in agroforestry systems. Soil biology and agriculture in the tropics. Soc Biol 2010; 21: 185-208.
[http://dx.doi.org/10.1007/978-3-642-05076-3_9]

[84] Pagano MC, Cabello MN, Scotti MR. Phosphorus response of three native Brazilian trees to inoculation with four arbuscular mycorrhizal fungi. Agric Technol Thail 2007; 3: 231-140.

[85] Pagano MC, Scotti MR. Arbuscular and ectomycorrhizal colonization of two *Eucalyptus* species in semiarid Brazil. Mycoscience 2008; 49(6): 379-84. a
[http://dx.doi.org/10.1007/S10267-008-0435-3]

[86] Pagano MC, Cabello MN, Bellote AF, Sá NM, Scotti MR. Intercropping system of tropical leguminous species and Eucalyptus camaldulensis, inoculated with rhizobia and/or mycorrhizal fungi in semiarid Brazil. Agroforestry Systems 2008b; 74: 231–242.

[87] Pagano MC, Cabello MN, Bellote AF, Sá NM, Scotti MR. Intercropping system of tropical leguminous species and *Eucalyptus camaldulensis*, inoculated with rhizobia and/or mycorrhizal fungi in semiarid Brazil. Agrofor Syst 2008; 74(3): 231-42. b
[http://dx.doi.org/10.1007/s10457-008-9177-7]

[88] Pagano MC, Scotti MR, Cabello MN. Effect of the inoculation and distribution of mycorrhizae in *Plathymenia reticulata* Benth under monoculture and mixed plantation in Brazil. New For 2009; 38(2): 197-214.
[http://dx.doi.org/10.1007/s11056-009-9140-0]

CHAPTER 4

Remediation of Wastewater Using Biotechnological Techniques

Sonu Sharma¹, Monu Sharma¹, Joginder Singh² and Raman Kumar¹,*

¹ Department of Biotechnology, Maharishi Markandeshwar (Deemed to be University), Mullana, Ambala 133207, India

² Department of Chemistry, Maharishi Markandeshwar (Deemed to be University), Mullana, Ambala 133207, India

Abstract: Wastewater contamination is increasing day by day because of increase in industrial operations and anthropogenic activities. Wastewater is a by product of industrial and domestic operations which is directly disposed into the environment and contain large amount of toxic materials harmful for human, animals as well as environment. Wastewater coming from industries is highly contaminated hence its recovery is a major concern. Developing countries and less developed countries generate large amount of wastewater in comparison to developed countries. Biotechnology provides best solution to get rid of this problem. Different technique/methods such as use of activated sludge, trickling filters, biosorption, bio-accumulation, use of nanoparticles play a major role in treatment of water. Role of microorganisms *via* microbial fuel cells and membrane biofilm bioreactors have also been used for removing metals present in wastewater. This chapter aims to provide complete information about biotechnological approaches for wastewater treatment in a cost- effective manner along with complete removal of sludge and toxic compounds.

Keywords: Bio-accumulation, Biosorption. trickling filters, Microbial fuel cells, Membrane biofilm reactors, Wastewater treatment.

INTRODUCTION

Increase in pollution due to anthropogenic activities is one of the biggest environmental problems of the century. Pollutants such as heavy metals, chemicals, dyes, pesticides released from various sources such as municipalities, industry and agriculture have contributed majorly to deterioration of environment. Industrialization and human growth have contributed to development of various environmental toxins and subsequent metal pollution [1]. Human activities have

* **Corresponding author Raman Kumar:** Department of Biotechnology, Maharishi Markandeshwar (Deemed to be University), Mullana, Ambala 133207, India; E-mail: ramankumar4@gmail.com

lead to release of large amounts of hazardous pollutants and effluents in soil, water, agricultural lands and air [2]. Industrial operations though contribute to change in human society *via* social and economic means but at the same time increase environmental concerns. Tremendous growth in industrial sector seen over the years has contributed to release and accumulation of several toxins such as heavy metals and other chemicals in air, water and land. Managing industrial water pollution due to industry, urbanization, and population growth is one of biggest challenges of today. Rapid urban growth as well industrial activities accelerate water pollution as huge amounts of heavy metals get disposed in the environment. Sewage is one of the major sources of water pollution. At present, water pollution is a global environmental issue as high concentration of toxic waste is getting added in the water bodies and their levels exceed the limits specified by World Health Organization and Environmental Protection agencies. Metals when disposed into environment prove toxic and directly or indirectly impact the human body as they interfere in natural reactions of the body. Excess of heavy metal exposure induces oxidative stress. According to recent reports, about 10-20 million people die every year because of waterborne diseases. It is predicted that billions of people on earth will not have it access to safe drinking water after few decades. The water supply will be reduced to one third of the current supply because the number water resources will decrease. Therefore, it is necessary that an effective wastewater treatment technology is developed. Both public and government need to look for the ways so that wastewater is treated before it is disposed. Water being a necessity for life needs special attention with respect to its use and treatment in way so that its sustainability can be maintained.

Industrial wastewater treatment is a complex process and requires huge set up and input [3, 4]. Pollution caused by industrial operations cause serious health problems. Natural potential of plants and microorganisms can be exploited to remove/treat pollutants present in wastewater. Remediation based on biological means have proved useful in degrading, altering or absorbing high levels of contaminants such as radionuclides, metals, medical waste, organic compounds *viz.* polyaromatic hydrocarbons (PAHs), biphenyls (PCBs). Catabolic functions of microbes have shown potential to degrade/treat pollutants. In bioremediation, different pollutants/toxins present in water are removed by bacteria, fungi, algae [5, 6]. Microbe-based remediation techniques for treatment of environmental pollutants have emerged as the latest technological methods with sustainable approach [7 - 10]. Both living and non-living microalgal biomass has been used as a biosorbent to get removal of metals [11, 12]. Cyanobacteria species such as *Oscillatoria*, *Phormidium*, *Spirogyra* and *Anabaena* have shown high growth even in contaminated water which could be due to their resistance to stress conditions. The microalgae species show specific methods of combating heavy metal toxicity. These include iron reduction, genetic control, isolation, chelation and role of enzymatic and non-enzymatic antioxidants which reduce heavy metals by redox reactions. *Bacillus* sp. showed an excellent capacity in salvation of iron stress. In addition, strains of *Aspergillus* such as *Aspergillus flavus* (FS4) and *Aspergillus fumigatus* (FS6) showed ability to disperse pollutants [13, 14]. Metagenomics are the most advanced tools used in removal of metals from contaminated water involving the microbial community. The technology has shown advancement at a faster pace. Nano-based technology has proved effective in treatment of wastewater to get clean safe drinking water [15].

WASTEWATER GENERATION IN INDIAN INDUSTRIES

Industrialization contributes to social sustainability but at the same time raises major issue such as pollution. Social growth should be done provide environmental sustainability is also focussed [16]. Pulp and paper, diesel and the tanning are some of them fast-growing industries that release wastewater that can pose high environmental risks [13, 17]. Indian tannery, pulp and paper, as well as the distillery industry release about 25,000 liters, 50,000 m^3 and 5-10 million liters of wastewater per day [18 - 20]. Release of such quantity of waste from industry is an issue of major concern for living beings. Being one of the most important natural resources that are important for all aspects of life, water is getting polluted due to rapid increase in population, urbanization as well as industrialization. Leather industry uses chromium-based agents at a large scale to turn raw leather into usable leather. many of the physic-chemical properties such as color, pH, TDS, TSS, BOD, COD, solid (TS), electrical conductivity (EC), heavy metals and metalloids *viz.* Mg, Cu, Zn, Cd, Fe, Cr, As, Pb, and Ni found in used wastewater released from paper and other heating industries are above the permissible levels. Aquatic life gets adversely affected by polluted water. Large areas of agricultural and aquaculture systems are polluted by toxic metals are extracted from glue and paper, tanning leather, and polish industries. About 2.9 billion urban dwellers release approximately 0.64 kg of the municipality solid waste (MSW) daily. Studies suggest that quality has been improved as now each person produces about 1.2 kg municipal solid wastes daily. It is expected that by 2025, 4.3 billion people will be producing about 1.42 billion kg/ capita /day approximately *via* 2.2 billion tons of MSW generated per year.

Heavy Metals Pollution and its Toxicity

Metals such as Ni, Cr, Ur, Hg, Zn, Cd, Se, Au, Ag, and As act as toxicants that pollute the environment. They adversely affect quality of soil, agriculture production and pose a major threat to human health [18, 21]. As a result of human activities, number of heavy metals contaminated water sources increase annually [22]. Heavy metals are toxic and unsafe [23]. The levels of heavy metals in water that are accepted under Robust Environmental Responsibility and Compensation Act, USA include Pb (0.015 mg L^{-1}), Cr (0.01 mg L^{-1}), Cd (0.05 mg L^{-1}), Hg (0.002 mg L^{-1}), Ar (0.01 mg L^{-1}), and Ag (0.05 mg L^{-1}) [24]. exposure to high concentrations of heavy metals cause degenerative disorders such as atherosclerosis, Parkinson's, Alzheimer's disease, cancer *etc.* in humans [25, 26]. Exposure of human body to high levels of Pb lead to serious health effects such as loss of balance and damage to the nervous system, while high levels of Cd affects body's immune system and functioning of vital organs such as heart, liver and kidneys [27]. Toxic metals lead to serious respiratory problems such as decreased lung function or lung cancer in adults and children [28]. Mercury, a disruptive neurotransmitter damages central nervous system by affecting hearing, speech and weakening of muscles. Mercury accumulates in aquatic biota found in the water bodies and gets converted to methylmercury which is very harmful. Transfer of toxic methylmercury to humans occurs *via* consumption of contaminated fish and other marine animals. Mineral mines, metal formation, paints, batteries, nuclear and nuclear power plants are some of the major sources of heavy metal pollution [29]. Accumulation of iron in lungs of humans residing in contaminated site has been reported [30]. Toxic effects of industrial wastewater contaminated with heavy metals have been reported in various aquatic organisms such as *Chlorella vulgaris, Selenastrum capricornutum* (green algae), *Daphnia magna* (crustacean), fish, few earthworm species such as *Oryzia slatipes*, *Eisenia fetida* and many others [31].

WASTEWATER TREATMENT

In present scenario, wastewater treatment is relatively advanced. Its challenge and sustainability remains a challenge. Activated sludge (AS), oxidation pools, droplets filters, biofilters, and anaerobic treatments are some of the well known wastewater treatment methods based on biotechnology. In addition, biotrickling filters, solid composting techniques, and biosorption are some other methods followed for treatment of wastewater. It is important in every way to find the right microorganisms that will damage the organisms and carry out therapeutic measures in the right conditions [32].

Activated Sludge

Activated sludge is a low cost wastewater treatment technique. Activated sludge (AS) floc mainly contains hydrophilic polar material. Zoologea found in AS floc show high levels of contaminants such as heavy metals. The capacity for removing pollutants is based on biological factors and attraction of pollutants towards living organisms Fig. (**1**). Both physical and chemical processes such as gravity, ion exchange, bonding, non-polar precipitation, oxide depletion, enzymatic pathway and electrical adsorption involved in removal of contaminants are based on the principle of adsorption. The presence of extracellular polymers in sludge cells increases the rate of microbial adsorption [33].

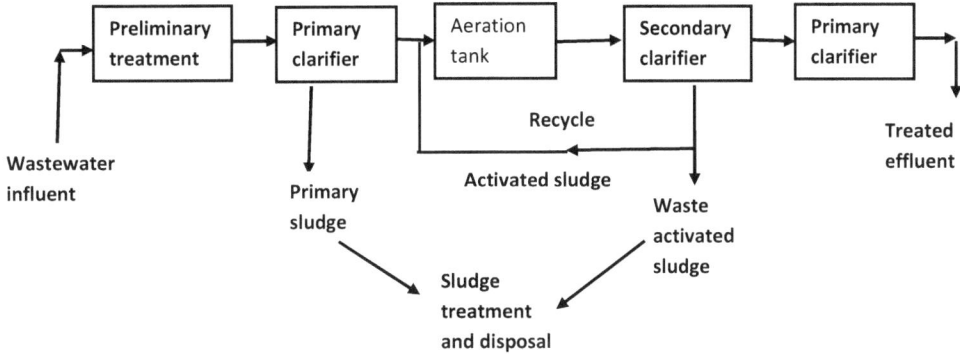

Fig. (1). Schematic representation of Activated Sludge Wastewater Treatment.

Anaerobic Treatment

In anaerobic treatment, contaminated water is subjected to treatment of N, P, and contagious propagules followed by removal of live carbon. The main advantages of anaerobic bioprocesses in comparison to physicochemical techniques are its cost effectiveness and eco-friendly nature. Anaerobic digestion is used mainly in the sewage treatment. A significant reduction in hardness, BOD, COD, nitrates, phosphates from wastewater has been noted after anaerobic treatment. The process also helps in production of biogas and bio-fertilizer. The technique of anaerobic sludge blanket (UASB) solutions is used to treat sulfur-contaminated water Fig. (2).

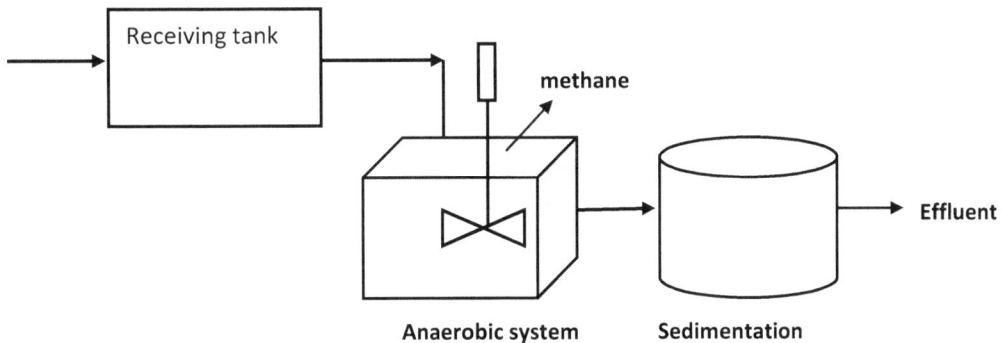

Fig. (2). Schematic representation of Anaerobic Wastewater Treatment.

This process produces a small amount of biological waste, shows high treatment capacity, no high cost investments, no oxygen requirements, methane production and low nutritional requirements for cultures. The process of anaerobic digestion does not work effectively at low temperature and needs to be linked to low-cost heat recovery process. Anaerobic digestion of swine fluid with sludge mud showed high efficacy of COD removal (about 85%) after a time period of ten days [34].

Membrane Bioreactor

This type of wastewater treatment system uses a combination of biological process along with membrane separation. Membrane bioreactor (MBR) has proven to be an effective strategy for resolving issues related to AS such as poor sludge quality, biological instability and low levels of mixed alcohol. Membrane filtration helps in retaining biomass and other components required for production of specific environment Fig. (3) [35]. Value of membrane technology has increased due to advantages such as ease of measurement, low tread, and easy to install with other processes.

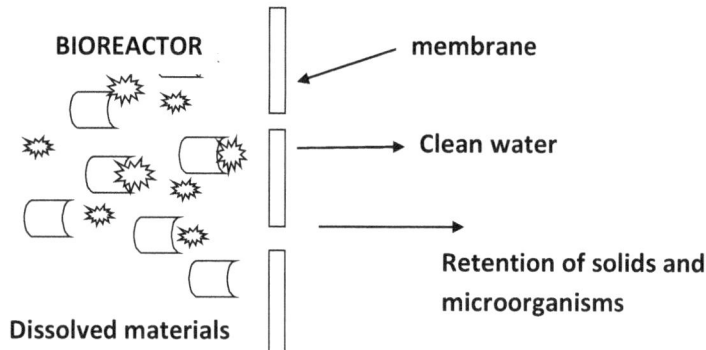

Fig. (3). Schematic representation of Membrane bioreactor.

MBR effectively removes COD and minor contaminants present in wastewater. This technique can provide an effective treatment of wastewater from various industrial and domestic sources. Although some new water treatment methods are cost effective and easy operating as compared to MBR technology. MBR system can prove to be a good technique for wastewater management in the near future provided the design of the bioreactor and efficiency of waste removal is improved [36, 37]. Studies proved that use of MBR in the treatment of surface contaminated water brought active mud to the MBR system. To avoid this deceptive membrane problem, carrier bacteria can aid in storing biomass in a reactor so that high rate of biodegradation is achieved [38]. Small batch membrane bioreactor (SMBR) sequences effectively remove antibiotics and organic materials from contaminated water, thus reducing the environmental risk [39].

Media filters (droplets filters)

Technique involving the use of biological trickling filter (BTF) has been used successfully in the treatment of domestic wastewater. These water treatment systems usually have a fixed bed of stone or plastic media where the contaminated water stays and a layer of microbial biofilm that helps in the treatment Fig. (**4**). The removal of impurities from a wastewater occurs by the process of absorption. Inorganic ions such as nitrite and nitrate ions can be removed by microbial biofilm layer under aerobic conditions. As thickness of biofilm increases, it falls into sewage and becomes part of the second sewage. The separation and removal of secondary mud of BTF contaminants in sediment tank is very specific [40].

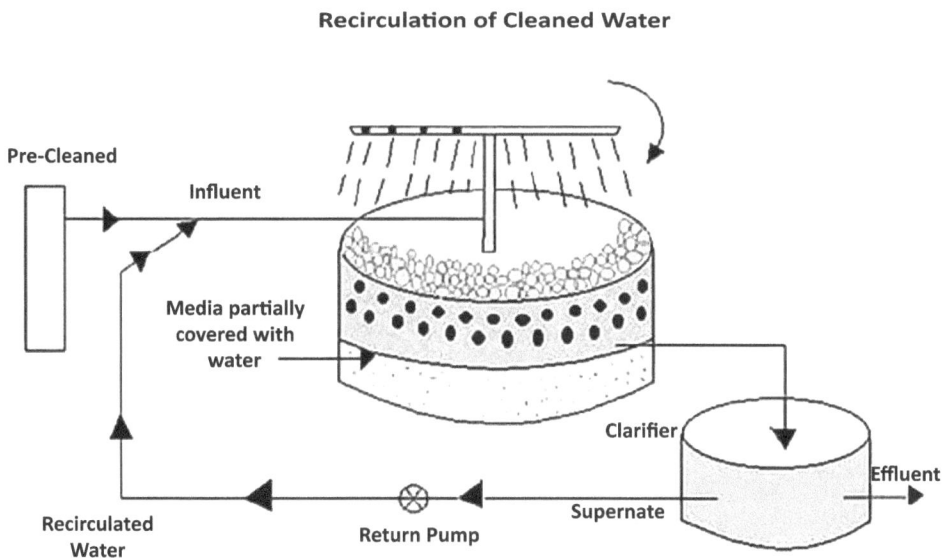

Fig. (4). Schematic representation of a Biological trickling filter (BTF) (Source: web.deu.edu.tr).

Biotrickling filters have proven to be relatively inexpensive and effective in removing dynamic molecules. Microorganisms are a major component of BTFs. The kind of bacteria present in these filters has a profound effect on their function. A two-story filter is generally used due to low-density desulfurization method for biogas purification and N-elimination [41].

Rotating Biological Contractor

Rotating biological contractor (RBCs) is a non-invasive aerobic film. It is used as a seconday treatment for wastewater. The set up for RBC contains glass container of different sizes and circular disc made up of polymer materials such as polyethylene, polystyrene, acrylic plastic, and polyvinyl chloride Fig. (**5**). In RBC, reduction of organic matter has been noted.

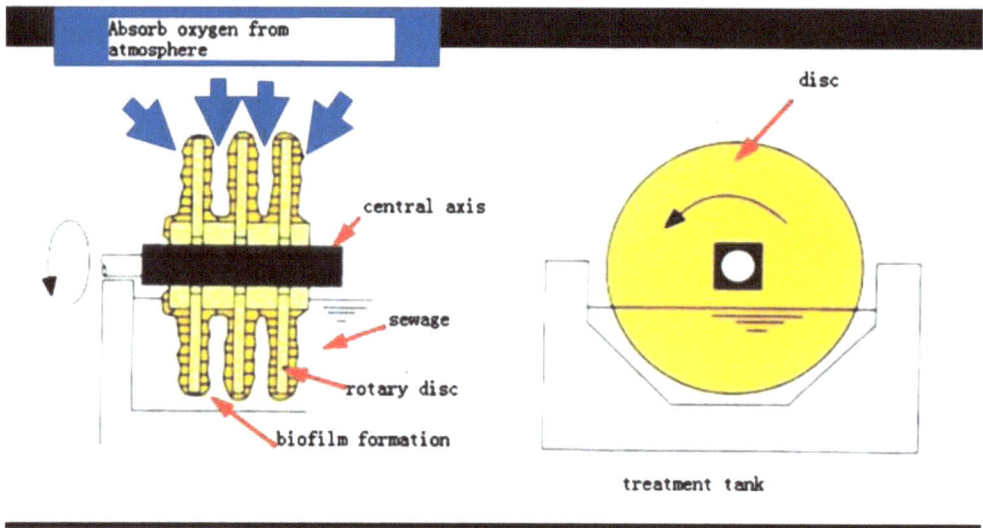

Fig. (5). Diagrammatic representation of Rotating biological contractor. (Source: m.yhenviro.com).

These disks are immersed in contaminated water and rotated in an electric motor with a variable speed. RBC exists in more than one category. The performance of RBCs is affected by various parameters such as wastewater heat, biological loading, biomass, hydraulic loading, RBC media, rotation speed, dissolved oxygen levels. varioyus advantages of RBC method include cost effectiveness, simple working, durability and eco-friendly performance. A high-density communication surface made of a rotating disk builds good communication between contaminants and germs. RBC system has been fouund effective in removing phenol at a higher rate (more than 56%) [42].

Bioleaching

Bioleaching or microbial leaching techniques for wastwewater removal are continously increasing mainly applies for low grade ores recovery and amount that cannot be processed by conventional methods. The bioleaching principle based on detachment of metals by oxidising agent which are produced by microorganisms [43]. Bioleaching is effective even at environmental temperature *i.e* below 30°C [44]. Researchers mainly focus on modeling stirrer tank reactor and bio-heap leaching in order to improve leaching efficiencies [45]. Bioleaching can also applies for electronic waste treatment or other type of wastegenerally consider for recycling or recovery [46 - 48].

Bioremediation

The process of removing toxins from the environment by the use of biological agents such as fungi, bacteria, algae and higher plants is called bioremediation. The technique is considered as a clean, cost-effective, and environment friendly technology that is very effective in removal of contaminants from wastewater [49]. The removal of contaminants such as metals can be achieved either on same site (*in situ*) or at different sites (*ex-situ*) *via* bioremediation [50]. Food, nutrients and microorganisms form key elements of the bioremediation process which are known as bioremediation triangles [51 - 53]. Microorganisms get their food from the contaminated water. They use chemical pollutants as a source of energy during their metabolic processes [48 - 50]. The growth of organisms gets stimulated by nutrients and chemicals present in the wastewater. The carbon source required for bacterial growth is provided by water. Moreover, microbes gain energy through oxidation while breaking the chemical bonds of pollutants.

The applicability of bioremediation processes has been improved by biotechnology. Genetic engineering technique also proves useful in burden of toxic/pollutants from the environment but is costly and efficiency technique. The potential of the organisms that play a vital role in bioremediation can be increased *via* technology of genetic engineering. Genetically modified microbes have proven to be effective in successfully remediating highly polluted waters. Microbial species such as *Bacillus thuringiensis* and *Deinococcusradio durans* have been used successfully to remove oil spills and ionic mercury from radiation waste [54]. Techniques such as phytoremediation, bioleaching, bio-stimulation have been used in removing contaminants from wastewater. Microbial species including *Nitrosomonas, Pseudomonas, Penicillium, Bacillus, Xanthobacter, Flavobacterium* and *Mycobacterium* form an important part of consortium and have shown to be very efficient in biodegradation of pollutants [54].

Biotransformation

The process of converting toxic waste/compounds from waster into elements of less toxicity with the aid of microorganisms such as bacteria, fungi is termed as biotransformation [55]. Microbe-mediated technology is getting importance all over the world because microbes covert toxic compounds to non-toxic metabolites by enzymatic reactions. Natural biotransformation is a natural process in which microbes present in water break and/or eliminate pollutants present in the water and hence play a role in restoring water to its original state [55]. The process of biotransformation is important for degradation (disposal or reduction) of chemicals. By the process of natural biotransformation, microorganisms turn polluted water into treated water by the action of microbes under natural conditions. The natural parameters help in achieving high rate of microbial growth and a fast rate of degeneration of specific substrate [56].

Bioaccumulation

The process of accumulating or increasing the chemical compounds in the body is defined as bioaccumulation. According to the International Union of Pure and Applied Chemistry, bioaccumulation of any chemical/compound in the body of any living organism depends upon the potential of natural matrix (bioconcentration) and acquirement by food (biomagnification). Chemicals generally get bioaccumulated in tissues. Bioaccumulation is influenced by several factors (physical, chemical, biological) and availability of chemicals.

Nanoparticles

Nanoparticles remove various types of pollutants found in contaminated water water. Nano particles are considered as modern and advance approach for removal of metal pollution from environment. Nanoparticles such as zero-valent iron or magnesium show high performance [57]. NZVI remove toxic substances from water [37, 58]. Iron nanoparticles (FeS-NPs) (~ 200 nm) degrade lindane, an organic compound and many other contaminants present in drinking water [59, 60]. The rate of degradation of lindane by NPs occurs at the rate of 5 mg/L per hour. The microbial treatment leads to complete reduction of lindane and its residues. Fabric dye can be effectively removed by nano zero-valent iron (nZVI). The removal of textile dyes present in water by ZVI-NPs depends on pH of water, volume of nanoparticles, concentration of pollutants and volume of dye. A recent study showed that 90% of dye can be removed at pH 9 within an hour if the dye concentration was 200 mL [61]. A similar study [62] showed that 99.75% of the dye could be removed in 6 hours if dye concentration was 25 mg/L at pH 8.3. Increase in concentration of nZVI improved the reaction kinetics between fabric dye and NPs. Natural dyes from water can be effectively treated with nano catalysts made of silver fibers and amidoxime [63]. Palladium-based (Pd) NPs proved useful in removing halogen contamination. Studies showed that trichloroethylene can be effectively treated using Pd-NPs and bimetallic alloys similar to Pd/Au *via* hydro dechlorination. Both free and immovable NPs such as metallo porphyrinogens show potential to degrade the chlorinated organic compounds. Degradation of persistent contaminants with the help of NPs can be considered a good option for the removal of these compounds. Iron nZVI showed capacity of degrading pesticides and herbicides such as chlorpyrifos, molinate, and atrazine. Studies proved that nZVI also assist in degradation of DDT. A significant decrease in concentration of DDT from water (92%) and groundwater (22.4%) has been reported.

Biosorption

Biosorption is defined as the uptake of metals as well as non-metals with the help of living organisms. This technique is not a standard wastewater treatment technique [64]. Huge amount of microbial biomass has shown good biosorption potential, hence formed a basis of treatment of metal containing wastewater [65, 66]. This technique proved effective in comparison to ion exchange or precipitation biosorption and do not produce sludge [67].

CONCLUDING REMARKS

Industrial wastewater needs immediate action and treatment to prevent environment contamination. Implementation of bioremediation and metagenomics is difficult in terms of their efficiency and economic aspects. The removal of metals from contaminated water through microorganisms works very well. Biodegradation using microbes has been successful in improving, cleaning, controlling, and revitalizing polluted areas. Microbes possess great adsorption ability and remediation techniques remove heavy metals in an effective manner. More research is needed to get large scale wastewater treatment. Incorporation of biological processes into advanced nanotechnology can provide a practical and effective treatment for wastewater treatment and purification. Nanoparticles like nanofibers, CNTs, TiO_2, and ZVIs provide effective wastewater treatment. Superior results have also been reported using integration of NPs with biological processes such as algal MBR, MFC and active mud process. Impact on the environment and health of people need to be considered when opting for nanotechnology-based water treatment. Safety and cost of operating of these technologies need to be addressed through research studies. Research studies should focus on addressing parameters such as safety and cost of operating these technologies with consideration such as impacts on the environment and health of people. Water being a precious natural resource need to be conserved and biotechnology can provide a good option for its treatment in a way so that it can be used in a sustainable way.

REFERENCES

[1]   Mustafa S, Bhatti HN, Maqbool M, Iqbal M. Microalgae biosorption, bioaccumulation and biodegradation efficiency for the remediation of wastewater and carbon dioxide mitigation: Prospects, challenges and opportunities. J Water Process Eng 2021; 41: 102009.

[http://dx.doi.org/10.1016/j.jwpe.2021.102009]

[2]   Raghunandan K, Kumar A, Kumar S, Permaul K, Singh S. Production of gellan gum, an exopolysaccharide, from biodiesel-derived waste glycerol by Sphingomonas spp. 3 Biotech 2018; 8(1): 1-13.

[3]   Mgbemene CA, Nnaji CC, Nwozor C. Industrialization and its backlash: focus on climate change and its consequences. J Environ Sci Technol 2016; 9(4): 301-16.
[http://dx.doi.org/10.3923/jest.2016.301.316]

[4]   Mandeep PS, Shukla P. Microbial nanotechnology for bioremediation of industrial wastewater. Front Microbiol 2020; 11: 590631.
[http://dx.doi.org/10.3389/fmicb.2020.590631] [PMID: 33224126]

[5]   Hauptmann AL, Sicheritz-Pontén T, Cameron KA, *et al.* Contamination of the Arctic reflected in microbial metagenomes from the Greenland ice sheet. Environ Res Lett 2017; 12(7): 074019.
[http://dx.doi.org/10.1088/1748-9326/aa7445]

[6]   Patra DK, Acharya S, Pradhan C, Patra HK. Poaceae plants as potential phytoremediators of heavy metals and eco-restoration in contaminated mining sites. Environ Technol Innovation 2020; p. 101293.

[7]   Renu K, Chakraborty R, Myakala H, *et al.* Molecular mechanism of heavy metals (Lead, Chromium, Arsenic, Mercury, Nickel and Cadmium) - induced hepatotoxicity – A review. Chemosphere 2021; 271: 129735.
[http://dx.doi.org/10.1016/j.chemosphere.2021.129735] [PMID: 33736223]

[8]   Ayangbenro A, Babalola O. (). A new strategy for heavy metal polluted environments: a review of microbial biosorbents. Int. Int J Environ Res Public Health 2017; 14(1): 94.
[http://dx.doi.org/10.3390/ijerph14010094] [PMID: 28106848]

[9]   Kumar A, Singh VK, Singh P, Mishra VK. Microbe Mediated Remediation of Environmental Contaminants. Woodhead Publishing 2020.

[10]   Leong YK, Chang JS. Bioremediation of heavy metals using microalgae: Recent advances and mechanisms. Bioresour Technol 2020; 303: 122886.
[http://dx.doi.org/10.1016/j.biortech.2020.122886] [PMID: 32046940]

[11]   Abinandan S, Subashchandrabose SR, Venkateswarlu K, Perera IA, Megharaj M. Acid-tolerant microalgae can withstand higher concentrations of invasive cadmium and produce sustainable biomass and biodiesel at pH 3.5. Bioresour Technol 2019; 281: 469-73.
[http://dx.doi.org/10.1016/j.biortech.2019.03.001] [PMID: 30850256]

[12]   Upadhyay AK, Mandotra SK, Kumar N, Singh NK, Singh L, Rai UN. Augmentation of arsenic enhances lipid yield and defense responses in alga Nannochloropsis sp. Bioresour Technol 2016; 221: 430-7.
[http://dx.doi.org/10.1016/j.biortech.2016.09.061] [PMID: 27665531]

[13]   Sharma P, Tripathi S, Chandra R. Highly efficient phytoremediation potential of metal and metalloids from the pulp paper industry waste employing Eclipta alba (L) and Alternanthera philoxeroide (L): Biosorption and pollution reduction. Bioresour Technol 2021; 319: 124147. a
[http://dx.doi.org/10.1016/j.biortech.2020.124147] [PMID: 32992272]

[14]   Talukdar D, Jasrotia T, Sharma R, *et al.* Evaluation of novel indigenous fungal consortium for enhanced bioremediation of heavy metals from contaminated sites. Environmental Technology & Innovation 2020; 20: 101050.
[http://dx.doi.org/10.1016/j.eti.2020.101050]

[15]   Sahu JN, Karri RR, Zabed HM, Shams S, Qi X. Current perspectives and future prospects of nano-biotechnology in wastewater treatment. Separ Purif Rev 2021; 50(2): 139-58.
[http://dx.doi.org/10.1080/15422119.2019.1630430]

[16]   Kassaye G, Gabbiye N, Alemu A. Phytoremediation of chromium from tannery wastewater using local plant species. Water Practice and Technology 2017; 12(4): 894-901.

[http://dx.doi.org/10.2166/wpt.2017.094]

[17]   Christopher JG, Kumar G, Tesema AF, Thi NBD, Kobayashi T, Xu K. Bioremediation for tanning industry: a future perspective for zero emission. Manage Hazard Wastes. INTECH 2016; pp. 91-102.

[18]   Abhijit G, Jyoti J, Aditya S, Ridhi S, Candy S, Gaganjot K. Microbes as potential tool for remediation of heavy metals: a review. J Microb Biochem Technol 2016; 8(4): 364-72.

[19]   Tare V, Gupta S, Bose P. Case studies on biological treatment of tannery effluents in India. J Air Waste Manag Assoc 2003; 53(8): 976-82.
         [http://dx.doi.org/10.1080/10473289.2003.10466250] [PMID: 12943317]

[20]   Tripathi S, Singh K, Singh A, Mishra A, Chandra R. Organo-metallic pollutants of distillery effluent and their toxicity on freshwater fish and germinating Zea mays seeds. Int J Environ Sci Technol 2021; 1-14.

[21]   Upadhyay N, Vishwakarma K, Singh J, *et al.* Tolerance and reduction of chromium (VI) by Bacillus sp. MNU16 isolated from contaminated coal mining soil. Front Plant Sci 2017; 8: 778.
         [http://dx.doi.org/10.3389/fpls.2017.00778] [PMID: 28588589]

[22]   Genevois N, Villandier N, Chaleix V, Poli E, Jauberty L, Gloaguen V. Removal of cesium ion from contaminated water: Improvement of Douglas fir bark biosorption by a combination of nickel hexacyanoferrate impregnation and TEMPO oxidation. Ecol Eng 2017; 100: 186-93.
         [http://dx.doi.org/10.1016/j.ecoleng.2016.12.012]

[23]   Esmaeili A, Aghababai Beni A. Optimization and design of a continuous biosorption process using brown algae and chitosan/PVA nano-fiber membrane for removal of nickel by a new biosorbent. Int J Environ Sci Technol 2018; 15(4): 765-78.
         [http://dx.doi.org/10.1007/s13762-017-1409-9]

[24]   Kushwaha A, Rani R, Kumar S, Gautam A. Heavy metal detoxification and tolerance mechanisms in plants: Implications for phytoremediation. Environ Rev 2016; 24(1): 39-51.
         [http://dx.doi.org/10.1139/er-2015-0010]

[25]   Birla H, Minocha T, Kumar G, Misra A, Singh SK. Role of oxidative stress and metal toxicity in the progression of Alzheimer's disease. Curr Neuropharmacol 2020; 18(7): 552-62.
         [http://dx.doi.org/10.2174/1570159X18666200122122512] [PMID: 31969104]

[26]   Muszynska E, Hanus-Fajerska E. Why are heavy metal hyperaccumulating plants so amazing? BioTechnologia. J Biotechnol Comput Biol Bionanotechnol 2015; 96(4)

[27]   Flora GJ. Arsenic toxicity and possible treatment strategies: Some recent advancement. Curr Trends Biotechnol Pharm 2012; 6(3): 280-9.

[28]   Dadzie ES. Assessment of heavy metal contamination of the Densu River, Weija from leachate, 2012.

[29]   Kamarudzaman AN, Chay TC, Amir A, Talib SA. Biosorption of Mn (II) ions from aqueous solution by Pleurotus spent mushroom compost in a fixed-bed column. Procedia Soc Behav Sci 2015; 195: 2709-16.
         [http://dx.doi.org/10.1016/j.sbspro.2015.06.379]

[30]   Huang Y, Chen Q, Deng M, *et al.* Heavy metal pollution and health risk assessment of agricultural soils in a typical peri-urban area in southeast China. J Environ Manage 2018; 207: 159-68.
         [http://dx.doi.org/10.1016/j.jenvman.2017.10.072] [PMID: 29174991]

[31]   Miyazaki A, Amano T, Saito H, Nakano Y. Acute toxicity of chlorophenols to earthworms using a simple paper contact method and comparison with toxicities to fresh water organisms. Chemosphere 2002; 47(1): 65-9.
         [http://dx.doi.org/10.1016/S0045-6535(01)00286-7] [PMID: 11996137]

[32]   Buyukgungor H, Gurel L. The role of biotechnology on the treatment of wastes. Afr J Biotechnol 2009; 8(25)

[33]   Darweesh M, Zedan AM, El-Banna A, Elbasiuny H, Elbehiry F. Biotechnology for Green Future of

Wastewater Treatment, 2021.
[http://dx.doi.org/10.1007/698_2021_788]

[34]　Zeng Z, Zhang M, Kang D, *et al.* Enhanced anaerobic treatment of swine wastewater with exogenous granular sludge: Performance and mechanism. Sci Total Environ 2019; 697: 134180.
[http://dx.doi.org/10.1016/j.scitotenv.2019.134180] [PMID: 32380626]

[35]　Bertanza G, Pedrazzani R. Removal of Trace Pollutants by Application of MBR Technology for Wastewater Treatment.Green Technologies for Wastewater Treatment. Springer 2012; pp. 31-43.
[http://dx.doi.org/10.1007/978-94-007-1430-4_3]

[36]　Khan MA, Ngo HH, Guo W, *et al.* Can membrane bioreactor be a smart option for water treatment? Bioresour Technol Rep 2018; 4: 80-7.
[http://dx.doi.org/10.1016/j.biteb.2018.09.002]

[37]　Li L, Suwanate S, Visvanathan C. Performance evaluation of attached growth membrane bioreactor for treating polluted surface water. Bioresour Technol 2017; 240: 3-8.
[http://dx.doi.org/10.1016/j.biortech.2017.01.043] [PMID: 28162925]

[38]　Xu Z, Song X, Li Y, Li G, Luo W. Removal of antibiotics by sequencing-batch membrane bioreactor for swine wastewater treatment. Sci Total Environ 2019; 684: 23-30.
[http://dx.doi.org/10.1016/j.scitotenv.2019.05.241] [PMID: 31150873]

[39]　Park JBK, Tanner CC, Craggs RJ. Assessment of sludge characteristics from a Biological Trickling Filter (BTF) system. J Water Process Eng 2018; 22: 172-9.
[http://dx.doi.org/10.1016/j.jwpe.2018.02.006]

[40]　Tanikawa D, Fujise R, Kondo Y, Fujihira T, Seo S. Elimination of hydrogen sulfide from biogas by a two-stage trickling filter system using effluent from anaerobic–aerobic wastewater treatment. Int Biodeterior Biodegradation 2018; 130: 98-101.
[http://dx.doi.org/10.1016/j.ibiod.2018.04.007]

[41]　Rana S, Gupta N, Rana RS. Removal of organic pollutant with the use of rotating biological contactor. Mater Today Proc 2018; 5(2): 4218-24.
[http://dx.doi.org/10.1016/j.matpr.2017.11.685]

[42]　Watling HR. The bioleaching of sulphide minerals with emphasis on copper sulphides — A review. Hydrometallurgy 2006; 84(1-2): 81-108.
[http://dx.doi.org/10.1016/j.hydromet.2006.05.001]

[43]　Riekkola-Vanhanen M. Talvivaara mining company – From a project to a mine. Miner Eng 2013; 48: 2-9.
[http://dx.doi.org/10.1016/j.mineng.2013.04.018]

[44]　Petersen J, Dixon DG. Modelling zinc heap bioleaching. Hydrometallurgy 2007; 85(2-4): 127-43.
[http://dx.doi.org/10.1016/j.hydromet.2006.09.001]

[45]　Borja D, Nguyen K, Silva R, *et al.* Experiences and future challenges of bioleaching research in South Korea. Minerals (Basel) 2016; 6(4): 128.
[http://dx.doi.org/10.3390/min6040128]

[46]　Jang YC. Waste electrical and electronic equipment (WEEE) management in Korea: generation, collection, and recycling systems. J Mater Cycles Waste Manag 2010; 12(4): 283-94.
[http://dx.doi.org/10.1007/s10163-010-0298-5]

[47]　Lim HS, Lee JS, Chon HT, Sager M. Heavy metal contamination and health risk assessment in the vicinity of the abandoned Songcheon Au–Ag mine in Korea. J Geochem Explor 2008; 96(2-3): 223-30.
[http://dx.doi.org/10.1016/j.gexplo.2007.04.008]

[48]　Darpito C, Shin WS, Jeon S, *et al.* Cultivation of Chlorella protothecoides in anaerobically treated brewery wastewater for cost-effective biodiesel production. Bioprocess Biosyst Eng 2015; 38(3): 523-30.

[http://dx.doi.org/10.1007/s00449-014-1292-4] [PMID: 25270406]

[49]   El-Sheekh MM, Farghl AA, Galal HR, Bayoumi HS. Bioremediation of different types of polluted water using microalgae. Rend Lincei Sci Fis Nat 2016; 27(2): 401-10.
[http://dx.doi.org/10.1007/s12210-015-0495-1]

[50]   Chakraborty R, Wu CH, Hazen TC. Systems biology approach to bioremediation. Curr Opin Biotechnol 2012; 23(3): 483-90.
[http://dx.doi.org/10.1016/j.copbio.2012.01.015] [PMID: 22342400]

[51]   Salgot M, Folch M. Wastewater treatment and water reuse. Curr Opin Environ Sci Health 2018; 2: 64-74.
[http://dx.doi.org/10.1016/j.coesh.2018.03.005]

[52]   Verma S, Kuila A. Bioremediation of heavy metals by microbial process. Environmental Technology & Innovation 2019; 14: 100369.
[http://dx.doi.org/10.1016/j.eti.2019.100369]

[53]   El-Sheekh MM, El-Abd MA, El-Diwany AI, Ismail AS, Omar TH. Poly-3-hydroxybutyrate (PHB) production by Bacillus flexus ME-77 using some industrial wastes. Rend Lincei Sci Fis Nat 2015; 26(2): 109-19.
[http://dx.doi.org/10.1007/s12210-014-0368-z]

[54]   Mamo FS, Marianne KW, Reidun DS, Inger A. The influence of estimated service life on the embodied emissions of zero emission buildings (ZEBs) when choosing low-carbon building products. XIV DBMC 14th International Conference on Durability of Building Materials and Components, 2017.

[55]   Alvarino T, Suarez S, Lema J, Omil F. Understanding the sorption and biotransformation of organic micropollutants in innovative biological wastewater treatment technologies. Sci Total Environ 2018; 615: 297-306.
[http://dx.doi.org/10.1016/j.scitotenv.2017.09.278] [PMID: 28982079]

[56]   Speight JG. El-GendyNS Introduction to petroleum biotechnology. Gulf Professional Publishing 2017.

[57]   Singh J, Yang JK, Chang YY, Koduru JR. Fenton-like degradation of methylene blue by ultrasonically dispersed nano zero-valent metals. Environ Process 2017; 4(1): 169-82.
[http://dx.doi.org/10.1007/s40710-016-0199-2]

[58]   Zou Y, Wang X, Khan A, *et al.* Environmental remediation and application of nanoscale zero-valent iron and its composites for the removal of heavy metal ions: a review. Environ Sci Technol 2016; 50(14): 7290-304.
[http://dx.doi.org/10.1021/acs.est.6b01897] [PMID: 27331413]

[59]   Paknikar KM, Nagpal V, Pethkar AV, Rajwade JM. Degradation of lindane from aqueous solutions using iron sulfide nanoparticles stabilized by biopolymers. Sci Technol Adv Mater 2005; 6(3-4): 370-4.
[http://dx.doi.org/10.1016/j.stam.2005.02.016]

[60]   Ramasamy K, Malik MA, Helliwell M, Tuna F, O'Brien P. Iron thiobiurets: single-source precursors for iron sulfide thin films. Inorg Chem 2010; 49(18): 8495-503.
[http://dx.doi.org/10.1021/ic1011204] [PMID: 20735018]

[61]   Fan J, Guo Y, Wang J, Fan M. Rapid decolorization of azo dye methyl orange in aqueous solution by nanoscale zerovalent iron particles. J Hazard Mater 2009; 166(2-3): 904-10.
[http://dx.doi.org/10.1016/j.jhazmat.2008.11.091] [PMID: 19128873]

[62]   Shu HY, Chang MC, Chen CC, Chen PE. Using resin supported nano zero-valent iron particles for decoloration of Acid Blue 113 azo dye solution. J Hazard Mater 2010; 184(1-3): 499-505.
[http://dx.doi.org/10.1016/j.jhazmat.2010.08.064] [PMID: 20833471]

[63]   Wu ZC, Zhang Y, Tao TX, Zhang L, Fong H. Silver nanoparticles on amidoxime fibers for photo-catalytic degradation of organic dyes in waste water. Appl Surf Sci 2010; 257(3): 1092-7.

[http://dx.doi.org/10.1016/j.apsusc.2010.08.022]

[64]   Yurtsever M, Şengil İA. Biosorption of Pb(II) ions by modified quebracho tannin resin. J Hazard Mater 2009; 163(1): 58-64.
[http://dx.doi.org/10.1016/j.jhazmat.2008.06.077] [PMID: 18667272]

[65]   Büyükgüngör H, Wilk M. Schubert H. Biosorption of Lead by Citrobacter freundii Immobilized on Hazelnut Shells, 1996.

[66]   Orhan Y, Hrenovič J, Büyükgüngör H. Biosorption of heavy metals from wastewater by biosolids. Eng Life Sci 2006; 6(4): 399-402.
[http://dx.doi.org/10.1002/elsc.200520135]

[67]   Pino GH, Souza de Mesquita LM, Torem ML, Saavedra Pinto GA. Biosorption of cadmium by green coconut shell powder. Miner Eng 2006; 19(5): 380-7.
[http://dx.doi.org/10.1016/j.mineng.2005.12.003]

CHAPTER 5

Soil Reclamation and Conservation Using Biotechnology Techniques

Bhupinder Dhir[1,*]

[1] *School of Sciences, Indira Gandhi National Open University, New Delhi, India*

Abstract: Pollution and unsustainable use of natural resources such as land and soil has resulted in their destruction. Restoration of degraded land and soil is essential for maintenance of essential ecosystem services such as preservation of biodiversity, nutrient/water cycling and meeting the food requirement for living beings. Bioremediation has appeared as technology with high potential for restoring damaged soil and degraded lands. Biotechnological techniques such as development of efficient microbial consortia with an enhanced capacity to remove various contaminants from soils and improvement in nutrient retention in soil have opened new prospects in bioremediation with an aim to recover productive capacity of soil. The techniques such as bioventing, bioaugmentation, biosparging have also proved useful in restoring degraded and non-productive soils to a great extent. The biotechnological techniques, thus can act as an ecofriendly method for remediation, restoration and reclamation of degraded/damaged soils.

Keywords: Biotechnology, Bioaugmentation, Biostimulation, Biofertilizers, Restoration, Soil.

INTRODUCTION

Soil is an important natural resource that supports basic functions of ecosystem and provides essential services to an ecosystem. It provides habitat to living organisms, plays an important role in cycling of nutrients and breakdown of organic matter. Soil contamination and deterioration has emerged as a major threats due to increase in anthropogenic activtities [1]. Overexploitaton of land resources for agriculture and construction purposes has resulted in decline in fertility of soils and affected their productive potential. According to studies, the productive capacity of about one-fourth of agricultural land throughout the world has been affected due to damage to soil. This has resulted in decline in agricultural production. Predictions suggest that restoration or reclamation of soil

* **Corresponding author Bhupinder Dhir:** School of Sciences, Indira Gandhi National Open University, New Delhi, India; E-mail: bhupdhir@gmail.com

is very important to increase the agricultural production (by about 50%) to meet the growing demand for food.

The presence of high amounts of heavy metals in soils has been emerged as a major problem related to soil degradation. About 35% of the soils in Europe have been contaminated with metals. Agricultural, mining, metallurgical activities lead to metal(oid) contamination in soil. Organic contaminants get added mainly from the industrial activities. Techniques such as excavation, incineration, chemical washing, and vitrification are some of the physicochemical methods that have been used to treat contaminated soils. Most methods of soil remediation show certain limitations. They are economically infeasible and not environment friendly. Studies have shown that physicochemical methods treat the contaminants effectively but in the process affect the soil processes, functions and biota [2]. Most of these techniques affect the ecological status of the remediated soil. Off late the need for restoration and reclamation of the degraded soils was realized and efforts were made to develop technologies that can help in getting the natural resource (soil/land) restored with an aim to maintain sustainability of the environment. Biological methods have emerged as sustainable remediation technologies that helped in restoring/ reclaming the productivity of the soil without inducing any negative effects on soil biota [3]. These techniques aim at reducing the concentration of soil contaminants and recovering of soil functionality. Biological methods of soil remediation thus provide help in reclamation of degraded soils [4].

The restoration process help in maintaining safe, clean environment and support plant growth. Reclamation process improves soil quality making it suitable for a sustainable use, while conservation process pertains to preservation, protection and planned use of soil [5]. The damaged soils (such as eroded, mined) can be treated in such a way that they return to a condition that occured prior to degradation. Biotechnology has provided a good potential tool that can help us in restoring and conserving the degraded soil [6, 7].

Some processes such as intrinsic bioremediation (bio-attenuation) help in recovering the soil contamination and degradation to a great extent. It is a natural reduction of organic pollutants by micro-organisms present in soil. This type of remediation depends upon potential of microbial population to reduce contaminant levels. It is cost effective than conventional engineered technologies, but remediation requires large time frame to accomplish the task.

The present chapter provides an overview of techniques of biotechnology that help in reclamation and conservation of soil.

VARIOUS TECHNOLOGIES USED IN RESTORATION OF DEGRADED SOILS

Biotechnological techniques have shown immense potential in restoration and retrieval of degraded soil [8].

Removal of Organic Contaminants Using Microbes

Microbes have shown immense potential to degrade organic contaminents present in the soil [9]. Microorganisms degrade organic pollutants under oxic conditions (presence of oxygen) by respiration or under anoxic conditions *via* processes such as denitrification, methanogenesis, and sulfidogenesis. Complete and fast degradation of the pollutants is noted under aerobic conditions. The pollutants present in the soil get trapped into soil pores and/or adsorbed to the soil matrix resulting in their immobilization. The microbial communities present in the soil change with change in the environmental conditions. Only the microbes which are resistant survive and thus play role in cleaning the polluted soil. The remediation of contaminated soil by microbes is regulated by physical factors such as temperature, pH of soil, soil moisture content, soil quality, soil nutrient content and concentration of oxygen. Change in any of these factors can alter the population of microbes affecting the bioremediation potential to a great extent.

Various types of hydrocarbons get degraded by different bacterial genera. Microbes present naturally in the contaminated soil break the complex hydrocarbons into simple form *via* their enzymatic systems. The degradation of hydrocarbons occurs under both aerobic and anaerobic conditions. In anaerobic condition, The bacteria present in the deep parts of the sediments use nitrates, sulfates and iron as electron acceptor to degrade the hydrocarbons under oxygen deprived conditions. Enzyme bacterial dioxygenase integrate oxygen into carbon molecule *via* series of enzyme catalyzed reactions under aerobic conditions. Oxygen gets added to alcohol groups to form aldehyde and further into carboxylic group by the action of other enzymes. This gets degraded to form acetyl co-A *via* beta oxidation. Bacteria such as *Desulfococcus, Thauera, Dechloromonas* and *Azoarcus* show hydrocarbon degradation ability under anaerobic conditons. Some bacteria such as *Alcaligenes, Sphingomonas, Pseudomonas, Bacillus, Nocardia, Acinetobacter, Micrococcus, Achromobacter, Rhodococcus, Alcaligenes, Moraxella, Mycobacterium, Aeromonas, Xanthomonas, Athrobacter, Flavobacterium, Micrococcus, Azospirillum* show ability to degrade crude oil.

Bacteria and fungi show capacity to degrade polyaromatic hydrocarbons (Table **1**). Microorganisms degrade PAHs by bringing oxidation in the aromatic ring of the compound followed by their breakdown to form metabolites [10, 11]. Enzymes such as laccase and manganese peroxidase play an important role in degradation of PAH in ligninolytic fungi [12]. These enzymes degrade PCB compounds by incorporating two hydroxyl groups into the aromatic ring. The reactivity of the PCBs increases because of the fission reaction in the enzymatic ring [15]. Surfactants produced by microorganisms increase desorption of PAHs from the soil matrix [13, 14]. Dechlorination of polychlorinated biphenyls (PCBs) by microbes removes *m-* and *p-* chlorines resulting in formation of *ortho* substituted mono tetrachlorobiphenyls [16]. Degradation of PCB *via* substitution of chlorine by *Phanerochaete chrysosporium* has been noted. The dechlorination of PCB produces mono and di-chlorinated biphenyls which easily get degraded by bacteria such as *Burkholderia*. Oxidation leads to breakage of ring and compound gets mineralized.

Table 1. Microorganisms biodegrading various organic pollutants [17].

Microorganism Used	Pollutants Degraded
Pseudomonas spp	Benzene, anthracene, hydrocarbons, PCBs
Alcaligenes spp	Halogenated hydrocarbons, linear alkylbenzene sulfonates, polycyclic aromatics, PCBs
Arthrobacter spp	Benzene, hydrocarbons, pentachlorophenol, phenoxyacetate, polycyclic aromatic compounds
Bacillus spp	Aromatics, long chain alkanes, phenol, cresol
Corynebacterium spp	Halogenated hydrocarbons, phenoxyacetates
Flavobacterium spp	Aromatics
Azotobacter spp	Aromatics
Rhodococcus spp	Naphthalene, biphenyl
Mycobacterium spp	Aromatics, branched hydrocarbons benzene, cycloparaffins
Nocardia spp	Hydrocarbons
Methanogens	Aromatics
Xanthomonas spp	Hydrocarbons, polycyclic hydrocarbons
Streptomyces spp	Phenoxyacetate, halogenated hydrocarbon diazinon
Cunniughamela elegans	PCBs, polycyclic aromatics, biphenyls

The degradation of organophosphates by microbes includes hydrolysis of P-O alkyl and P-O-aryl bond. Enzymes, phosphotriesterases (PTEs) have been found responsible for degradation of OP in microorganisms. *Burkholderia cocovenenas* has been found to degrade phenanthrene [10].

Remediation of contaminated soil by microbes is influenced by various factors including amount of water, temperature and pH of soil, oxygen concentration, quality of soil and nutrient content. If any of these factor changes, the population of microbes decreases which in turn affects bioremediation.

Microbial activity can be accelerated by using bioaugmentation and bio stimulation strategies. In bioaugmentation exogenous oil degrading bacteria are supplemented to enhance soil microbiota.

Biosurfactants are the surfactants produced by microorganisms. These compounds increase the bioavailability of hydrocarbons for microbes hence supporting its degradation. The microbes that produce biosurfactants show high remediation potential. Biosurfactants produced by various microbes such as *Acinetobacter sp., Bacillus subtilis A1, Pseudomonas aeruginosa, Rhodococcus erythropolis M-25, Rhodococcus ruber Em1, Pseudomonas stutzeri* has been found to play a major role in remediation.

The mycelium of fungus is very helpful in the degradation of hydrocarbons. The fungi degrade hydrocarbons by oxidation carried out with the help of enzymes laccase, lignin peroxidase and manganese peroxidase. Fungi *Candida, Stropharia, Rhodotorula, Pleurotus, Penicillium, Phanerochaete, Fusarium* have shown capacity to degrade oil.

Bioaugmentation

Bioaugmentation method involves application of wild or genetically modified microorganisms to polluted/contaminated soils. This proves helpful in removing/remediating undesired compounds present in soil [18]. The presence of microbes increases the rate of degradation of the complex pollutants and removes them from the soil at a faster rate. The process increases the inherent capacity and genetic diversity of soil. Individual microbial strain or a consortium is inoculated. Inoculation of mixture (cocktail) of microbial strains proves efficient in removing contaminants from soil. Biodegradation of the target organic contaminants get stimulated. This is because in a consortium, different metabolic activities of microorganisms combine and complement each other thus treating various compounds. The efflux pumps and surface-active compounds produced by microbes help in removing toxic compounds present inside the cells (Fig. **1**). If the microbial consortia are well adapted to the soil/site that needs to be restored/reclaimed, bioaugmentation is successful [19].

Bioaugmentation has played a important role in removal of wide range of pollutants including pesticides such as DDT, lindane, endosulfan, pentachlorophenol (PCP), organic compounds such as polyaromatic hydrocarbons (PAHs) and petroleum hydrocarbons from soil [11, 20 - 28] (Table **2**).

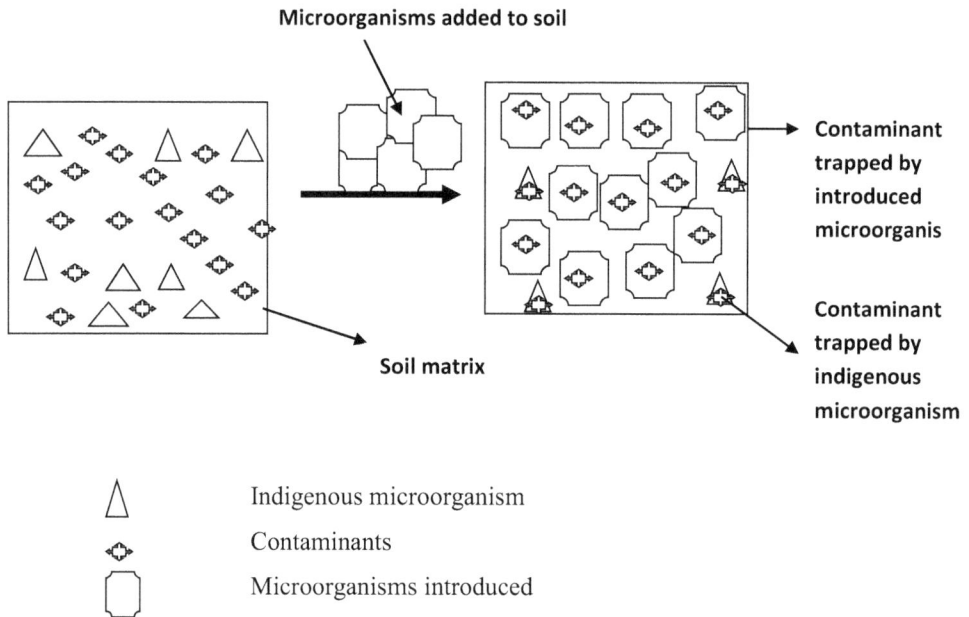

Fig. (1). Diagram showing the process of Bioaugmentation.

Table 2. Microorganisms used in removal of various contaminats by bioaugmentation.

Microorganism Used	Pollutants Degraded
Pseudomonas putida	4 chlorobenzoic acid, Napthalene, Phenol, trichloroethane
Pseudomonas fluorescens	2,4-dinitrotoluene, Biphenyl, polychlorinated biphenyl
Pseudomonas sp.	3 phenoxybenzoic acid
Burkholderia xenovorans	Arochlor
Cupriavidus necator	Chlorobenzoates, Arochlor
Rhodococcus sp.	4-chlorobenzoate, petroleum products
Eschirichia coli	Atrazine
Arthrobacter, Burkholderia, Pseudomonas, Rhodococcus	Petroleum hydrocarbons

The rate of bioaugmentation get affected by various factors including pH, temperature, moisture, amount of organic matter and nutrients, aeration and type of soil [29]. The rate of removal of contaminants by bioaugmentation varies depending on the physical conditions. *Burkholderia* sp. FDS-1 present in the soil showed capacity to degrade nitrophenolic pesticides at temperature of 30°C and alkaline pH. The enzymatic reactions of cells contribute to catabolic activities. Soil rich in organic matter bring out of degradation of 2, 4-dichlorophenoxyacetic acid at high rate. Moisture rich soil supplemented with *Achromobacter piechaundii* TBPZ showed rapid degradation of tribromophenol. The degradation activity of bacteria gets affected adversely under moisture stress conditions of soil. Degradation of quinoline by *Burkholderia pickettii* and chlorinated benzenes by *Pseudomonas* sp. has been reported [30, 31].

The technique showed restricted use and acceptability because of certain limitations. Te microbial strains used for bioaugmentation ideally (i) Possess high potential to degrade target contaminants, (ii) Show high growth rate, (iii) Show tolerance to high concentration of contaminants and (iv) Can survive in different environmental conditions. Besides these, microbes also show ability to resist predation, competition and toxins present in the soils.

Biostimulation

A remediation technique in nutrients such as phosphorus, nitrogen, oxygen is added to highly contaminated sites. The nutrients added to the soil improve the efficiency of the microbes for degradation of contaminants. The degradation of hazardous and toxic contaminants occurs at a faster rate [32] (Table **3**). Remedation of contaminants such as petroleum hydrocarbons, sulphate and polyester polyurethanes have been successfully achieved through biostimulation. Microbes attack the intramolecular bonds found in these polymers. The rate of degradation of contaminants by microbes depend availability of carbon (C) source and environmental factors such as pH, moisture content and temperature. These factors affect the rate of metabolic activities.

Table 3. Some of the contaminants removed by biostimulation.

Nutrients Added to Soil	Target Pollutants
Dairy manure	Atrazine
Maize straw	Methabenzthiazuron
Activated sludge	Atrazine and simazine
Animal manure and sewage sludge	Atrazine and alachlor
Cellulose, straw and compost	Atrazine
Cornmeal, ryegrass and poultry litter	Cyanazine and fluometuron
Plant residues, ground seed, or commercial meal	Alchlor, metolachlor, atrazine, trifluralin

Microbes play a major role in bioaugmentation and biostimulation. The properties such as nutrient unavailability, abiotic factors (pH, temperature or oxygen) affect the number of bacteria present in the soil, thereby altering their bistimulation as well as bioaugmentation potential responsible for degrading pollutants such as oil hydrocarbons [8].

Bioventing

In the process of bioventing, oxygen availability in soil brings *in situ* biodegradation of compounds by existing soil microorganisms. Low rates of air flow provide enough oxygen to sustain microbial activity. The technique of bioventing increases the activity of indigenous bacteria, archaea by inducing air or oxygen flow into the unsaturated zone. This stimulates biodegradation of hydrocarbon derivatives (crude oil or crude oil products) and volatile compounds under natural *in situ* conditions [33]. In contaminated soil, oxygen is supplied directly through injection (Fig. **2**). Successfull remediation of compounds such as petroleum hydrocarbons, non-chlorinated solvents, some pesticides, wood preservatives and other organic chemicals present in the contaminated soils has been done using this technique.

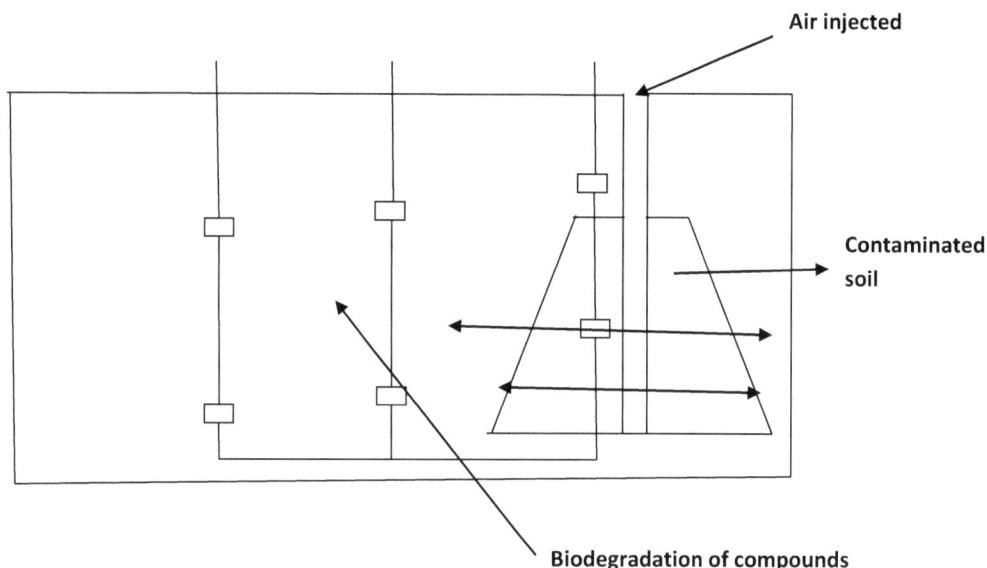

Fig. (2). Diagram showing Bioventing system.

Biosparging

Biosparging is a technique in which indigenous microorganisms degrade organic constituents present in the saturated zone under *in situ* conditions. The biological activity of the indigenous microorganisms is increased by supplying air (or oxygen) and nutrients (if required) in the saturated zone of soil. The concentrations of petroleum constitutents that are dissolved in groundwater, adsorbed to soil below the water table, and within the capillary fringe are reduced. The technique of biodegradation is mainly involved in the removal of constitutents. The microorganisms present in the soil degrade organic contaminants present in the soil. The supply of oxygen by air or pure oxygen sparging into the soil layer increases the biological activity of the microbes.

Biosparging can be combined with soil vapor extraction or bioventing (also referred to as vapor extraction), and/or can be used with other remedial technologies if volatile constituents are present. Oxygen increases the biodegradation rates. Petroleum residues present in the soil can be removed by this technique. Light and medium weight petroleum products volatilise and get removed easily *via* this technique [34].

Techniques for Enhancing Nutrient Content of Soils

Biofertilizers

Microbes as biofertilizers restore the damaged and unproductive soils. Bacteria, fungi, arbuscular mycorrhiza fungi (AMF) act as growth promoting biological agents in plants. Rhizosphere bacteria including N_2-fixing cyanobacteria, phosphorus, potassium solubilising microorganisms, mycorrhiza, plant disease suppressive bacteria, stress tolerant endophytes, microbes witrh biodegradation potential that help in growth of plants are known as plant growth promoting rhizobacteria (PGPR) [35]. Plant growth promoting rhizobacteria (PGPR) help in cycling of nutrients *via* nitrogen fixation, phosphate mobilization and the release of other nutrients to soil solution [36, 37].

Addition of PGPR to soil improves its physicochemical properties and microbial biodiversity. Application of biofertilizers helps in nutrient cycling thereby increasing soil productive capacity. The processes such as nitrogen fixation, phosphate and potassium solubilisation/mineralization, release of plant growth regulating substances, production of antibiotics and breakdown of organic matter in the soil increase micro- and macro-nutrients content of soil. The improvement in growth of plants is due to increased production of growth regulators (such as auxins, gibberellins and cytokinins), increase in metabolic activities of roots and enhancement in biological nitrogen fixation is attributed to biofertilizers [38]. PGPR, thus directly affect the plant growth by improving properties such as germination of seed, development of root, uptake of nutrient, water and indirectly by biocontrol activity such as suppression of bacterial, fungal and nematode pathogens.

Azotobacter, *Azospirillum*, *Azoarcus*, *Klebsiella*, *Bacillus*, *Pseudomonas*, *Arthrobacter*, *Enterobacter* are the major PGPR strains identified so far [39]. Soil can be restored when *Rhizobium* is used in association with leguminous trees and *Frankia* association occurs with non leguminous species. *Rhizobium* and *Frankia* assist in reclamation of soil by inducing nodule formation, high biomass production in plants without input of nitrogen fertilizers. The root morphology in plants can be improved *via* production of plant growth regulating substances (indole acetic acid, gibberellins and cytokinins) and siderophore if *Azotobacter*, *Azospirillum* and *Rhizobium* are inoculated in soil. Absorption of nutrients increased wth increase in number of lateral roots and root hairs that provide more surface area. The improvement in water status and nutrient profile supports growth and development in plants. Exposure to biofertilizers induce an increase in plant height, number of leaves, diameter of stem, seed filling and dry weight of seeds. Wheat, mustard, barley, oat, sunflower, seasum, rice, linseeds, castor, sorghum, cotton, maize, jute, sugar beet, tea, coffee, tobacco, rubber and coconut are some of the crops that have shown improvement in growth after biofertilizer application [40].

Biotechnological methods have been used to develop non-leguminous plants that devloped the ability to survive under adverse conditions such as low nutrient supply. Inoculation with *Rhizobium* increased yield of grain crops. *Rhizobium* application improved growth and development in various crop species such as pea, alfalfa, Bengal gram, lentil, sugar beet, berseem, groundnut and soybean. These fertilizers play an important role in integrated nutrient management system that helps in sustaining agricultural productivity.

Use of Mycorrhiza

The symbiotic associations between plant roots and fungi form mycorrhizae. Arbuscular mycorrhiza fungi establish symbiotic association with the roots of plant species [41, 42]. Increase in uptake of nutrients and water improve growth and survival of seedling. The absorption of phosphate from the soil is facilitated by inorganic phosphate (Pi) transporters present on the hyphae. They increase the life of root and provide protection to plants against the pathogens. Application of fungal inocula to roots of seedlings allows formation of mycorrhizae. The micropropagated plants that are infected during rooting show increase in their survival rate in the field and plantations on degraded lands [43].

Plant-associated microorganisms *i.e.* AM fungi and bacteria help in removal /treatment of contaminants such as heavy metals, radionuclides, organic xenobiotics (including volatile organic compounds, oil derived alkanes or polycyclic aromatic hydrocarbons) by enhancing the abilities of plant for the remediation (phytoremediation).

In legume plants, mutualistic symbiotic relationship of arbuscular mycorrhizal (AM) fungi and N_2-fixing bacteria fix atmospheric N_2 is found [44 - 46]. The formation of AM increases the plant ability to capture nutrients, cycling them in soils with low nutrient availability. AM symbiosis helps plant in maintaining productive capacity under adverse conditions [47]. The colonization of AM alters the chemical composition of root exudates, mycelium and brings modifications into the root environment which affects the microbial structure and diversity. Mycorrhizal colonization is also supposed to up-regulate genes related to auxin biosynthesis and root morphogenesis.

Heavy metals (HMs) get irreversibly sequestered by AM fungi and hence contribute to metal stabilization in the soil [48, 49].

Techniques for Protecting Soil from use of Harmful Chemicals

Biopesticides

Bacteria, fungi, viruses, plants, animals and minerals produce certain chemicals that help in controlling pests. They are also referred as biological pesticides or biocontrol agents. Biocontrol agents generally include living organisms such as bacteria, viruses, fungi, nematodes and protozoa [50]. The mode of action varies with each type of biopesticide. Hyperparasitism, competition, secretion of volatile compounds, antibiosis and parasitism are some of the major properties of biopesticides. Biopesticides act as an alternate to synthetic chemicals used in pest management [51, 52]. Biopesticides thus play a significant role in integrated pest management for sustainable agriculture [53].

Biopesticides mainly include bacterial species such as *Bacillus, Pseudomonas, Burkholderia, Xanthomonas, Enterobacter, Streptomyces* and *Serratia. Pseudomonas, Bacillus thuringiensis* and fungi like *Trichoderma viride* colonise the rhizosphere of plants and thus play an important role in controlling pests and diseases [54]. Some of the fungal species that act as biopesticides include *Trichoderma, Beauveria, Metarhizium, Paecilomyces, Fusarium, Pythium, Penicillim* and *Verticillium*. Nematode species such as *Steinernama and Heterarhabditis* act as biopesticides. *Bacillus cereus* produces multiple compounds that suppress more than two pathogens and thus prove effective in disease management. *Lysobacter* and *Myxobacteria* suppress pathogens by producing lytic enzymes which hydrolyze compounds. Rhizospheric and PGPR strains such as *Trichoderma* spp. protect plants against pathogens by competitiing for space nutrients and antibiosis [44]. Siderophore produced by PGPR strains chelate iron in the rhizosphere region. This leads to iron deficiency in the pathogenic microbes.

Priming also provides protection to plants against both roots and foliar pathogens and induces systemic resistance (ISR) [55]. The symbiotic nature of arbascular mycorrhizae (AM) provides protection to plants against pathogens and insects.

ROLE OF BIOTECHNOLOGY IN SOIL RESTORATION

Various methods and procedures using biotechnological approach are been developed to check/restrict the contamination of soil and help in restoration of contaminated soils. These methods mainly focus on development of microorganisms or plants that efficiently remove contaminants from soil and hence prevent it from degradation.

Use of Bioengineered Microbes

Microorganisms (such as algal, fungal and bacterial species) possess capacity to remove/degrade various contaminants from soil. They follow different mechanisms to remove contaminats such as heavy metals. Microorganisms accumulate metals in their cell walls and or show the capacity to transport the metals into free space (intracellular and intercellular) and organelles. Metal ions bind to the cell surface of microorganisms *via* adsorption. Carboxyl, sulphate, phosphate and amino groups present on the cell wall of microbes help in binding metals ions. The removal of metal ions occurs *via* exchange of bivalent metal ions with the counter ions of the polysaccharides present in the cell wall of microbes. Bacteria produce extracellular polymeric substances (EPS) that help in removal of heavy metal ions. Oxalic acid, citric acid excreted by microbes help in the chelation of metal ions. Metal uptake in microbes mainly occurs *via* biosorption which is physico-chemical adasorption process. Some bacteria such as *Desulfovibrio* and *Desulfotomaculum* show capacity to precipitate metals as hydroxides or sulfates. Microorganisms transform toxic metal ions to non–toxic form *via* mechanism of biomethylation. Bacteria, filamentous fungi and yeasts carry out methylation of metals such as Hg, As, Cd, Se, Sn, Te and Pb. Metal removal or remediation by fungal and algal species also occurs *via* synthesis of metal binding proteins or peptides.

Biotechnology has proved useful in improving the potential of microbes for achieving high removal and biodegradation for contaminants such as heavy metals by technique known as genetic modification [56]. The genetic engineering technique has proved useful in improving the performance of bacteria by producing enzymes and modifying metabolic pathways that lead to high degradation of the target contaminants. Biotechnological techniques have been used to develop new strains/varieties of microbes having high bioremediation potential. *Arabidopsis thaliana* developed by introducing gene glutathione-S-transferase from tobacco showed high tolerance to metals such as Al, Cu, and Na [57].

Genetic engineering has proved useful in improve or redesign the potential of microorganisms that possess intrinsic capability to sequester heavy metals in high quantities and greater resistance to environmental conditions [58, 59]. Microbes with superior potential for removal and recovery of metal contaminants from polluted soil have been developed using biotechnological techniques [60]. A genetically engineered *E.coli* showed capacity to transport metals, synthesize metallothionein (a metal binding protein) and accumulate high levels of metals. Metal accumulation in bacterial cells can increase by expressing metallothioneins (MTs) in cells. This can help in development of microbial-based biosorbents that can be used for remediation of metal contamination [61].

Microbial strains such as *Pseudomonas pseudoalkali* have also been developed with show enhanced capacity for degrading organic compounds such as poly chlorinated biphenyls (PCBs). Genes respoonsibel for bph-making enzymes responsible for degradation of PCBs has been isolated and introduced in bacterial strains. Technique known as 'DNA shuffling' that involves mixing of DNA from two different PCB degrading bacterial strains has been developed. The chimeric bph genes formed as a result produce enzymes that are capable of degrading wide range of PCBs. Tod-genes responsible for degradation of toluene have also been introduced in bacterials strains mainly *Pseudomonas putida*. Genetically engineered bacteria have the capacity to degrade different ranges of chlorinated compounds have also been developed. *Desulfitlobacterium* sp. Y51 has shown potential to dechlorinate PCE (Poly chloroethylene) and convert it into cw-1--dichloroethylene (cDCE).

Mer genes (genes for mercury reistance) responsible for degradation of organic mercurial compounds have been isolated from bacteria. A hybrid enzyme has been created by exchanging the subunits of the enzymes. Hybrid deoxygenase was found capable of degrading trichloroethylene (TCE) at as faster rate. The todCl gene isolated from toluene degrading bacteria has been successfully introduced in the bacterial strain KF707. The hybrid strain formed as aresult show high degradation of TCE.

A microbe developed by genetic enginerring also proves useful in increasing fertility of soil [62].

Use of Bioengineered Plant Species

Genetically engineered plants developed *via* biotechnology have their genome modified in such a way that they can tolerate high concentrations of contaminants and disintegrate them [63, 64]. The development of pollutant resistant plant varieties also has proved beneficial in restoring the degraded lands. The development of plant species with superior potential to grow on degraded land is another approach followed by researchers to reclaim and revegetate the degraded lands. Large scale multiplication of plants has been achieved through culture of shoot apical meristems, buds, rooting of shoots and micropropagaton. Genes involved in the uptake, translocation and sequestration of metals has been identified, isolated and introduced in plants [65, 66]. The transfer of genes responsible for these traits in plants will improve their phytoremediation potential. The genes that promote the synthesis of metal chelators which increase metal uptake capacity have been introduced in plants [67 - 69] (Table **4**).

Table 4. Some examples of transgenic plants for metal tolerance/phytoremediation.

Plant	Metal Accumulated	Gene	Reference
tobacco	Cd, Ca, Mn	gene *CAX-2* (vacuolar transporters) from *A. thaliana*	73
tobacco	Ni, Pb	gene *Nt CBP4* from tobacco	74
tobacco	Fe	gene *Ferretin* from soybean;	75
rice	Fe	gene *Ferretin* from soybean;	75, 76
A. thaliana, tobacco	Hg	gene *merA* from bacteria	69, 77, 78,
Indian mustard	Se	glutathione reductase from bacteria	79

Overexpression of glutamyl cysteine synthetase improves metal accumulation *in Populus angustifolia, Nicotaiana tobacum, Silene cucubalis* [70]. Genetically modified *Arabidopsis thaliana* showing high As accumulation and tolerance to As express SRSIp/ArsC and ACT 2p/γ-ECS have been developed [66, 71]. Overexpression of bacterial genes *viz.arsC* and *gECS1* that code for arsenate reductase and glutamylcycteine synthase in *Arabidopsis thaliana* increased the plants capacity to capture and degrade arsenate [72].

The selection of the plant species with high capacity to tolerate, accumulate, and/or remove environmental pollutants such as TNT, PCBs, pharmaceuticals, dyes from textile, phenolics, radionuclides, and heavy metals have been done using hairy roots (HRs) culture technique. The metabolic and catalytic pathways involved in the uptake of pollutants can be explored using this technique.

Enzymes such as peroxidases and laccases released by plant roots help in degrading pollutants.

CONCLUSION AND FUTURE PROSPECTS

Pollutants can be successfully removed from the environment mainly through mechanism of bioremediation and phytoremediation. Biotechnological techniques have also been developed involving use of microorganisms that show immense potential to remove/degrade pollutants, and plant species with high capacity to remove pollutants from soil. Mass scale propagation of plant species can be developed using biotechnological procedures. Use of bioengineered microbes and plants with high capacity to tolerate high level of pollutants has also proved useful in reclamation and conservation of soil to a great extent. Application of plant growth promoting rhizobacteria (PGPRs), endo- and ectomycorrhizal fungi play a important role in improving soil nutrient content by increasing nutrient uptake. Biopesticides act as a viable substitute for synthetic pesticides. They show low toxicity, and biodegradability. Further studies are required to validate the use of the bioengineered microbial and plant species so that soil reclamation and restoration can be achieved at a larger scale and faster pace.

REFERENCES

[1]   Goudie AS. The human impact on the natural environment: past, present and future. John Wiley and Sons 2013; p. 424.

[2]   Kim JM, Le NT, Chung BS, *et al.* Influence of soil components on the biodegradation of benzene, toluene, ethylbenzene, and o-, m-, and p-xylenes by the newly isolated bacterium *Pseudoxanthomonas spadix* BD-a59. Appl Environ Microbiol 2008; 74(23): 7313-20.
[http://dx.doi.org/10.1128/AEM.01695-08] [PMID: 18835999]

[3]   Rafael GL, Becerril JM, Garbisu C. Biological methods of polluted soil remediation for an effective economically-optimal recovery of soil health and ecosystem services. J Environ Sci Public Health 2020; 4: 112-33.

[4]   Tripathi V, Adil Edrisi S, Chen B, *et al.* 2017.

[5]   Maestre FT, Solé R, Singh BK. Microbial biotechnology as a tool to restore degraded drylands. Microb Biotechnol 2017; 10(5): 1250-3.
[http://dx.doi.org/10.1111/1751-7915.12832] [PMID: 28834240]

[6]   Sinha RK, Valani D, Sinha S, *et al.* Bioremediation of contaminated sites: a low-cost nature's biotechnology for environmental clean up by versatile microbes, plants & earthworms. Nova Publishers 2010; p. 73.

[7]   Blanco-Canqui HN, Lal R. Restoration of eroded and degraded soils principles of soil conservation and management. Trends Biotechnol 2017; 35(9): 847-59.
[PMID: 28606405]

[8]   Adams GO, Fufeyin PT, Okoro SE, *et al.* Bioremediation, biostimulation and bioaugmentation: a review. Int Environ Bioremed Biodeg 2015; 3(1): 28-39.

[9]   Prescott MI, Harle JD, Klein DA. Microbiology of Food. 5th ed. New York, USA: McGraw-Hill Ltd 2002; pp. 964-76.

[10]   Bamforth SM, Singleton I. Bioremediation of polycyclic aromatic hydrocarbons: current knowledge and future directions. J Chem Technol Biotechnol 2005; 80(7): 723-36.

[11]   García-Sánchez M, Košnář Z, Mercl F, Aranda E, Tlustoš P. A comparative study to evaluate natural attenuation, mycoaugmentation, phytoremediation, and microbial-assisted phytoremediation strategies for the bioremediation of an aged PAH-polluted soil. Ecotoxicol Environ Saf 2018; 147: 165-74.
[http://dx.doi.org/10.1016/j.ecoenv.2017.08.012] [PMID: 28843188]

[12]   Lau KL, Tsang YY, Chiu SW. Use of spent mushroom compost to bioremediate PAH-contaminated samples. Chemosphere 2003; 52(9): 1539-46.
[http://dx.doi.org/10.1016/S0045-6535(03)00493-4] [PMID: 12867186]

[13]   Makkar RS, Rockne KJ. Comparison of synthetic surfactants and biosurfactants in enhancing biodegradation of polycyclic aromatic hydrocarbons. Environ Toxicol Chem 2003; 22(10): 2280 -92.
[http://dx.doi.org/10.1897/02-472] [PMID: 14551990]

[14]   Chebbi A, Hentati D, Zaghden H, *et al.* Polycyclic aromatic hydrocarbon degradation and biosurfactant production by a newly isolated *Pseudomonas sp.* strain from used motor oil-contaminated soil. Int Biodeterior Biodegradation 2017; 122: 128-40.
[http://dx.doi.org/10.1016/j.ibiod.2017.05.006]

[15]   Pieper DH, Seeger M. Bacterial metabolism of polychlorinated biphenyls. J Mol Microbiol Biotechnol 2008; 15(2-3): 121-38.
[PMID: 18685266]

[16]   Wiegel J, Wu Q. Microbial reductive dehalogenation of polychlorinated biphenyls. FEMS Microbiol Ecol 2000; 32(1): 1-15.
[http://dx.doi.org/10.1111/j.1574-6941.2000.tb00693.x] [PMID: 10779614]

[17]   Chatterjee S, Chattopadhyay P, Roy S, Sen SK. Bioremediation: a tool for cleaning polluted environments. J Appl Biosci 2008; 11: 594-601.

[18]   Mrozik A, Piotrowska-Seget Z. Bioaugmentation as a strategy for cleaning up of soils contaminated with aromatic compounds. Microbiol Res 2010; 165(5): 363-75.
[http://dx.doi.org/10.1016/j.micres.2009.08.001] [PMID: 19735995]

[19]   Safdari MS, Kariminia HR, Rahmati M, *et al.* Development of bioreactors for comparative study of natural attenuation, biostimulation, and bioaugmentation of petroleum-hydrocarbon contaminated soil. J Hazard Mater 2018; 342: 270-8.
[http://dx.doi.org/10.1016/j.jhazmat.2017.08.044] [PMID: 28843796]

[20]   Abhilash PC, Srivastava S, Singh N. Comparative bioremediation potential of four rhizospheric microbial species against lindane. Chemosphere 2011; 82(1): 56-63.
[http://dx.doi.org/10.1016/j.chemosphere.2010.10.009] [PMID: 21044795]

[21]   Das N, Chandran P. Microbial degradation of petroleum hydrocarbon contaminants: an overview. Biotechnol Res Int 2011; 2011: 1-13.
[http://dx.doi.org/10.4061/2011/941810] [PMID: 21350672]

[22]   Saez JM, Álvarez A, Benimeli CS, Amoroso MJ. Enhanced lindane removal from soil slurry by immobilized Streptomyces consortium. Int Biodeterior Biodegradation 2014; 93: 63-9.
[http://dx.doi.org/10.1016/j.ibiod .2014.05.013]

[23]   Wang L, Chi XQ, Zhang JJ, Sun DL, Zhou NY. Bioaugmentation of a methyl parathion contaminated soil with *Pseudomonas* sp. strain WBC-3. Int Biodeterior Biodegradation 2014; 87 : 116-21.
[http://dx.doi.org/10.1016/j.ibiod.2013.11.008]

[24]   Chen M, Xu P, Zeng G, Yang C, Huang D, Zhang J. Bioremediation of soils contaminated with polycyclic aromatic hydrocarbons, petroleum, pesticides, chlorophenols and heavy metals by composting: Applications, microbes and future research needs. Biotechnol Adv 2015; 33(6): 745-55.
[http://dx.doi.org/10.1016/j.biotechadv.2015.05.003] [PMID: 26008965]

[25]   Kuppusamy S, Palanisami T, Megharaj M, Venkateswarlu K, Naidu R. *Ex situ* remediation technologies for environmental pollutants: a critical perspective. Rev Environ Contam Toxicol 2016; 236: 117-92. a
[PMID: 26423074]

[26]   Kuppusamy S, Palanisami T, Megharaj M, Venkateswarlu K, Naidu R. *In situ* remediation approaches for the management of contaminated sites: a comprehensive overview'. Rev Environ Contam Toxicol 2016; 236: 1-115. b
[PMID: 26423073]

[27]   Rabus R, Boll M, Heider J, *et al.* Anaerobic Microbial Degradation of Hydrocarbons: From Enzymatic Reactions to the Environment. J Mol Microbiol Biotechnol 2016; 26(1-3): 5-28.
[PMID: 26960061]

[28]   dos Santos JJ, Maranho LT. Rhizospheric microorganisms as a solution for the recovery of soils contaminated by petroleum: A review. J Environ Manage 2018; 210: 104-13.
[http://dx.doi.org/10.1016/j.jenvman.2018 .01.015] [PMID: 29331851]

[29]   Tyagi M, da Fonseca MMR, de Carvalho CCCR. Bioaugmentation and biostimulation strategies to improve the effectiveness of bioremediation processes. Biodegradation 2011; 22(2): 231-41.
[http://dx.doi.org/10. 1007/s10532-010-9394-4] [PMID: 20680666]

[30]   Tchelet R, Levanon D, Mingelgrin U, Henis Y. Parathion degradation by a *Pseudomonas sp.* and a *Xanthomonas sp.* and by their crude enzyme extracts as affected by some cations. Soil Biol Biochem 1993; 25(12): 1665-71.
[http://dx.doi.org/10.1016/0038-0717(93)90168-B]

[31]   Vidali M. Bioremediation. An overview. Pure Appl Chem 2001; 73(7): 1163-72.
[http://dx.doi.org/10.1351/pac 200173071163]

[32]   Goswami M, Chakraborty P, Mukherjee K, *et al.* Bioaugmentation and biostimulation: a potential strategy for environmental remediation. J Microbiol Exp 2018; 6(5): 223-31.

[33]   Parthipan P, Preetham E, Machuca LL, Rahman PKSM, Murugan K, Rajasekar A. Biosurfactant and degradative enzymes mediated crude oil degradation by bacterium *Bacillus subtilis* A1. Front Microbiol 2017; 8: 193.
[http://dx.doi.org/10.3389/fmicb.2017.00193] [PMID: 28232826]

[34]   Ghazali FM, Rahman RNZA, Salleh AB, Basri M. Biodegradation of hydrocarbons in soil by microbial consortium. Int Biodeterior Biodegradation 2004; 54(1): 61-7.
[http://dx.doi.org/10.1016/j.ibiod.2004.02.002]

[35]   Bhardwaj D, Ansari MW, Sahoo RK, Tuteja N. Biofertilizers function as key player in sustainable agriculture by improving soil fertility, plant tolerance and crop productivity. Microb Cell Fact 2014; 13(1): 66.
[http://dx.doi.org/10.1186/1475-2859-13-66] [PMID: 24885352]

[36]   Barea JM, Richardson AE. Phosphate mobilisation by soil microorganisms.Principles of Plant-Microbe Interactions. Heidelberg: Springer International Publishing Switzerland 2015; pp. 225-34.
[http://dx.doi.org/10.1007/978-3-319-08575-3_24]

[37]   Richardson AE, Barea JM, McNeill AM, Prigent-Combaret C. Acquisition of phosphorus and nitrogen in the rhizosphere and plant growth promotion by microorganisms. Plant Soil 2009; 321(1-2): 305-39.
[http://dx.doi.org/10.1007/s11104-009-9895-2]

[38]   Bhardwaj D, Ansari MW, Sahoo RK, Tuteja N. Biofertilizers function as key player in sustainable agriculture by improving soil fertility, plant tolerance and crop productivity. Microb Cell Fact 2014; 13(1): 66.
[http://dx.doi.org/10.1186/1475-2859-13-66] [PMID: 24885352]

[39]   Souza R, Ambrosini A, Passaglia LMP. Plant growth-promoting bacteria as inoculants in agricultural soils. Genet Mol Biol 2015; 38(4): 401-19.

[http://dx.doi.org/10.1590/S1415-475738420150053] [PMID: 26537605]

[40]   Ramasamy M, Geetha T, Yuvaraj M. Role of biofertilizers in plant growth and soil health, nitrogen fixation In: Rigobelo Everlon Cid, Serra Ademar Pereira, Eds. IntechOpen. 2020.
[http://dx.doi.org/10.5772/intechopen.87429]

[41]   Smith SE, Read DJ. Mycorrhizal Symbiosis. 3rd ed., New York: Elsevier, Academic Press 2008.

[42]   Heijden MGA, Martin FM, Selosse MA, Sanders IR. Mycorrhizal ecology and evolution: the past, the present, and the future. New Phytol 2015; 205(4): 1406-23.
[http://dx.doi.org/10.1111/nph.13288] [PMID: 25639293]

[43]   Bonfante P, Genre A. Mechanisms underlying beneficial plant–fungus interactions in mycorrhizal symbiosis. Nat Commun 2010; 1(1): 48.
[http://dx.doi.org/10.1038/ncomms1046] [PMID: 20975705]

[44]   Barea JM, Pozo MJ, Azcón R, Azcón-Aguilar C. Microbial interactions in the rhizosphere.Molecular Microbial Ecology of the Rhizosphere. Hoboken, New Jersey, USA: Wiley Blackwell 2013; Vol. 1: pp. 29-44. a
[http://dx.doi.org/10.1002/9781118297674.ch4]

[45]   Olivares J, Bedmar EJ, Sanjuán J. Biological nitrogen fixation in the context of global change. Mol Plant Microbe Interact 2013; 26(5): 486-94.
[http://dx.doi.org/10.1094/MPMI-12-12-0293-CR] [PMID: 23360457]

[46]   de Bruijn FJ. Biological nitrogen fixation.Principles of Plant-Microbe Interactions. Heidelberg: Springer International Publishing Switzerland 2015; pp. 215-24.
[http://dx.doi.org/10.1007/978-3-319-08575-3_23]

[47]   Jeffries P, Barea JM. Arbuscular Mycorrhiza – a key component of sustainable plant-soil ecosystems.The Mycota, a comprehensive treatise on fungi as experimental systems for basic and applied research. Berlin, Heidelberg, Berlin, Heidelberg: Springer 2012; pp. 51-75.
[http://dx.doi.org/10.1007/978-3-642-30826-0_4]

[48]   Cornejo P, Meier S, Borie G, Rillig MC, Borie F. Glomalin-related soil protein in a Mediterranean ecosystem affected by a copper smelter and its contribution to Cu and Zn sequestration. Sci Total Environ 2008; 406(1-2): 154-60.
[http://dx.doi.org/10.1016/j.scitotenv.2008.07.045] [PMID: 18762323]

[49]   Barea JM, Pozo MJ, López-Ráez JA, *et al.* Arbuscular Mycorrhizas and their significance in promoting soil-plant systems sustainability against environmental stresses. 2013. b
[http://dx.doi.org/10.1201/b15251-16]

[50]   Geraldin MW. Lengai, James W, Muthomi Biopesticides and their role in sustainable agricultural production. J Biosci Med 2018; 6: 7-41.

[51]   Nawaz M, Mabubu JI, Hua H. Current status and advancement of biopesticides: microbial and botanical pesticides. J Entomol Zool Stud 2016; 2: 241-6.

[52]   Kachhawa D. Microorganisms as a Biopesticides. J Entomol Zool Stud 2017; 3: 468-73.

[53]   Srijita D. Biopesticides: An eco-friendly approach for pest control. World J Pharm Pharm Sci 2015; 6: 250-65.

[54]   Suprapta DN. Potential of microbial antagonists as biocontrol agents against plant fungal pathogens. Int Soc Southeast Asian Agric Sci J 2012; 2: 1-8.

[55]   Selosse MA, Bessis A, Pozo MJ. Microbial priming of plant and animal immunity: symbionts as developmental signals. Trends Microbiol 2014; 22(11): 607-13.
[http://dx.doi.org/10.1016/j.tim.2014.07.003] [PMID: 25124464]

[56]   Perpetuo EA, Souza CB, Nascimento CAO. Engineering bacteria for bioremediation, Progress in molecular and environmental bioengineering - From snalysis and modeling to technology applications. Angelo Carpi, IntechOpen 2011.

[http://dx.doi.org/10.5772/19546]

[57] Ezaki B, Gardner RC, Ezaki Y, Matsumoto H. Expression of aluminum-induced genes in transgenic arabidopsis plants can ameliorate aluminum stress and/or oxidative stress. Plant Physiol 2000; 122(3): 657-66.
[http://dx.doi.org/10.1104/pp.122.3.657] [PMID: 10712528]

[58] Bae W, Chen W, Mulchandani A, Mehra RK. Enhanced bioaccumulation of heavy metals by bacterial cells displaying synthetic phytochelatins. Biotechnol Bioeng 2000; 70(5): 518-24.
[http://dx.doi.org/10.1002/1097-0290(20001205)70:5<518::AID-BIT6>3.0.CO;2-5] [PMID: 11042548]

[59] Mejáre M, Bülow L. Metal-binding proteins and peptides in bioremediation and phytoremediation of heavy metals. Trends Biotechnol 2001; 19(2): 67-73.
[http://dx.doi.org/10.1016/S0167-7799(00)01534-1] [PMID: 11164556]

[60] Leng P, Zhang Z, Pan G, Zhao M. Applications and development trends in biopesticides. Afr J Biotechnol 2014; 86: 19864-73.

[61] Bae W, Mehra RK, Mulchandani A, Chen W. Genetic engineering of Escherichia coli for enhanced uptake and bioaccumulation of mercury. Appl Environ Microbiol 2001; 67(11): 5335-8.
[http://dx.doi.org/10.1128/AEM .67.11.5335-5338.2001] [PMID: 11679366]

[62] Kumar BL, Gopal DVRS. Effective role of indigenous microorganisms for sustainable environment. 3 Biotech 2015; 5(6): 867-76.
[http://dx.doi.org/10.1007/s13205-015-0293-6] [PMID: 28324402]

[67] Clemens S, Palmgren MG, Krämer U. A long way ahead: understanding and engineering plant metal accumulation. Trends Plant Sci 2002; 7(7): 309-15.
[http://dx.doi.org/10.1016/S1360-1385(02)02295-1] [PMID: 12119168]

[63] Samanta SK, Singh OV, Jain RK. Polycyclic aromatic hydrocarbons: environmental pollution and bioremediation. Trends Biotechnol 2002; 20(6): 243-8.
[http://dx.doi.org/10.1016/S0167-7799(02)01943-1] [PMID: 12007492]

[64] Buhari ML, Sulaiman BR, Vyas NL, Sulaiman B, Harisu Umar Y. Role of Biotechnology in Phytoremediation. J Bioremediat Biodegrad 2016; 7: 330.

[65] Yang X, Feng Y, He Z, Stoffella PJ. Molecular mechanisms of heavy metal hyperaccumulation and phytoremediation. J Trace Elem Med Biol 2005; 18(4): 339-53.
[http://dx.doi.org/10.1016/j.jtemb.2005.02.007] [PMID: 16028496]

[66] Mello-Farias PC, Chaves ALS. Biochemical and molecular aspects of toxic metals phytoremediation using transgenic plants.Transgenic approach in Plant Biochemistry and Physiology. Kerala, India: Research Singpost 2008; pp. 253-66.

[67] Sharma HC, Crouch JH, Sharma KK, Seetharama N, Hash CT. Applications of biotechnology for crop improvement: prospects and constraints. Plant Sci 2002; 163(3): 381-95.
[http://dx.doi.org/10.1016/S0168-9452(02)00133-4]

[68] Pilon-Smits E, Pilon M. Phytoremediation of metals using transgenic plants. Crit Rev Plant Sci 2002; 21(5): 439-56.
[http://dx.doi.org/10.1080/0735-260291044313]

[69] Eapen S, D'Souza SF. Prospects of genetic engineering of plants for phytoremediation of toxic metals. Biotechnol Adv 2005; 23(2): 97-114.
[http://dx.doi.org/10.1016/j.biotechadv.2004.10.001] [PMID: 15694122]

[70] Fulekar MH, Singh A, Bhaduri AM. Genetic engineering strategies for enhancing phytoremediation of heavy metals. Afr J Biotechnol 2008; 8: 529-35.

[71] Dhankher OP, Li Y, Rosen BP, et al. Engineering tolerance and hyperaccumulation of arsenic in plants by combining arsenate reductase and γ-glutamylcysteine synthetase expression. Nat Biotechnol

2002; 20(11): 1140-5.
[http://dx.doi.org/10.1038/nbt747] [PMID: 123 68812]

[72]   Muhammad Aslam M, K Karanja J, Bin Gias Uddin J, Zaynab M, Hassanyar A. Use of Biotechnological Tools for Environmental Cleanup. Acta Scientific Agriculture 2019; 3(12): 46-8.
[http://dx.doi.org/10. 31080/ASAG.2019.03.0721]

[73]   Hirschi KD, Korenkov VD, Wilganowski NL, Wagner GJ. Expression of *arabidopsis* CAX2 in tobacco. Altered metal accumulation and increased manganese tolerance. Plant Physiol 20 00; 124(1): 125-34.
[http://dx.doi.org/10.1104/pp.124.1.125] [PMID: 10982428]

[74]   Arazi T, Sunkar R, Kaplan B, Fromm H. A tobacco plasma membrane calmodulin-binding transporter confers Ni^{2+} tolerance and Pb^{2+} hypersensitivity in transgenic plants. Plant J 1999; 20(2): 171-82.
[http://dx.doi.org/10.1046/j.1365-313x.1999.00588.x] [PMID: 10571877]

[75]   Goto F, Yoshihara T, Saiki H. Iron accumulation in tobacco plants expressing syabean ferritin gene. Transgenic Res 1998; 7(3): 173-80.
[http://dx.doi.org/10.1023/A:1008836812714]

[76]   Goto F, Yoshihara T, Shigemoto N, Toki S, Takaiwa F. Iron fortification of rice seed by the soybean ferritin gene. Nat Biotechnol 1999; 17(3): 282-6.
[http://dx.doi.org/10.1038/7029] [PMID: 10096297]

[77]   Rugh CL, Bizily SP, Meagher RB. Phytoremediation of environmental mercury pollution. 2000.

[78]   Bizily SP, Rugh CL, Meagher RB. Phytodetoxification of hazardous organomercurials by genetically engineered plants. Nat Biotechnol 2000; 18(2): 213-7.
[http://dx.doi.org/10.1038/72678] [PMID: 10657131]

[79]   de Souza MP, Pickering IJ, Walla M, Terry N. Selenium assimilation and volatilization from selenocyanate-treated Indian mustard and muskgrass. Plant Physiol 2002; 128(2): 625-33.
[http://dx.doi.org/10.1104 /pp.010686] [PMID: 11842165]

CHAPTER 6

Remediation of Environmental Contaminants using Nanoparticles

Bhupinder Dhir[1,*]

[1] *School of Sciences, Indira Gandhi National Open University, New Delhi, India*

Abstract: Nanotechnology plays an important role in monitoring, preventing, and remediating environmental pollution. Nanomaterials are used in the detection and removal of contaminants such as heavy metals, organic pollutants (aliphatic and aromatic hydrocarbons), and biological agents such as viruses, bacteria, and parasites. Nanomaterials act as good adsorbents, catalysts, and sensors due to their large specific surface areas and high reactivities. Physicochemical properties, such as large surface area, facilitate easier biodegradation/remediation of environmental contaminants. Carbon nanomaterials, namely carbon nanotubes, graphene, graphene oxide, and zero-valent iron nanoparticles, have shown great potential for the removal of heavy metals and organic contaminants from water and soil. Hence, nanoremediation represents an innovative approach to safe and sustainable remediation of environmental contamination.

Keywords: Carbon nanotubes, Nanomaterials, Nanotechnology, Soil, Water.

INTRODUCTION

Environmental pollution has emerged as one of the major problems all over the world. The major source of pollution is the presence of contaminants that are non-biodegradable in nature. They directly exert toxic effects on living beings but indirectly affect the interactive relationships in ecological communities. Toxic environmental pollutants such as heavy metals and organic compounds (pesticides, dyes) pose risks to human health [1, 2]. The remediation/removal of pollutants from the environment has been done using various methods. Most of the treatment methods employed are difficult, time-consuming, expensive, require pretreatment, and are less effective. Nanotechnology has emerged as a technology with the potential to remediate/treat a wide range of contaminants and hence plays a role in improving the quality of the environment [3, 4]. Major applications of nanotechnology include

---

* **Corresponding author Bhupinder Dhir:** School of Sciences, Indira Gandhi National Open University, New Delhi, India; E-mail: bhupdhir@gmail.com

1. Remediation and removal of environmental contaminants,
2. Detection and prevention of pollution.

Nanoparticles (NPs) have been used to remediate various forms of environmental contamination [5, 6]. Nanomaterials have shown the potential to remove/treat various pollutants such as organic compounds, metal ions, and many biological materials from the environment. Carbon nanotubes (CNTs), graphene, graphene oxide, activated carbon, nanoscale zeolites, nanofibers, and titanium dioxide are some of the nanomaterials used in the remediation of environmental contaminants. The properties of nanoparticles, such as big surface area, high surface-to-volume ratio, and reactivity, contribute to their high remediation capacity.

Soil and groundwater contaminated with inorganic pollutants such as heavy metals, organic contaminants, and pharmaceutical and personal care products ion has been successfully treated using nanoremediation [7 - 10]. Remediation of contaminants occurs either by sequestration or degradation. Both *in situ* (removal of contaminants at their source site) and *ex-situ* (treatment of contaminants at a site different from their source) treatment of contaminants can be done using nanotechnology. *In situ* remediation helps in the treatment of contaminants at the source (such as crevices and aquifers) and eliminates costly operations such as the transport of contaminants. Nanoscience helps in effective remediation for a certain pollutant (such as metal), depending upon their affinity and selectivity.

NANOMATERIALS USED IN ENVIRONMENTAL REMEDIATION

Nanomaterials have been used successfully for remediating contaminants present in water and land areas of hazardous waste sites [10]. The use of nanomaterials in environmental remediation has proved better than conventional techniques. Nanosorbents, nanoclays, nano-aerogels, nano-iron oxides, nanoscale zerovalent iron (nZVI), nano-metal oxides, dendrimers and nanofibres are some of the materials that have shown great potential to remove/treat contaminants Table (**1**). Major nanomaterials that have been exploited in the removal of contaminants include carbon-based nanomaterials, dendrimers, ferritin and metallopor phyrinogens. Treatment of surface water, groundwater, wastewater, soil and sediments contaminated with heavy metals, microorganisms, and organic and inorganic solutes has been done using nanoscale particles [11 - 13].

Oxide-based Nanoparticles

Oxide-based nano-particles include inorganic nanoparticles prepared by using non-metals and metals. These include titanium oxides, titanium oxide/dendrimers composites, zinc oxides, magnesium oxide, manganese oxides, and ferric oxides. They help in removing hazardous pollutants from wastewater.

Table 1. Nanomaterials used in the remediation of pollutants.

Type of Nanomaterial	Pollutant	Area of application
Amphiphilic polyurethane nanoparticles	Phenanthrene	Soil
Nanoscale zero-valent iron	Heavy metals, hydrocarbons, oil	Water
Nanocellulose	Heavy metals, dyes	Water
Polymer nanocomposites	Hydrocarbons, heavy metals	Water and soil
Carbon nanotubes	Ethylbenzene, copper, nickel ions, cationic dye, oil	Water
Dendrimer nanoparticle composite	Organic pollutants, metal ions	Water
Titanium dioxide	Organic pollutants	Water, Soil
Bimetallic nanoparticles	Polybrominated diphenyl ethers, chlorine	Water, Soil
Magnetic nanoparticles	Heavy metals	Water, Soil
Graphene-based nanomaterials	Cationic compounds	Water, Air
Plasmonic-based nanomaterials	Organic pollutants, pathogenic microorganisms	Water

Manganese oxide (MnO), Magnesium oxide (MgO) and iron-based nano-particles such as Fe_2O_3 have been successfully used in the removal of heavy metals from wastewater. Their high surface area and polymorphic structure contribute to their adsorption capacity [14]. Nano-adsorbents such as nano assemblies, nanoplates, nano-sheets, and ZnO nano-rods are used in the removal of heavy metals from wastewater [15, 16]. High removal of heavy metals such as Pb and Cd (II), from wastewater using mesoporous hierarchical ZnO nano-rods, has been reported [16].

Factors such as pH, temperature, adsorbent dose, and incubation time regulate the adsorption of heavy metals [17]. Research showed that modification of the surface of MnOs increased the adsorption capacity of Fe_2O_3 [18]. The affinity for the removal of different pollutants such as Cr, Co, Ni, Cu, Cd, Pb, and As from wastewater increased when the surface of nano-adsorbents was modified [17].

Zero-Valent Iron Nanoparticles

Nanoscale iron particles remediate and transform a variety of environmental contaminants. These nanoparticles have been proven useful in treating environmental contaminants such as toxic metals, and organic and inorganic compounds. These nanoparticles exhibit high reactivity in remediating compounds from aqueous media [19]. These particles effectively remove contaminants from the environment. Studies have shown that nanoscale zero-valent iron (NZVI) removes arsenic (V) from groundwater. Zero-valent Fe-NPs supported by hydrophilic carbon (Fe/C) and poly (acrylic acid) (Fe/PAA) act as reactive materials and assist in the dehalogenation of chlorinated hydrocarbons present in groundwater and soils. Dehalogenation of trichloroethylene (TCE) has been achieved using nickel-iron NPs. Zero-valent metals (ZVMs) remove pentachlorophenol (PCP) from aqueous solutions. This occurs due to dechlorination or sorption to the surface of ZVM. Transformation of DDT, DDD [1,1-dichloro-2,2-bi's(p-chlorophenyl)ethane], and DDE [2,2-bi's(p-chlorophe nyl)-1, 1-dichloroethylene] by zero-valent iron has also been noted [20].

CARBON-BASED MATERIALS

Nanocrystals and carbon nanotube(s) are some Carbon-based nanomaterials that act as sorbents, high-flux membranes, and depth filters and hence play a role in pollution prevention strategies [20]. Carbon nanotubes (CNTs) act as efficient nanoadsorbents and remove water pollutants such as heavy metals, organic pollutants,*etc.* CNT-modified equivalents that remove organic contaminants from water have been developed [21, 22]. CNTs act as favorable nanoadsorbents for pharmaceuticals and personal care products because of their unique characteristics and high adsorption capacity. Abatement of organic compounds is supposed to occur *via* π -π and n–π interactions. Adsorption of organic compounds is influenced by hydrogen bonding and Lewis acid-base interactions.

Carbon Nanotubes (CNTs)

Carbon nanotubes remove heavy metals and various organic contaminants from wastewater through adsorption [23 - 26]. Therefore carbon nanomaterials are utilized in the purification of heavy metal-contaminated water. High surface area helps in the retention of high adsorption capacity. The potential of both single-walled carbon nanotubes (SWCNTs), multiwalled carbon nanotubes (MWCNTs), and hybrid carbon nanotubes (HCNTs) for the removal of contaminants from aqueous solution has been established.

Both oxidized and composite forms of carbon nanotubes show tremendous potential to adsorb heavy metal ions. Mesopores present in the carbon nanotubes assist in the removal of heavy metal ions. Removal of heavy metals depends on the properties such as the surface of nanoparticles, their electrochemical potential, and ion exchangeability. Large specific surface area contributes to high quantum efficiency, chemical stability, good adsorption capacity, and catalytic activity of nanoparticles. Adsorption of various organic molecules by carbon nanotubes is also reported.

The modification in the surface of CNT increases their adsorption activity. Modification techniques, such as acid treatment, metal impregnation, and functional molecules/group grafting, have been used [24, 27 - 29]. The characteristics of CNTs surface, like BET surface area, surface charge, dispersion, and hydrophobicity, get altered after using these techniques. New functional groups get added on the surface of CNTs, and this results in an increase in their adsorption capacity [17].

SWCNTs

The porous structure, high surface area, easy surface functionalization, and nanosize establish SWCNTs as a material that is widely used for the control/remediation of environmental pollution. These properties make SWCNTs a very promising material for application in water treatment.

MWCNTs

Carbon nanotubes that contain multiple rolled layers of graphene are called multiwall carbon nanotubes (MWCNTs). High surface area, electrical and thermal conductivity, and tensile strength are some unique properties of MWCNTs that contribute to their potential for application in water treatment, especially in the removal of heavy metal ions. The chemical interactions with functional groups present in MWCNTs help in the adsorption of heavy metal ions. Nanocomposites such as MWCNTs-Fe_2O_3, MWCNTs-ZrO_2, MWCNTs-Fe_3O_4, MWCNTs-Al_2O_3, and MWCNTs-MnO_2-Fe_2O_3 have been successfully used for removing heavy metal ions (Cr, As, Ni, Pb, and Cu) from water. High number of oxygenated functional groups present on the surface of the carbon nanotubes contribute to this. Exceptionally high sorption capacity for heavy metals has been shown by oxidized MWCNTs. Significant removal of Mn (71.5%) and Fe (52%) has been reported when a concentration of 50 ppm of these metal ions is present in an aqueous solution [30]. CNT-based nanocomposites have also shown the capacity to remove Fe and Mn from water.

Photocatalysts

The photocatalysts contain semiconductor metals that help in the degradation of a variety of persistent organic pollutants such as dyes, detergents, pesticides, and volatile organic compounds present in wastewater [31]. Photocatalytic reactions are based on the interaction of light energy with metallic nanoparticles. Nanoparticles show high photocatalytic activities [32]. Photocatalysis technique has shown high potential for water purification and wastewater treatment. Semiconductor nano-catalysts effectively degrade halogenated and non-halogenated organic compounds, persistent compounds like pharmaceutical compounds and personal care products (PCPPs), and heavy metals [33]. High reactivity and chemical stability of TiO_2 increase its application in photocatalysis [34, 35]. TiO_2 possesses high antimicrobial power and shows an efficient ability to inactive the pathogenic organism such as bacteria [36]. ZnO also found its role due to its photocatalytic action.

Photocatalytic Nanotubes

Nanotubes made up of carbon molecules are highly electronegative, electrically insulating, and easily polymerizable. They easily degrade organic compounds and chlorinated chemicals. Titanium dioxide nanotubes, in particular, have been proven to be photocatalytic degraders of chlorinated compounds.

SAMMS Particles

Self-assembled monolayers on mesoporous silica (SAMMS) particles include a nanoporous ceramic substrate with a functional group monolayer. Mercury, radionuclides, arsenate, chromate, selenite, and pertechnate are some contaminants that easily get sorbed to SAAMS. Functional molecules /contaminants bind to the silica surface of SAMMS *via* covalent bonding.

Graphene

Graphene, an allotrope of carbon, possesses certain special features that make it useful for various environmental applications. Graphene oxide (GO) is produced by the chemical oxidation of graphite. Heavy metals get adsorbed on the surface of GO *via* hydroxyl and carboxyl groups [37 - 39]. Graphene oxide specifically removes organic dyes from water. The properties such as high surface area, mechanical strength, lightweight, flexibility, and chemical stability establish it as an ideal adsorbent for the removal of heavy metals from wastewater [40 - 42]. Graphene and its other composite also show high efficiency for the removal of heavy metals from wastewater. They also show potential for the adsorption of gaseous contaminants. Graphene-based photocatalytic materials help in water decontamination [43].

Other nanomaterials that have been used in environmental remediation include:

Ferritin – It is an iron storage protein and has shown the capacity to remove toxic contaminants such as technetium and chromium from ground and surface water.

Dendrimers – Dendrimers are highly branched macromolecules. They consist of a central core, interior cells, and a terminal or peripheral cell group. The void spaces present in dendrimers help in interaction with other substances. A dendrimers-nanoparticle composite that shows high catalytic activity and high reactivity can be prepared. These composites are used in the treatment of water and dye.

Metalloporphyrinogens – These are metal complexes that possess organic porphyrin molecules. They have been shown to reduce chlorinated hydrocarbons like PCE, TCE, and carbon tetrachloride present in contaminated groundwater and soil.

Polymeric NPs

Polymer nanomaterials show high potential for use in environmental applications. They possess properties such as high chemical and thermal stability and show good adsorption capacity for the removal of toxic metal ions, dyes, and microorganisms from water/wastewater [44, 45]. Polymer nanoparticles show a high sorption capacity for hydrophobic organic contaminants such as polycyclic aromatic hydrocarbons (PAHs), and phenanthrene (PHEN) [46]. Polymeric nanoparticles (NPs), namely urethane acrylate (PMUA) modified using poly(ethylene)glycol, show enhancement in the capacity for mineralization of phenanthrene (PHEN) present in water.

Amphiphilic polyurethane (APU) nanoparticles for the remediation of polynuclear aromatic hydrocarbons (PAHs) from soils have been developed. The hydrophilic surface of these nanoparticles supports mobility in the soil, while the hydrophobic interior provides affinity for the removal of hydrophobic organic contaminants. Reports suggest that APU NPs remove phenanthrene from contaminated aquifer sand.

Poly (amidoamine) or dendrimers (PAMAM) are nanopolymers containing functional groups such as primary amines, carboxylates, and hydroxamates. These polymers trap various contaminants such as metal ions (such as Cu, Ag, Au, Fe, Ni, Zn). Hence find their use in the remediation of water contaminated with metal ions. They act as chelating agents and bind with metal ions, assisting in water purification in ultrafilters. They have also been used as antibacterial/antivirus agents.

NANOMATERIALS IN WATER TREATMENT

Nanotechnology has proved to be helpful in improving the quality of water. Generally, methods such as separation, filtration, bioremediation, and disinfection are used to treat water. Nanomaterials that are used in the remediation of water mainly include zeolites, carbon nanotubes (CNTs), single-enzyme nanoparticles, zerovalent iron (ZVI), and biopolymers [47 - 49].

Nanosorbents that are composed of nanoscale particles of inorganic or organic materials also show a capacity for contaminant absorption. Nanosorbents find their use in the treatment of groundwater and wastewater [50 - 52].

In the e*x-situ treatment* method of remediation, groundwater is removed through wells and trenches and treated above ground. Air stripping, carbon adsorption, biological reactors, or chemical precipitation are the processes applied earlier but produced highly contaminated waste that needs proper disposal. In nanotechnologies, remediation of contaminants in water occurs by degradation. The remediation of organic contaminants is done by photo-oxidation catalyzed by metal oxide nanoparticles such as TiO_2. In another method, Fe nanoparticles are injected into groundwater through application wells.

In another method, *i.e.*, *in situ*, treatment is done by creating a reactive zone with immobile nanoparticles or reactive nanoparticles. These nanoparticles migrate to contaminated zones. The permeable reactive barrier (PRB) is the most commonly used *in situ* method of remediation. It is used for cleaning up contaminated groundwater. These barriers are composed of materials that degrade or immobilize contaminants as the groundwater passes through them. The barriers are installed within the flow path of a contaminant plume. Depending upon the type of contaminant, the material of the barrier is chosen. PRBs remediate only those contaminant plumes that pass through them but not those which are present beyond the barrier. PRB treatment is used to remove pollutants such as chlorinated hydrocarbons, aromatic nitro compounds, polychlorinated biphenyls (PCBs), pesticides, and chromate compounds. The only limitation for the applicability of the technique is cost.

Various zero-valent metal (nZV) nanoparticles play a significant role in *in situ* remediation. There are two types of ZVI, known as nanoscale ZVI (nZVI) and reactive nanoscale iron product (RNIP). Zero-valent iron (ZVI) nanoparticles are composed of iron (Fe) and have a diameter of 100–200 nm, whereas RNIP particles consist of Fe and Fe_3O_4 (50/50 wt%). Zero-valent iron (ZVI) is commonly used as a filter material in PBR. It showed a significant reduction in contaminants.

ZVI shows a high rate of reactivity and hence is able to remove a large number of contaminants, such as Cu and chlorinated hydrocarbons.

Nano-iron can be used *via* direct injection into the soil or sediment. Nanoparticles can be mixed with water to form a slurry. The injected particles remain in the form of a suspension, and a treatment zone is formed. In other methods, nanoparticles attach to a solid matrix, such as activated carbon. Nano-iron gets substituted with other metals such as Pd, Ag, Pt, Co, Cu, Au, and Al. Metals show the ability to reduce contaminants. Iron and iron–nickel–copper have been shown to degrade trichloroethene and trichloroethane.

Ferritin, an iron-containing protein, controls the formation of mineralized structures. Ferritin is composed of 24 structurally similar polypeptides connected to each other to form a cage-like protein structure. Research studies have shown that ferritin can resist photoreduction, is more stable, and can easily remediate toxic metals and chlorocarbon. It converts Cr(VI) into Cr(III).

Polymer nanoparticles are amphiphilic, *i.e.*, they possess both hydrophobic and hydrophilic parts. The polymer forms a polymer cell that has a big diameter (nanometers) inside the hydrophobic part, while the hydrophilic part is outside in conditions of water availability. The polymer nanoparticles are kept stable *via* a crosslink that occurs before the aggregation of particles. They act as surfactants and improve the remediation of hydrophobic organic contaminants.

Nanotechnology helps in cleaning water by removing germs. Nanomaterials show strong antimicrobial properties as (i) They produce reactive oxygen species (ROS) by photocatalytic reactions. The ROS produced damaged cell components and viruses. Examples of such nanomaterials are TiO_2, ZnO, and fullerol, (ii) They damage the cell envelope of bacteria. Examples of such nanomaterials are carboxyfullerene, CNTs, ZnO, and silver nanoparticles, (iii) They disturb energy transduction. Examples of such nanomaterials are Ag and aqueous fullerene nanoparticles and (iv) They inhibit the activity of enzymes and the synthesis of DNA. An example of such nanomaterials is chitosan. Damage to the cell membrane of bacterial cells by CNTs is the major inactivation mechanism while oxidation stress is suggested as the other possible mechanism of action [53 - 55]. TiO_2 has been proposed as the best material because it is stable in water, is non-toxic, and is cost-effective.

Nanofibres used in filtration membranes and nanobiocides improve the quality of water. Nanofibres in which the surface has been modified reduce membrane fouling caused by bacteria by inhibiting bacteria, thereby improving the quality of water. Silver nanoparticles present in polyvinyl alcohol (PVA) and polyacrylonitrile (PAN) nanofibres have shown excellent antimicrobial activity. A significant reduction (91% to 99%) in bacteria in a contaminated water sample treated with PVA nanofibres has been noted. PAN nanofibres showed 100% mortality in bacteria in the same contaminated water sample. Both PVA and PAN nanofibres are non-toxic and biodegradable synthetic polymers that show excellent antimicrobial activity.

Photocatalyst has been used in water remediation in the United States Environmental Protection Agent (US EPA) SITE program. The photocatalyst

showed the ability to remove contaminants such as 1,1-dichloroethane, cis-1,-
-dichloroethane, 1,1,1-trichloroethane, xylene, and toluene from groundwater.
Pilot scale studies showed that benzene, toluene, ethylbenzene, and xylene
(BTEX) present in groundwater could easily be removed by TiO_2.

Other nanotechnologies for water remediation include

- Self-assembled monolayers on mesoporous silica (SAMMS);
- Dendrimers or dendritic polymers;
- Single nanoparticle enzyme (SEN);
- Nanocrystalline zeolites.

Nanofiltration

Nanofiltration has emerged as one of the leading technology in water treatment. In
this technology, pressure acts as the driving force for filtration. Membranes used
in nanofiltration remove multivalent ions, pesticides, and heavy metals at a large
scale. In nanofiltration membranes, different molecules are targeted based on their
molecular weight. For example, Dow Filmtec, nanofiltration membranes show the
capacity to remove molecules of weight high than 90, 200, or 270 g/mol.

REMEDIATION OF AIR POLLUTION USING NANOMATERIALS

Nanotechnology has also proved its potential in the remediation of air pollution.
Nanotechnology has shown its potential to clean the air by removing toxic gases
such as CO and VOCs. Nanomaterials such as CNTs, gold particles, and other
adsorbents play an important role in the adsorption of toxic gases [56]. CNTs are
composed of hexagonally arranged carbon atoms in graphene sheets that surround
the tube axis. Both single-walled nanotubes (SWNTs) and multi-walled nanotubes
(MWNTs) are unique macromolecules that have a one-dimensional structure,
thermal stability, and exceptional chemical properties. These nanomaterials act as
good adsorbents that remove organic and inorganic pollutants both in air and
aqueous environment.

The adsorption capacity of CNTs is because of pore structure and the existence of
a broad spectrum of surface functional groups of CNTs that can be modified by
chemical or thermal treatment to get optimal performance. Dioxin molecules
interact with the surface of nanotubes and overlapping events increase the
adsorption potential inside the pores. Studies showed that CNTs have good
potential of removing dioxins [57]. The curved surface of the nanotube gives
stronger interaction forces between dioxin and CNTs. CNTs could be used as an
adsorbent for the removal/uptake of NO_x, SO_2, and CO_2 [58].

Unique properties and structures contribute to enhancing the potential applications of SWNTs and MWNTs. SWNTs are reported to act as chemical sensors for NO_2 and NH_3. The electrical resistance of SWNTs changes significantly after exposure to NO_2 or NH_3 gas.

Nano-catalysts with the increased surface area are proven useful in treating/removing gaseous reactions. They speed up the rate of chemical reactions that transform harmful vapors from automobile exhausts and industrial operations into harmless gases. Nanofiber catalysts composed of manganese oxide remove volatile organic compounds from industrial smokestacks. The small pores of the nanostructured membrane separate methane or carbon dioxide from the exhaust.

Photocatalysts *viz.* titanium dioxide (TiO_2), zinc oxide (ZnO), iron (III) oxide (Fe_2O_3), and tungsten oxide (WO_3) oxidize organic pollutants into nontoxic materials. TiO_2, due to its non-toxic nature, high photoconductivity, high photostability, easy availability, and inexpensive nature, is used in advanced methods of photochemical oxidation used in the remediation of water. ZnO as photocatalysts are supposed to help in the detection and remediation of contaminants. Laboratory experiments showed that ZnO had been successfully used to detect and eliminate 4-chlorocatechol.

REMEDIATION OF SOILS USING NANOMATERIALS

Soil contamination is a serious problem all over the world. The quality of soil has deteriorated due to increased levels of pesticides, herbicides, fertilizers, salinity, and many other compounds. Nanoparticles play an important role in the removal of contaminants present in the soil. Nanoparticles convert heavy metals to less toxic forms and degrade organic contaminants such as pesticides (DDT, carbamates) and chlorinated organic solvents [59 - 61]. Magnetic nanoparticles have been proven to be useful in removing salts from the soil. Nanotubes used to develop composite help in the detection of chemicals such as organophosphorus pesticides and other pathogenic substances present in low concentrations.

Amphiphilic polyurethane (APU) nanoparticles find their use in the remediation of soil contaminated with hydrophobic contaminants such as polyaromatic hydrocarbons (PAHs). These particles show the ability to desorb PAH. The APU nanoparticles are stable and independent of their concentration in the aqueous phase. The APU particles have hydrophobic interior regions that show a high affinity for phenanthrene (PHEN). The affinity of APU nanoparticles for PHEN can be controlled by changing the size of the hydrophobic segment.

Nanoparticles such as nZVI, TiO_2, BNPs, and magnetite nanoparticles help in the degradation of several organic pollutants including diuron and phenanthrene [62 - 64]. Metal nanoparticles can be used as biocatalysts for reductive dechlorination.

NANOPHYTOREMEDIATION

Phytoremediation is the process in which plants and their associated microorganisms help in the treatment of contaminants present in various components of the environment such as water, soil, and air. Plants degrade contaminants present in the soil, sludge, sediments, and water. Remediation of contaminants using nanoparticles is known as Nanoremediation. Nanophytore mediation is a technique in which techniques such as nanotechnology and phytoremediation, are collectively used for the remediation of environmental pollutants or cleaning of the environment [65]. Naturally occurring or genetically modified plants play a major role in nanophytoremediation. Nanophytore mediation removes contaminants in an eco-friendly manner and has been found well-suited for the treatment/remediation of moderately contaminated sites.

The success of nanophytoremediation depends upon the properties (physical and chemical) of the contaminants, environmental factors, and characteristics of the plant. The uptake of pollutants by plant species depends mainly on the size of the nanoparticle. The nanoparticles need to be taken up from the roots and carried to other parts of the plant. The process is regulated by the type and chemical composition of the nanoparticle [66]. The selection of the plant used in nanoremediation depends upon factors such as fast growth rate, large biomass production, well-developed root systems with increased surface area, easy harvesting, and better tolerance towards pollutants/contaminants.

Interaction between nanoparticles and plants results in physiological and morphological changes. Silver nanoparticles have been shown to induce abscisic acid and gibberellin production. These growth regulators help the plant to tolerate stress as well as increase the uptake of nutrients and water for improved growth.

Increased surface area per unit mass of nanomaterials helps in bringing a large amount of material into contact with surrounding materials and affects the reactivity positively. Nanomaterials directly catalyze the degradation of waste and toxic materials but also help in enhancing the efficiency of microorganisms that degrade waste and toxic materials.

Though nanotechnology is proved to be very useful in the remediation of environmental components, it can induce effects on human health and the environment. Some of the materials can be proven highly toxic at the nanoscale [67, 68], hence studies focusing on toxicity and health risks associated with nanomaterials also needs to be analyzed. Studies related to the fate and behaviour of nanoparticles in humans and the environment need to be conducted [69]. Research studies related to the impact of nanoparticles on human health can help in potential risk assessment [70, 71].

CONCLUSION

Nanoparticles and nanofibers function as catalysts and adsorbent,s and hence play a significant role in the removal of harmful gases, inorganic and organic pollutants and biological substances. Iron nanoparticles, ferritin, polymeric nanoparticles, nanofibres, nanobiocides and nanoenzymes have been commonly used in the process of water purification or nanofiltration techniques. Nanotechnology has also been applied in the removal of toxic gases such as CO, VOCs, and dioxins from the air by using CNTs, gold nanoparticles, and other adsorbents. NZVPs, due to their high reactivity, help in bringing effective remediation. A better understanding of the mobility, bioavailability, stability, longevity, toxicity/risks of nanoparticles needs to be conducted. The ecological impact of NPs remains unreported. The application of nanotechnology in the abatement of environmental pollution can help in maintaining environmental sustainability.

REFERENCES

[1]   Morgan R. Soil, Heavy Metals, and Human Health.Soils and Human Health (Brevik EC, Burgess LC). Boca Raton: CRC Press 2013; pp. 59-80.

[2]   Schwarzenbach RP, Escher BI, Fenner K, *et al.* The challenge of micropollutants in aquatic systems. Science 2006; 313(5790): 1072-7.
[http://dx.doi.org/10.1126/science.1127291] [PMID: 16931750]

[3]   Tratnyek PG, Johnson RL. Nanotechnologies for environmental cleanup. Nano Today 2006; 1(2): 44-8.
[http://dx.doi.org/10.1016/S1748-0132(06)70048-2]

[4]   Bhushan B. Springer Handbook of Nanotechnology. New York: Springer 2010; p. 3.
[http://dx.doi.org/10.1007/978-3-642-02525-9]

[5]   Khin MM, Nair AS, Babu VJ, Murugan R, Ramakrishna S. A review on nanomaterials for environmental remediation. Energy Environ Sci 2012; 5(8): 8075-109.
[http://dx.doi.org/10.1039/c2ee21818f]

[6]   Taran M, Safaei M, Karimi N, Almasi A. Benefits and Application of Nanotechnology in Environmental Science: an Overview. Biointerface Res Appld Chem 2021; 11: 7860 – 7870.

[7]   Pandey B, Fulekar MH. Nanotechnology: Remediation Technologies to Clean Up the Environmental Pollutants. Res J Chem Sci 2012; 2: 90-6.

[8]   Das S, Sen B, Debnath N. Recent trends in nanomaterials applications in environmental monitoring and remediation. Environ Sci Pollut Res Int 2015; 22(23): 18333-44.
[http://dx.doi.org/10.1007/s11356-015-5491-6] [PMID: 26490920]

[9]   Kyzas GZ, Matis KA. Nanoadsorbents for pollutants removal: A review. J Mol Liq 2015; 203: 159-68.
[http://dx.doi.org/10.1016/j.molliq.2015.01.004]

[10] Araújo R, Meira Castro AC, Fiúza A. Use of Nanoparticles in Soil and Water Remediation Processes 2015; 2: 315-20.

[11] Baby R, Saifullah B, Hussein MZ. Carbon Nanomaterials for the Treatment of Heavy Metal-Contaminated Water and Environmental Remediation. Nanoscale Res Lett 2019; 14(1): 341.
[http://dx.doi.org/10.1186/s11671-019-3167-8] [PMID: 31712991]

[12] Ganie AS, Bano S, Khan N, *et al.* Nanoremediation technologies for sustainable remediation of contaminated environments: Recent advances and challenges. Chemosphere 2021; 275: 130065.
[http://dx.doi.org/10.1016/j.chemosphere.2021.130065] [PMID: 33652279]

[13] Guerra F, Attia M, Whitehead D, Alexis F. Nanotechnology for Environmental Remediation: Materials and Applications. Molecules 2018; 23(7): 1760.
[http://dx.doi.org/10.3390/molecules23071760] [PMID: 30021974]

[14] Luo T, Cui J, Hu S, Huang Y, Jing C. Arsenic removal and recovery from copper smelting wastewater using TiO_2. Environ Sci Technol 2010; 44(23): 9094-8.
[http://dx.doi.org/10.1021/es1024355] [PMID: 21053910]

[15] Ge F, Li MM, Ye H, Zhao BX. Effective removal of heavy metal ions Cd^{2+}, Zn^{2+}, Pb^{2+}, Cu^{2+} from aqueous solution by polymer-modified magnetic nanoparticles. J Hazard Mater 2012; 211-212: 366-72.
[http://dx.doi.org/10.1016/j.jhazmat.2011.12.013] [PMID: 22209322]

[16] Kumar KY, Muralidhara HB, Nayaka YA, Balasubramanyam J, Hanumanthappa H. Hierarchically assembled mesoporous ZnO nanorods for the removal of lead and cadmium by using differential pulse anodic stripping voltammetric method. Powder Technol 2013; 239: 208-16.
[http://dx.doi.org/10.1016/j.powtec.2013.02.009]

[17] Gupta VK, Tyagi I, Sadegh H, Ghoshekand RS, Makhlouf ASH, Maazinejad B. Nanoparticles as adsorbent; a positive approach for removal of noxious metal ions: a review. Science, Technology and Development 2015; 34(3): 195-214.
[http://dx.doi.org/10.3923/std.2015.195.214]

[18] Wang H, Yuan X, Wu Y, *et al.* Facile synthesis of amino-functionalized titanium metal-organic frameworks and their superior visible-light photocatalytic activity for Cr(VI) reduction. J Hazard Mater 2015; 286: 187-94.
[http://dx.doi.org/10.1016/j.jhazmat.2014.11.039] [PMID: 25585267]

[19] Ma L, Zhang W. Enhanced biological treatment of industrial wastewater with bimetallic zero-valent iron. Environ Sci Technol 2008; 42(15): 5384-9.
[http://dx.doi.org/10.1021/es801743s] [PMID: 18754450]

[20] Rahmani A, Ghaffari H, Samadi M. A comparative study on arsenic (III) removal from aqueous solution using nano and micro sized zero-valent iron. J Environ Health Sci Eng 2011; 8(2): 157-66.

[21] Mauter MS, Elimelech M. Environmental applications of carbon-based nanomaterials. Environ Sci Technol 2008; 42(16): 5843-59.
[http://dx.doi.org/10.1021/es8006904] [PMID: 18767635]

[22] Li Y, Liu F, Xia B, *et al.* Removal of copper from aqueous solution by carbon nanotube/calcium alginate composites. J Hazard Mater 2010; 177(1-3): 876-80.
[http://dx.doi.org/10.1016/j.jhazmat.2009.12.114] [PMID: 20083351]

[23] Kandah MI, Meunier JL. Removal of nickel ions from water by multi-walled carbon nanotubes. J Hazard Mater 2007; 146(1-2): 283-8.
[http://dx.doi.org/10.1016/j.jhazmat.2006.12.019] [PMID: 17196328]

[24] Gong JL, Wang B, Zeng GM, *et al.* Removal of cationic dyes from aqueous solution using magnetic multi-wall carbon nanotube nanocomposite as adsorbent. J Hazard Mater 2009; 164(2-3): 1517-22.
[http://dx.doi.org/10.1016/j.jhazmat.2008.09.072] [PMID: 18977077]

[25]   Ren X, Chen C, Nagatsu M, Wang X. Carbon nanotubes as adsorbents in environmental pollution management: A review. Chem Eng J 2011; 170(2-3): 395-410.
[http://dx.doi.org/10.1016/j.cej.2010.08.045]

[26]   Lu C, Chiu H. Adsorption of zinc(II) from water with purified carbon nanotubes. Chem Eng Sci 2006; 61(4): 1138-45.
[http://dx.doi.org/10.1016/j.ces.2005.08.007]

[27]   Huang Z, Wang X, Yang D. Adsorption of Cr(VI) in wastewater using magnetic multi-wall carbon nanotubes. Water Sci Eng 2015; 8(3): 226-32.
[http://dx.doi.org/10.1016/j.wse.2015.01.009]

[28]   Tawabini BS, Al-Khaldi SF, Khaled MM, Atieh MA. Removal of arsenic from water by iron oxide nanoparticles impregnated on carbon nanotubes. J Environ Sci Health Part A Tox Hazard Subst Environ Eng 2011; 46(3): 215-23.
[http://dx.doi.org/10.1080/10934529.2011.535389] [PMID: 21279891]

[29]   Zhang C, Sui J, Li J, Tang Y, Cai W. Efficient removal of heavy metal ions by thiol-functionalized superparamagnetic carbon nanotubes. Chem Eng J 2012; 210: 45-52.
[http://dx.doi.org/10.1016/j.cej.2012.08.062]

[30]   Ihsanullah , Al-Khaldi FA, Abusharkh B, *et al.* Adsorptive removal of cadmium(II) ions from liquid phase using acid modified carbon-based adsorbents. J Mol Liq 2015; 204: 255-63.
[http://dx.doi.org/10.1016/j.molliq.2015.01.033]

[31]   Elsehly EMI, Chechenin NG, Bukunov KA, *et al.* Removal of iron and manganese from aqueous solutions using carbon nanotube filters. Water Sci Technol Water Supply 2016; 16(2): 347-53.
[http://dx.doi.org/10.2166/ws.2015.143]

[32]   Lin ST, Thirumavalavan M, Jiang TY, Lee JF. Synthesis of ZnO/Zn nano photocatalyst using modified polysaccharides for photodegradation of dyes. Carbohydr Polym 2014; 105: 1-9.
[http://dx.doi.org/10.1016/j.carbpol.2014.01.017] [PMID: 24708945]

[33]   Akhavan O. Lasting antibacterial activities of Ag–TiO_2/Ag/a-TiO_2 nanocomposite thin film photocatalysts under solar light irradiation. J Colloid Interface Sci 2009; 336(1): 117-24.
[http://dx.doi.org/10.1016/j.jcis.2009.03.018] [PMID: 19394952]

[34]   Adeleye AS, Conway JR, Garner K, Huang Y, Su Y, Keller AA. Engineered nanomaterials for water treatment and remediation: Costs, benefits, and applicability. Chem Eng J 2016; 286: 640-62.
[http://dx.doi.org/10.1016/j.cej.2015.10.105]

[35]   Yu JC, Yu J, Ho W, Zhang L. Preparation of highly photocatalytic active nano-sized TiO_2 particles *via* ultrasonic irradiation. Chem Commun (Camb) 2001; 19(19): 1942-3.
[http://dx.doi.org/10.1039/b105471f] [PMID: 12240230]

[36]   Yu JC, Ho W, Lin J, Yip H, Wong PK. Photocatalytic activity, antibacterial effect, and photoinduced hydrophilicity of TiO_2 films coated on a stainless steel substrate. Environ Sci Technol 2003; 37(10): 2296-301.
[http://dx.doi.org/10.1021/es0259483] [PMID: 12785540]

[37]   Li J, Guo S, Zhai Y, Wang E. Nafion–graphene nanocomposite film as enhanced sensing platform for ultrasensitive determination of cadmium. Electrochem Commun 2009; 11(5): 1085-8.
[http://dx.doi.org/10.1016/j.elecom.2009.03.025]

[38]   Lingamdinne LP, Koduru JR, Choi YL, Chang YY, Yang JK. Studies on removal of Pb(II) and Cr(III) using graphene oxide based inverse spinel nickel ferrite nano-composite as sorbent. Hydrometallurgy 2016; 165: 64-72. a
[http://dx.doi.org/10.1016/j.hydromet.2015.11.005]

[39]   Lingamdinne LP, Koduru JR, Roh H, Choi YL, Chang YY, Yang JK. Adsorption removal of Co(II) from waste-water using graphene oxide. Hydrometallurgy 2016; 165: 90-6. b
[http://dx.doi.org/10.1016/j.hydromet.2015.10.021]

[40] Azamat J, Sattary BS, Khatee A, Joo SW. Removal of a hazardous heavy metal from aqueous solution using functionalized graphene and boron nitride nanosheets: Insights from simulations. J Mol Graph Model 2015; 61: 13-20.
[http://dx.doi.org/10.1016/j.jmgm.2015.06.012] [PMID: 26186492]

[41] Dong Z, Zhang F, Wang D, Liu X, Jin J. Polydopamine-mediated surface-functionalization of graphene oxide for heavy metal ions removal. J Solid State Chem 2015; 224: 88-93.
[http://dx.doi.org/10.1016/j.jssc.2014.06.030]

[42] Zare-Dorabei R, Ferdowsi SM, Barzin A, Tadjarodi A. Highly efficient simultaneous ultrasonic-assisted adsorption of Pb(II), Cd(II), Ni(II) and Cu (II) ions from aqueous solutions by graphene oxide modified with 2,2′-dipyridylamine: Central composite design optimization. Ultrason Sonochem 2016; 32: 265-76.
[http://dx.doi.org/10.1016/j.ultsonch.2016.03.020] [PMID: 27150770]

[43] Chowdhury S, Balasubramanian R. Recent advances in the use of graphene-family nanoadsorbents for removal of toxic pollutants from wastewater. Adv Colloid Interface Sci 2014; 204: 35-56.
[http://dx.doi.org/10.1016/j.cis.2013.12.005] [PMID: 24412086]

[44] Tungittiplakorn W, Lion LW, Cohen C, Kim JY. Engineered polymeric nanoparticles for soil remediation. Environ Sci Technol 2004; 38(5): 1605-10.
[http://dx.doi.org/10.1021/es0348997] [PMID: 15046367]

[45] Pan B, Pan B, Zhang W, Lv L, Zhang Q, Zheng S. Development of polymeric and polymer-based hybrid adsorbents for pollutants removal from waters. Chem Eng J 2009; 151(1-3): 19-29.
[http://dx.doi.org/10.1016/j.cej.2009.02.036]

[46] Tungittiplakorn W, Cohen C, Lion LW. Engineered polymeric nanoparticles for bioremediation of hydrophobic contaminants. Environ Sci Technol 2005; 39(5): 1354-8.
[http://dx.doi.org/10.1021/es049031a] [PMID: 15787377]

[47] Botes M, Eugene Cloete T. The potential of nanofibers and nanobiocides in water purification. Crit Rev Microbiol 2010; 36(1): 68-81.
[http://dx.doi.org/10.3109/10408410903397332] [PMID: 20088684]

[48] Noubactep C, Caré S. On nanoscale metallic iron for groundwater remediation. J Hazard Mater 2010; 182(1-3): 923-7.
[http://dx.doi.org/10.1016/j.jhazmat.2010.06.009] [PMID: 20594643]

[49] Rao LN. Nanotechnological methodology for treatment of wastewater. Int J Chemtech Res 2014; 6: 2529.

[50] Amin MT, Alazba AA, Manzoor U. A review of removal of pollutants from water/wastewater using different types of nanomaterials. Adv Mater Sci Eng 2014; 2014: 1-24.
[http://dx.doi.org/10.1155/2014/825910]

[51] Scott TB, Popescu IC, Crane RA, Noubactep C. Nano-scale metallic iron for the treatment of solutions containing multiple inorganic contaminants. J Hazard Mater 2011; 186(1): 280-7.
[http://dx.doi.org/10.1016/j.jhazmat.2010.10.113] [PMID: 21115222]

[52] Theron J, Walker JA, Cloete TE. Nanotechnology and water treatment: applications and emerging opportunities. Crit Rev Microbiol 2008; 34(1): 43-69.
[http://dx.doi.org/10.1080/10408410701710442] [PMID: 18259980]

[53] Bina B, Pourzamani H, Rashidi A, Amin MM. Ethylbenzene removal by carbon nanotubes from aqueous solution. J Environ Public Health 2012; 2012: 1-8.
[http://dx.doi.org/10.1155/2012/817187] [PMID: 22187576]

[54] Liu Y, Chen X, Li J, Burda C. Photocatalytic degradation of azo dyes by nitrogen-doped TiO_2 nanocatalysts. Chemosphere 2005; 61(1): 11-8.
[http://dx.doi.org/10.1016/j.chemosphere.2005.03.069] [PMID: 15878606]

[55]   Anjum M, Miandad R. Waqas Muhammad, Gehany F, Barakat MA. Remediation of wastewater using various nanomaterials. Arab J Chem 2016; 9(12): 4897-919.

[56]   Yunus IS, Harwin , Kurniawan A, Adityawarman D, Indarto A. Nanotechnologies in water and air pollution treatment. Environ Technol Rev 2012; 1(1): 136-48.
[http://dx.doi.org/10.1080/21622515.2012.733966]

[57]   Binh ND, Oanh NTK, Parkpian P. Photodegradation of dioxin in contaminated soil in the presence of solvents and nanoscale TiO $_2$ particles. Environ Technol 2014; 35(9): 1121-32.
[http://dx.doi.org/10.1080/09593330.2013.861873] [PMID: 24701907]

[58]   Long RQ, Yang RT. Carbon nanotubes as superior sorbent for dioxin removal. J Am Chem Soc 2001; 123(9): 2058-9.
[http://dx.doi.org/10.1021/ja003830l] [PMID: 11456830]

[59]   Kristanti RA, Liong RMY, Hadibarata T. Soil Remediation Applications of Nanotechnology. Tropical Aquatic and Soil Pollution 2021; 1(1): 35-45.
[http://dx.doi.org/10.53623/tasp.v1i1.12]

[60]   Zhu N, Luan H, Yuan S, Chen J, Wu X, Wang L. Effective dechlorination of HCB by nanoscale Cu/Fe particles. J Hazard Mater 2010; 176(1-3): 1101-5.
[http://dx.doi.org/10.1016/j.jhazmat.2009.11.092] [PMID: 19969417]

[61]   Wang CB, Zhang W. Synthesizing nanoscale iron particles for rapid and complete dechlorination of TCE and PCBs. Environ Sci Technol 1997; 31(7): 2154-6.
[http://dx.doi.org/10.1021/es970039c]

[62]   Schrick B, Blough JL, Jones AD, Mallouk TE. Hydrodechlorination of trichloroethylene to hydrocarbons using bimetallic nickel-iron nanoparticles. Chem Mater 2002; 14(12): 5140-7.
[http://dx.doi.org/10.1021/cm020737i]

[63]   Kim YH, Carraway ER. Dechlorination of pentachlorophenol by zero valent iron and modified zero valent irons. Environ Sci Technol 2000; 34(10): 2014-7.
[http://dx.doi.org/10.1021/es991129f]

[64]   Sayles GD, You G, Wang M, Kupferle MJ. DDT, DDD, and DDE dechlorination by zero-valent iron. Environ Sci Technol 1997; 31(12): 3448-54.
[http://dx.doi.org/10.1021/es9701669]

[65]   Rizwan M, Singh M, Mitra CK, Morve RK. Ecofriendly Application of Nanomaterials: Nanobioremediation 2014; 2014: Article ID 431787.

[66]   Srivastav A, Yadav KK, Yadav S, Gupta N, Singh JK, Katiyar R, Kumar V. Nano-phytoremediation of Pollutants from Contaminated Soil Environment: Current Scenario and Future Prospects. In: Phytoremediation (Ansari A, Gill S, Gill R, Lanza GR, Newman L. eds.), Springer, Cham. 2018; pp 383-401.
[http://dx.doi.org/10.1007/978-3-319-99651-6_16]

[67]   Bystrzejewska-Piotrowska G, Golimowski J, Urban PL. Nanoparticles: Their potential toxicity, waste and environmental management. Waste Manag 2009; 29(9): 2587-95.
[http://dx.doi.org/10.1016/j.wasman.2009.04.001] [PMID: 19427190]

[68]   Grieger KD, Fjordbøge A, Hartmann NB, Eriksson E, Bjerg PL, Baun A. Environmental benefits and risks of zero-valent iron nanoparticles (nZVI) for *in situ* remediation: Risk mitigation or trade-off? J Contam Hydrol 2010; 118(3-4): 165-83.
[http://dx.doi.org/10.1016/j.jconhyd.2010.07.011] [PMID: 20813426]

[69]   Karn B, Kuiken T, Otto M. Nanotechnology and in situ remediation: a review of the benefits and potential risks. Environ Health Perspect 2009; 117(12): 1813-31.
[http://dx.doi.org/10.1289/ehp.0900793] [PMID: 20049198]

[70]   Singh S, Barick KC, Bahadur D. Fe3O4 embedded ZnO nanocomposites for the removal of toxic

metal ions, organic dyes and bacterial pathogens. J Mater Chem A Mater Energy Sustain 2013; 1(10): 3325-33.
[http://dx.doi.org/10.1039/c2ta01045c]

[71] Li Q, Mahendra S, Lyon DY, *et al.* Antimicrobial nanomaterials for water disinfection and microbial control: Potential applications and implications. Water Res 2008; 42(18): 4591-602.
[http://dx.doi.org/10.1016/j.watres.2008.08.015] [PMID: 18804836]

CHAPTER 7

Application of Nanoparticles in Environmental Monitoring

Bhoirob Gogoi[1,*], Neehasri Kumar Chowdhury[2], Suprity Shyam[3], Reshma Choudhury[4] and Hemen Sarma[5]

[1] *Department of Microbiology, Assam University, Silchar, Assam, India*

[2] *Department of Zoology, Gauhati University, Guwahati, Assam, India*

[3] *Department of Life Sciences, Dibrugarh University, Dibrugarh, Assam, India*

[4] *Department of Biotechnology, Royal Global University, Guwahati, Assam, India*

[5] *Department of Botany, N Saikia College, Titabar, Assam, India*

Abstract: The planet is dealing with a major problem of environmental pollution. Year after year, this problem worsens, causing harm to our planet. To combat the major environmental issues, various technologies have been developed over the years. The use of nanomaterials in environmental management is becoming more common. Nanomaterials are increasingly being used to clean the air, purify water, decontaminate soil, and detect pollution. Nanotechnology has emerged as a technique for cleaning up pollution and monitoring degradation of environmental sectors such as air, water and soil. Hence nanotechnology can contribute to the sustainability of the environment. This chapter discusses the use of nanomaterials in the monitoring of air pollutants, organic contaminants and other environmental pollutants, as well as the various methods involved in the production of nanoparticles.

Keywords: Environmental monitoring, Nanomaterials, Nanoparticles, Pollutants.

INTRODUCTION

Environmental contamination and energy deficiency are two major problems recognized all over the world. The advancement of nanotechnology in the last twenty years has focused on improving interaction in the plan, disclosure, creation, and novel use of counterfeit nanoscale materials. A variety of nano materials like iron, titanium dioxide, silica, zinc oxide, carbon nanotube, dendrimers, polymers, and have been constantly used to make the air clean, to refine water, and to purify soil [1].

* **Corresponding author Bhoirob Gogoi:** Department of Microbiology, Assam University, Silchar, Assam, India; E-mail: bhoirobgogoi1998@gmail.com

Interesting and conceivably valuable properties of nanomaterials include expanded surface zones and reactivity, improved strength-weight proportions, expanded electrical conductivity, and changes in shading and mistiness. Materials with these properties found their application in a various areas such as medication and environmental insurance [2]. Nanoscale materials discover use in a variety of various territories, like electronic, attractive and optoelectronic, biomedical, drug, corrective, energy, environmental, reactant, and materials applications. Nanomaterials and nanotechnology give an incredible technique for recognition and treatment of toxins found in the environment. Small sized particles have been utilized for achieving various tasks by humanity and there has been a new resurgence because of its capacity to combine and control such materials. In view of the capability of this innovation, there has been an overall expansion in interest in nanotechnology innovative work [3].

The subject of nanoparticles applications is extremely expansive. Nanotechnology has acquired a ton of consideration in the previous a very long time because of the one of a kind actual property of nanoscale materials. Nanomaterials present upgraded reactivity and in this way better *via*bility when contrasted with their bulkier partners because of their higher surface-to volume proportion. Nanotechnology is additionally being utilized to make sustainable power less expensive and more efficient. The utilization of nanotechnology in agribusiness area will lessen the aimless utilization of agrochemicals and in this manner will diminish the heap of substance toxin [3].The utilization of nanoparticles in environmental applications will definitely prompt the arrival of nanoparticles into the climate. Surveying their dangers in the climate requires a comprehension of their portability, bioavailability, harmfulness, and constancy. Though air-borne particles and inward breath of nanoparticles have pulled in a great deal of consideration significantly less is thought about the conceivable openness of amphibian and earth bound life to nanoparticles in water and soils. A growing body of evidence demonstrates that nanoparticles can be taken up by a wide variety of mammalian cell types, cross the cell membrane, and become disguised. Nanoparticles are also toxic to amphibian species, both unicellular and multicellular (e.g. microscopic organisms or protozoa) and creatures (e.g. *Daphnia* or fish) [4].

These outcomes from considers show that certain nanoparticles will have impacts on creatures, on the environment, at any rate at raised focuses. Nanomaterials present improved reactivity and in this way better *via*bility when contrasted with their bulkier partners because of their higher surface-to volume proportion. The following stage towards an appraisal of the dangers of nanoparticles in the environment will, in this way, be to gauge the openness to the diverse nanoparticles benefits are augmented while limiting the probability of accidental unfavorable results [4].

These outcomes from contemplates show that certain nanoparticles will have consequences for living beings and the environment, in any event at raised focuses. It is significant that the materials utilized for the remediation of contamination are not another toxin themselves after they have been utilized. The subsequent stage towards an evaluation of the dangers of nanoparticles to the environment, consequently, be to gauge the openness to the distinctive nanoparticles [4].

CLASSIFICATION OF NANOPARTICLES

Based on their physical and chemical features, nanoparticles are classified into different types Fig. (**1**). The major classes of NPs are briefly defined in the following section [5].

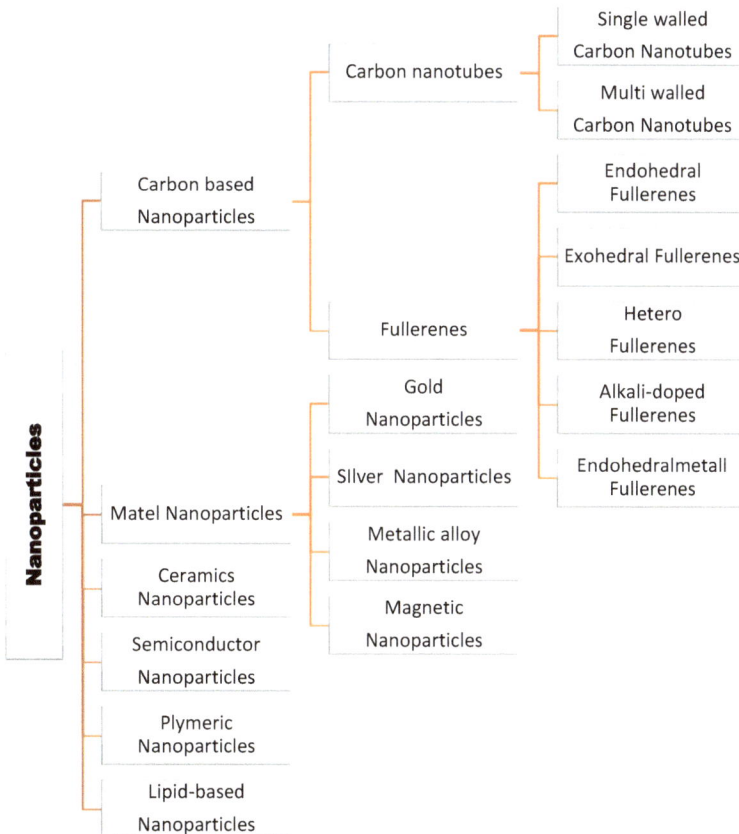

Fig. (1). Classification of Nanoparticles.

Carbon Based NPs

Carbon-based nanoparticles are divided into two major classes, *i.e.* carbon nanotubes and fullerenes [5].

Carbon Nanotubes

In the year 1991, carbon nanotubes were discovered by Iijma [6]. Carbon nanotubes are long, slim fullerenes where the mass of the cylinders is hexagonal carbon (graphite form) and frequently covered at each end [99]. Recently, two morphological forms of carbon nanotubes have gotten a lot of consideration: single-walled carbon nanotubes and multiwall carbon nanotubes. These types of results were caused by morphological defects in carbon nanotubes [7].

Fullerenes

In 1985, Kroto et al discovered third types of carbon allotrope [8] where 60 carbon atoms are arranged in truncated icosahedrons shape and diameter of approximately 7 Armstrong [9]. Fullerenes involve a wide scope of isomers also, homologous arrangements from the most examined C60 or C70 and containing 12 hexagonal and 20 pentagonal carbon rings [10, 11]. There are different sorts of fullerenes. For example, endohedral fullerenes, exohedral fullerenes, heterofullerenes, Alkali-doped fullerenes, and endohedralmetall fullerenes have been reported in recent times [7].

Metal Nanoparticles

Metallic nanoparticles (MNPs) have an inorganic metal or metal oxide core that is typically bounded by an organic or inorganic material or metal oxide shell. MNPs have a broad array of applications in our everyday life. The introduction of pilot-scale production of MNPs which have gained market in various creams, clothes, shampoos, footwear, and plastic containers has resulted from the creation of a new economically feasible method for MNPs production [12]. Broadly used MNPs are the following types:

Gold Nanoparticles (AuNPs)

Gold nanoparticles or colloidal gold are a well-known and widely used compound made up of tiny gold particles with a diameter ranging from 1 to 100 nm. AuNPs are used in immunochemical studies, used in drug delivery, protein delivery, and gene delivery process. They can also detect aminoglycoside antibodies. Gold nanorods are being used to detect cancer stem cells [13].

Silver Nanoparticles (AgNPs)

AgNPs are the most widely used metal nanoparticles due to their antimicrobial properties. In the textile industries, AgNPs are used as antimicrobial agents. It is also used in a variety of fields, such as biomedicine [14] agriculture [15] drug delivery [16] water treatment [17] and so on.

Metallic Alloy Nanoparticles

Metalic alloy nanoparticles are formed by fusing two or more metals. Many studies have shown that bimetallic or trimetallic alloy nanoparticles have a synergistic characteristic which makes them more efficient than other monometallic nanoparticles. Mixed alloyed nanoparticles, sub-cluster segregated alloyed nanoparticles, core-shell alloyed nanoparticles, and multiple core-shell alloyed nanoparticles are some well-known nanoparticles. Alloyed nanoparticles are widely used in imaging, diagnosis, and therapies [18].

Magnetic Nanoparticles

Magnetic nanoparticles are a multilayer structure containing small particles with a nanoscale dimension that can be regulated by the magnetic field and their magnetic properties differ from their bulk materials [19]. Magnetic nanoparticles are employed in a wide range of potential therapeutic uses such as hyperthermia, drug delivery, magnetic resonance imaging (MRI), sorting and manipulation of stem cells, DNA analysis, and gene therapy [20].

Ceramics Nanoparticles

Ceramics nanoparticles are heat resistant, non-metallic solids composed of metallic and non –metallic compounds. The sol-gel process was used to create ceramic nanoparticles in 1980 [100].Due to their high heat resistance and chemical inertness, they may be used for a variety of purposes, including caring for drugs, gens, protein, *etc.* They are also used in the delivery of drugs [21].

Semiconductor Nanoparticles

Semiconductor nanoparticles or quantum dots are microscopic crystalline particles and their diameter ranges from 2-6 nm. They have unique characteristics such as optoelectronic magnetic effect, high surface area, and quantum size effects. Because of these properties, they have a plethora of applications in a wide range of fields such as LEDs, diodes, solar cells, transistors, lasers, quantum computing, and medical imaging [22].

Polymeric Nanoparticles

Polymeric nanoparticles are colloidal, solid particles and filled with some active compounds that are either entrapped inside or surface adsorbed onto the polymeric center. The size of the polymeric nanoparticles ranges between 1-1000nm [23]. Polymeric nanoparticles are used for a variety of purposes, including drug delivery, bio-imaging, and theranostics [24].

Lipid-based Nanoparticles

Lipid-based nanoparticles have a lipid-based rigid center and a matrix of soluble molecules. Surfactants or emulsifiers helped to keep those nanoparticles' exterior cores stable and their size ranges from 10-1000nm [25]. Lipid-based nanoparticles are divided into two main groups, namely solid lipid nanoparticles (SLC) and nonstructural lipid carriers (NLC) [26]. Lipid-based nanoparticles are used in drug delivery presses. Recently, some lipid-based nanoparticles have been approved for BreC treatment [26].

SYNTHESIS OF NANOPARTICLES

Two main perspectives that are extensively used for the Nanoparticle synthesis are brought into context, *viz.* top down and bottom up methods [27] which are briefly described in the following sections

Top Down Methods

In the top down method, different physical and chemical treatments are used to convert the bulk materials into nanosized particles [28]. Some processes such as mechanical milling, thermal and laser ablation, *etc.* come under the top-down approaches which are widely used for preparation of nanoparticles. A short description of these processes is given below:

Mechanical Milling

Mechanical milling involves the use of a high-energy ball mill to grind appropriated batches of materials and grinding medium together. The main reason for processing is to diminish the size of the particles and blend the particles in to another stage. This technique was invented by John Benjamin in 1970 [29]. Mechanical milling is employed for the preparation of Al-based nanomaterials; Mg based nanomaterials, Co-based nanomaterials, Ni-based nanomaterials, Ti-based, Cu-based, C-based and Fe_2O_3 nanoparticles [30].

Thermal Decomposition

One of the most common methods for producing stable monodisperse suspension with self-assembly is thermal decomposition. In this process, nucleation happens as the metal precursor is applied to a hot solution containing a surfactant while growth occurs at a higher reaction temperature [31]. Thermal decomposition methods are used to create carbon and metal oxide-based nanoparticles [32].

Laser Ablation

Laser ablation is a process where laser irradiation is used to minimize the size of the robust goal materials which are submerged under a thin layer of liquids [33]. Mainly, three types of laser are used in the laser ablation process to create nanoparticles, namely Neodymium dopedyltrium aluminium garnet (Nd.YAG) laser at 106µm, Titanium doped sapphire (Ti. Sapphire) laser and copper vapor laser [34]. The Laser ablation process is employed for the synthesis of Si nanoparticles [35], Ag nanoparticles and Au nanoparticles [34].

Bottom Up Method

In the bottom-up or building up approach, nanostructured nanoparticles building blocks are first formed. After that, those building blocks are assembled and produce the desired nanoparticles. In the following sections, some of the most widely used strategies are described briefly:

Sol-gel Method

The sol-gel is a method which is employed for the production of nanoparticles. In sol-gel methods, a colloidal suspension of a solid in a liquid form is termed as sol and a stable macro molecule that has been immersed in a solvent termed as gel. For synthesis of colloids, two types of precursor metal alkoxides and alkoxysilanes are used. As a form of mutual solvent, alcohol is used in sol-gel methods. For creation of sol-gel, four major steps are involved: hydrolysis, condensation, particle growth and particle agglomeration [34].

Chemical Vapor Deposition

In chemical vapor deposition methods, various reducing agents such as sodium borohydride, glucose, ethanol, and citrate of sodium are used to remove ionic salt in a suitable medium in the presence of a surfactant [36]. This process is used for the synthesis of silver nanoparticles, copper nanoparticles [37], and Ultrafine FePt nanoparticles [38].

Physical Vapor Deposition

(PVD) is a widely used technique for the fabrication of thin films and surface coatings [3]. The material is evaporated inside the vaccum in order to avoid the impurities. Vapour particles travel towards the cold target (substrate) and gets deposited there, which results in the condensation to a solid state [39]. PVD can be achieved mainly by two methods; *i.e.,* Thermal evaporation and sputtering method [40]. In thermal evaporation method, high resistance coil is used for producing the heat which is used to evaporate the materials [41]. In Sputtering method, bombardment of the solid target takes place by energetic electrons resulting in the ejection of atoms [42].

By a simple physical vapor deposition method fabrication of ZnO nanowire on the NiO nanoparticle, deposited Al_2O_3 can be obtained [43]. Formation of Yttria-stabilized zirconia [44], thin-film solar cells, Cu (In, Ga) Se_2 thin film are possible by this method [45].

It is a convenient method for the formation of thin metal firms [40], but it also have some disadvantages like it is expensive to use and production of low volume of materials is also reported [46].

Biosynthesis

Biosynthesis is a method to produce nanoparticles that are biodegradable [47] and free from the use of toxic chemicals as by products [48] Fig. (**2**). Biological synthesis provides an alternative way for the physical and chemical methods as they are very expensive to use [49] than the biosynthesis method which infact, is high yielding as such [50]. Large number of microorganisms is associated in the biosynthesis of Nanoparticles as they can produce intracellular and extracellular inorganic NPs [51].

Plants, fungi, yeast and bacteria are the most common sources in this regard that produces NPs like Au, Ag from *Azadirachta indica* [52], Au from *Aloe vera* [53], Cds from *Schizosaccharomyces pombe* [54], Magnetite from *Fusariu moxysporum* and *Verticillium* sp., Ag from yeast strain MKY3 [54], Cds from *Schizosaccharomyces pombe* [55], Ag, Au from *Lactobacillus* strains [56], Cds from *Escherichia coli* [57]. It aims to synthesize the NPs at physiological pH, temperature and pressure by removing the harsh processing conditions in a very convenient way.

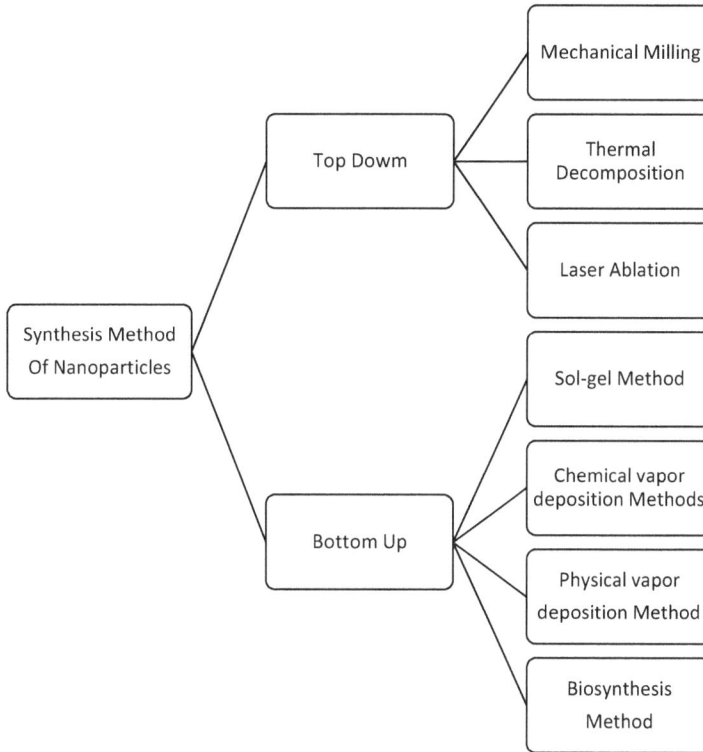

Fig. (2). Synthesis method of nanoparticles.

APPLICATION OF NANOMATERIALS IN ENVIRONMENTAL MONITORING

The nanoparticles based technologies find applications in several environmental areas. The applications of nanoparticles which are used in environmental technologies are collectively referred to as "Environmental nanotechnology".

Nanoparticles have well defined physical and chemical properties which are unique and can be used for desired applications by suitably manipulation them [58]. Some of the mostly used applications in environmental monitoring are cited below:

As Antimicrobials

The metal based nanoparticles comprises of an effective microbial agent which inhabits the growth of some common pathogenic Micro organisms. Therefore some of the nanoparticles such as silver, titanium dioxide and zinc oxide are much used as antimicrobials and additives in consumer industrial products and health related issues [59].

Silver nanoparticles are also incorporated into various fields to widen their applicability in materials and in bio medicals [60].

The zinc oxide nanoparticles have been found to be highly toxic and their stability is relatively low under harsh processing conditions. It can be combined with the potent antimicrobial properties which favours their application as antimicrobials [61].

Magnesium oxide nanoparticles are also used as antimicrobials that are prepared through an aerogel procedure (Ap-Mgo) which yields square and polyhedral shaped nanoparticles which are able to adsorb and retain significant amounts of elemental chlorine and bromine for a long time [62, 63] Ap-Mg O / X2 NPs exhibition - biocidal activity against gram positive and gram negative bacteria and spores that are desirable for a potent disinfectant [64, 65].

In Waste Water Treatment

Contamination include depositing of metal ions or increasing amount of unwanted metal ions, rock salts, gasoline, oils, chemical *etc.* in the ground water. The ground water when entering into the water streams shows negative impact on the health issues if consumed. It also causes damage to the ecosystem. It also causes damage to the ecosystem as well.

Therefore, researchers and scientists have come up with advanced nanotechno logies that treat waste or contaminated water in different ways. Advanced nanotechnology mainly includes four classes: Dendrimers, metal containing nanoparticles, Zeolites and carbonaceous Nanomaterials.

All these classes primarily aim to treat water that are contaminated with pollutants and germs.

Dendrimers: Dendrimers are usually branched 3-dimensional based structures which are synthesized by monodispersal of macromolecules. It is a tree like fashion structure which was first discovered by Fritz Vogtle in 1978 by Donald Tomalia and co-workers in the early 1980s [66].

The structure of dendrimers consists of three components, namely; a core, interior branch cells and terminal branch cells. Core is composed of atoms or groups of atoms from which the other branch of atoms called "dendrons" emerges out. This dendron reacts with the core of atoms and yields a dendrimer [67]. Today dendrimers, have several waste water treatment applications. *e.g.,* Dendron-enhanced ultrafiltration (DEUF) and poly (amidoamine) (PAMAM) dendrimers are extensively used dendrimers in this regard. Dendritic polymers are "soft" nanoparticles having sizes in the range of 1-20nm, which can be used to treat toxic metal ions, radionuclide and inorganic anions from recyclable water soluble ligands [68].

PAMAM dendrimers were the first commercially used dendrimers, since 1984 [69, 70]. It is stable and soluble in water having a low tendency [71]. PAMAM based silver complexes includes silver that remains conjugated to the dendrimer in the form of ions, stable metallic silver clusters or silver compounds. As PAMAM dendrimers are soluble, they release immobilized silver on the agar medium by self-diffusion causing the protection of silver and silver compounds which displays high anti-microbial activity without the loss of solubility.

• *Metal Nanoparticle*:

Metallic nanoparticles (MNPs) are most widely used to treat contaminated water against pesticides, industrial solvents and germs. It is found that the unique properties of MNPs to develop high capacity and selective sorbents for metal ions and anions.

Due to its high surface area of NP, they not only enhance the adsorption capacities of sorbent materials but also have unique adsorption properties due to different distribution of reactive sites and disordered surface region. Magnesium oxide (MgO) NPs and magnesium (Mg) NPs are effective against gram positive and gram negative bacteria and bacteria spores present in water; Stoimenov et al [72]. Ag (I) and silver compounds have been used to treat coliform bacteria found in waste water [73]. Ag NPs are now extensively used as active biocides against gram positive and gram negative bacteria including *E. coli*, *Staphylococcus aureus*, *Klebsiella pneumoniae* and *Pseudomonas aeruginosa* [74, 75].

Gold (Au) coating NP associated with palladium is very effective in removing tri-chloroethane (TCE) from the ground water. Zno NPs is used to remove arsenic from water. Iron ferrite and magnetite also helps in removal of metal ion.

• *Zeolite*: Natural zeolites removes heavy metals from acid mine wastewater. It
works with the principal of ion-exchange media for metal ions. It is low-cost and highly effective method. Clinoptilohte and chabazite are two natural zeolites subjected for the treatment of effluents contaminated with mixed heavy metals (Pb, Cu, Cd, Zn, Cr, Ni and Co) [76]. Clinoptilohte also removes ammonium ions from drinking water [77]. NH_4^+ is the common cation in water which affects human as well as animals. It can be removed by exchanging with biologically acceptable cations such as Na^+, K^+, Mg^{2+}, Ca^{2+} or H^+ residing on the exchange sites of the zeolites [78]. Modification of natural zeolites can also be done to

increase its effectiveness. Some of the known methods are acid treatment and surfactant functionalisation which helps in achieving higher adsorption capacity for different particles [79].

- *Carbonaceous Nanoparticles*:

Carbonaceous NPs are high capacity and selective sorbents servers for the organic solutes in aqueous solution. The hydrophilic carbon NP (HNPs) removes cd (II), Ni (II) and Pb (II) ions that are supported over silica beads to enhance the separation from treated water [80]. Some polymers shows antibacterial properties were developed for soluble and insoluble pyridinium-type of polymers involved in various surface coatings [81], azidated poly (vinyl chloride) [82] used to prevent various bacterial adhesions. In the presence and absence of coagulation, the nanoscale carbon black takes part in removing Natural Organic Matter (NMO) from water. Studies show that a H-ion concentration (PH) of (3-5) was favourable for NOM removal [83].

Air Pollutants

Air Pollution is a common threat to our environment in the day-to-day life which can be caused by several factors, either naturally or man-made. Nitrous oxide (NOx), fine suspended particulate matter (PM), carbon monoxide (CO), volatile organic compounds (VOCs), and ozone (O_3), are some of the air pollutants that brought concern into the environment and the ecosystem [84]. Nowadays measurements in pollutants are carried out by the commonly used analytical instruments such as gas chromatography, mass spectrometry, chemiluminescence, optical spectroscopy and infra-red spectrometry. But, these instruments are expensive and time consuming which is a major drawback for the tests to be done continuously and simultaneously in a huge scale [85]. Therefore nanotechnology provides the best way to handle the situation in a systematic manner. Nanotechnologies clean the air in the following ways:

- *By nanocatalysts:* They transform harmful vapors from vehicles and industries and turn it into harmless gases. Nanocatalysts are nanofiber catalysts made up of magnesium oxide (MgO) NP that excludes volatile organic compounds making it less harmful for the environment. Gold NPs plays an important role in controlling highly toxic Carbon monoxide to less toxic Carbon dioxide gas [86].
- *By nanostructured membranes*: Nanostructured membranes have small pores that separate methane of CO_2 from exhaust. It is highly selective and composed of single-walled carbon nanotubes (SWCNTs) which can be utilized as nanoscale vessels. Since it is selective, it can enclose the tetrafluromethane at 300k by maintaining a pressure of 1 bar [87]. Carbon nanotubes are extensively

used for the purification of air because CNT changes in shape after absorption of gas molecules on the surface of CNTs. This change in shape triggers redistribution of electrons which leads to the macroscopic changes in resistance [88].

- *By Nanocomposites*: The mercury vapors coming out from the sources of combustion are removed by silica- titania nanocomposites. For the improved mercury absorption, high surface area of nanosilica and titania molecules has been amalgamated for making the nanocomposites. Moreover, it is shown that Superior mercury removal efficiency was ensured with significant reduction of contact angle up to 10° by this silica-titanianano composite [89].

FUTURE ASPECTS OF NANOPARTICLES

The recent advances in nanotechnologies are improving our daily life by engaging it in different objects and applications which enhances its performance and efficiency. Due to its environmental friendly property, it can provide a clean environment by providing safer air, water and clean renewable energy for sustainable future [90, 91]. Since NPs are more preferable than any other analytical instruments, for its low cost and time efficiency. The future will able to use it for better improvement of the environment in no time. Future challenges may include large-scale production, enhancement of stability, reduced time to obtain desirable shape and size and possible applications in various fields [92].

It is revealed that AgNPs can be used as a desirable nanoweapon to develop friendly environment as there has been a hope to be used in water disinfectant, removal of phytopathogens and also as a plasma resonance enhancers [93]. The structure and function of NPs and the interaction between Fe and organic carbon, especially in seawaters, where there is an impact on climate change issues are the areas of interest to be carried out in further research works. Colloids are also an important NP to go through where the further important developments on its structure, trace elements and pathogen binding and the effects of binding on the bioavailability of pollutants and pathogens are underway with the objective of methodological developments, collection of good quality data on colloidal structure and colloidal interactions with contaminants and effects of colloid pollutant interactions on transport. Other aspects can also be predicted in engineered nanoparticles. Engineered nanoparticles are those materials which are <100 nm in size includes fullerenes, carbon, and other nanotubes, metals and their oxides, and sulfides and other materials are used in a variety of processes such as for the remediation of contaminated land [94]. Gold NPs are seen to be highly remarkable because of its optical, electronic, physicochemical and surface plasmon resonance (SPR) properties. Recently, the electron transfer regime between molecules and macroelectrodes has also become possible only because of the development of thiolate-protected gold clusters [95].

On the other hand, long term usage of some NPs may also cause impact on the living beings and the environment in the future. For *e.g.,* Silver nanoparticles (AgNPs) that are associated with textiles, industries and other sources, ultimately distributes in the water streams and causes threat to the ecosystem. According to Blaster et al, this AgNPs were bounded by mud and other sewages that may harm agricultural fields in the future leading to bioaccumulation and toxicological threat. AuNPs, TiO2NPs, CNTsNPs are examples of some of the NPs that damages causing oxidative stress [96], phytotoxic effects [97], phytotoxicity respectively [98]. Exposure of NPs in the nature and its consequences to the human as well is still limited [99]. NPs are replacing many toxic materials [100 - 102] but their exposure in the environment and their effects are less known.

CONCLUSION

Nanotechnology is an emerging science with broad applications and possible advantages. Nanomaterials have properties that empower both synthetic decrease and catalysis to relieve the poisons of concern.While nanomaterials have helpful applications, they additionally raise worries over expected ramifications for human wellbeing and the climate. Bioaccumulation potential, harmfulness, specialist and local area openness, and extreme destiny are among the worries that merit thought. The systems through which the various nanotechnologies are applied are notable, what befalls these materials after they have been applied for foreign substance catch or debasement is underexplored.Nanoparticles are likewise discovering use in elective energy creation, stockpiling, and transformation, and this can incredibly lessen the dependence on petroleum product ignition and generally speaking discharges. Nanotechnology gives an abundance of methodologies that can be utilized to address environmental pollution. The quest for contamination anticipation applications of nanotechnology ought to be embraced with thought of the likely effects across the whole lifecycle of the nanomaterials, including creation, use, and end-of-life mien.

AUTHOR'S CONTRIBUTION

The authors have equally contributed to this chapter. The authors declare that they have no known competing financial interests or personal relationships that could have appeared to influence the work reported in this chapter.

REFERENCES

[1]   Biswas P, Wu CY. Nanoparticles and the Environment. J Air Waste Manag Assoc 2005; 55(6): 708-46.
       [http://dx.doi.org/10.1080/10473289.2005.10464656] [PMID: 16022411]

[2]   Sarma H, Joshi S, Prasad R. Jampilek. Biobased Nanotechnology for Green Applications, 978-3-00-61985-5, Series ISSN2523-8027, 2021; p 668.

[3]   Aliofkhazraei M, Ali N. PVD Technology in Fabrication of Micro- and Nanostructured Coatings. Comprehensive Materials Processing 2014; 7: 49-84.
       [http://dx.doi.org/10.1016/B978-0-08-096532-1.00705-6]

[4]   Ibrahim RK, Hayyan M, AlSaadi MA, Hayyan A, Ibrahim S. Environmental application of nanotechnology: air, soil, and water. Environ Sci Pollut Res Int 2016; 23(14): 13754-88.
       [http://dx.doi.org/10.1007/s11356-016-6457-z] [PMID: 27074929]

[5]   Khan I, Saeed K, Khan I. Nanoparticles: Properties, applications and toxicities. Arab J Chem 2019; 12(7): 908-31.
       [http://dx.doi.org/10.1016/j.arabjc.2017.05.011]

[6]   Iijima S. Helical microtubules of graphitic carbon. Nature 1991; 354(6348): 56-8.
       [http://dx.doi.org/10.1038/354056a0]

[7]   Bhatia S. Nanoparticles Types, Classification, Characterization, Fabrication Methods and Drug Delivery Applications. Natural Polymer Drug Delivery Systems 2016; pp. 33-93.
       [http://dx.doi.org/10.1007/978-3-319-41129-3_2]

[8]   Kroto HW, Heath JR, O'Brien SC, Curl RF, Smalley RE. C60: Buckminsterfullerene. Nature 1985; 318(6042): 162-3.
       [http://dx.doi.org/10.1038/318162a0]

[9]   Krätschmer W, Lamb LD, Fostiropoulos K, Huffman DR. Solid C60: a new form of carbon. Nature 1990; 347(6291): 354-8.
       [http://dx.doi.org/10.1038/347354a0]

[10]   Ballesteros E, Gallego M, Valcárcel M. Analytical potential of fullerene as adsorbent for organic and organometallic compounds from aqueous solutions. J Chromatogr A 2000; 869(1-2): 101-10.
       [http://dx.doi.org/10.1016/S0021-9673(99)01050-X] [PMID: 10720229]

[11]   Taylor R, Hare JP, Abdul-Sada AK, Kroto HW. Isolation, separation and characterisation of the fullerenes C60 and C70: the third form of carbon. J Chem Soc Chem Commun 1990; 20(20): 1423-5.
       [http://dx.doi.org/10.1039/c39900001423]

[12]   Diegoli S, Manciulea AL, Begum S, Jones IP, Lead JR, Preece JA. Interaction between manufactured gold nanoparticles and naturally occurring organic macromolecules. Sci Total Environ 2008; 402(1): 51-61.
       [http://dx.doi.org/10.1016/j.scitotenv.2008.04.023] [PMID: 18534664]

[13]   Tomar A, Garg G. Short Review on Application of Gold Nanoparticles. Glob J Pharmacol 2013; 7(1): 34-8.

[14]   Chaloupka K, Malam Y, Seifalian AM. Nanosilver as a new generation of nanoproduct in biomedical applications. Trends Biotechnol 2010; 28(11): 580-8.
       [http://dx.doi.org/10.1016/j.tibtech.2010.07.006] [PMID: 20724010]

[15]   Nair R, Varghese SH, Nair BG, Maekawa T, Yoshida Y, Kumar DS. Nanoparticulate material delivery to plants. Plant Sci 2010; 179(3): 154-63.
       [http://dx.doi.org/10.1016/j.plantsci.2010.04.012]

[16]   Sarma H, Gupta S, Narayan M, Prasad R, Krishnan A. Engineered Nanomaterials for Innovative Therapies and Biomedicine, 2021.
       [http://dx.doi.org/10.1007/978-3-030-82918-6]

[17]   Dankovich TA, Gray DG. Bactericidal paper impregnated with silver nanoparticles for point-of-use water treatment. Environ Sci Technol 2011; 45(5): 1992-8.
[http://dx.doi.org/10.1021/es103302t] [PMID: 21314116]

[18]   Huynh KH, Pham XH, Kim J, *et al.* Synthesis, Properties, and Biological Applications of Metallic Alloy Nanoparticles. Int J Mol Sci 2020; 21(14): 5174.
[http://dx.doi.org/10.3390/ijms21145174] [PMID: 32708351]

[19]   Mørup S, Hansen MF, Frandsen C. Magnetic Nanoparticles.Comprehensive Nanoscience and Technology. Elsevier 2011; Vol. 1: pp. 437-91.
[http://dx.doi.org/10.1016/B978-0-12-374396-1.00036-2]

[20]   Fan TX, Chow SK, Zhang D. Biomorphic mineralization: From biology to materials. Prog Mater Sci 2009; 54(5): 542-659.
[http://dx.doi.org/10.1016/j.pmatsci.2009.02.001]

[21]   Thomas S, Harshita BSP, Mishra P, Talegaonkar S. Ceramic Nanoparticles: Fabrication Methods and Applications in Drug Delivery. Curr Pharm Des 2015; 21(42): 6165-88.
[http://dx.doi.org/10.2174/1381612821666151027153246] [PMID: 26503144]

[22]   Frechet JMJ, Tomalia DA. Dendrimers and other dendritic polymers. NewYork: Wiley and Sons 2001.
[http://dx.doi.org/10.1002/0470845821]

[23]   Zielińska A, Carreiró F, Oliveira AM, *et al.* Polymeric nanoparticles: Production, characterization, toxicology and ecotoxicology. Molecules 2020; 25(16): 3731.
[http://dx.doi.org/10.3390/molecules25163731] [PMID: 32824172]

[24]   Szczęch M, Szczepanowicz K. Polymeric core shell nanoparticles prepared by spontaneous emulisification solvent, evaporation and funtionalized by the layer by lay method. Nanomaterials (Basel) 2020; 10(3): 496.
[http://dx.doi.org/10.3390/nano10030496] [PMID: 32164194]

[25]   Rawt MK, Jain A, Singh S. Studies on Binary lipid nanoparticle of repaglinid:in vitro and in vivo evaluation. J Pharm Sci 2011; 100(6): 2366-78.
[http://dx.doi.org/10.1002/jps.22435] [PMID: 21491449]

[26]   Alavi M, Hamidi M. Passive and active targeting in cancer therapy by liposomes and lipid nanoparticles. Drug Metab Pers Ther 2019; 34(1): 1-8.
[http://dx.doi.org/10.1515/dmpt-2018-0032] [PMID: 30707682]

[27]   Ahmed S, Ahmad M, Swami BL, Ikram S. A review on plants extract mediated synthesis of silver nanoparticles for antimicrobial applications: A green expertise. J Adv Res 2016; 7(1): 17-28.
[http://dx.doi.org/10.1016/j.jare.2015.02.007] [PMID: 26843966]

[28]   Meyers ME, Mishra A, Benson DJ. Mechanical properties of nanocrystallin materials progress in material science. 2006; 51(4): 427-556.

[29]   Saravanan P, Deepika D, Hsu JH, Vinod VTP, Černík M, Kamat SV. A surfactant-assisted high energy ball milling technique to produce colloidal nanoparticles and nanocrystalline flakes in Mn–Al alloys. RSC Advances 2015; 5(112): 92406-17.
[http://dx.doi.org/10.1039/C5RA16550D]

[30]   Prasad Yadav T, Manohar Yadav R, Pratap Singh D. Mechanical Milling: a Top Down Approach for the Synthesis of Nanomaterials and Nanocomposites. Nanoscience and Nanotechnology 2012; 2(3): 22-48.
[http://dx.doi.org/10.5923/j.nn.20120203.01]

[31]   Salavati-Niasari M, Davar F, Mir N. Synthesis and characterization of metallic copper nanoparticles *via* thermal decomposition. Polyhedron 2008; 27(17): 3514-8.
[http://dx.doi.org/10.1016/j.poly.2008.08.020]

[32]   Bharde A, Rautaray D, Bansal V, *et al.* Extracellular biosynthesis of magnetite using fungi. Small

2006; 2(1): 135-41.
[http://dx.doi.org/10.1002/smll.200500180] [PMID: 17193569]

[33] Simakin AV, Voronov VV, Kirichenko NA, Shafeev GA. Nanoparticles produced by laser ablation of solids in liquid environment. Appl Phys, A Mater Sci Process 2004; 79(4-6): 1127-32.
[http://dx.doi.org/10.1007/s00339-004-2660-8]

[34] Jamkhande PG, Ghule NW, Bamer AH, Kalaskar MG. Metal nanoparticles synthesis: An overview on methods of preparation, advantages and disadvantages, and applications. J Drug Deliv Sci Technol 2019; 53: 101174.
[http://dx.doi.org/10.1016/j.jddst.2019.101174]

[35] Levoska J, Tyunina M, Leppävuori S. Laser ablation deposition of silicon nanostructures. Nanostruct Mater 1999; 12(1-4): 101-6.
[http://dx.doi.org/10.1016/S0965-9773(99)00074-4]

[36] Guzman M, Dille J, Godet S. Synthesis and antibacterial activity of silver nanoparticles against gram-positive and gram-negative bacteria. Nanomedicine 2012; 8(1): 37-45.
[http://dx.doi.org/10.1016/j.nano.2011.05.007] [PMID: 21703988]

[37] Dang TD, Tuyet T, Le T, Blanc EF, Dang MC. Synthesis and optical properties of copper nanoparticles prepared by a chemical reduction method. Adv Nat Sci Nanosci Nanotechnol 2011; p. 2015009.

[38] Elkins KE, Vedantam TS, Liu JP, *et al.* Ultrafine FePt Nanoparticles Prepared by the Chemical Reduction Method. Nano Lett 2003; 3(12): 1647-9.
[http://dx.doi.org/10.1021/nl034734w]

[39] Pandey PA, Bell GR, Rourke JP, *et al.* Physical vapor deposition of metal nanoparticles on chemically modified graphene: observations on metal-graphene interactions. Small 2011; 7(22): 3202-10.
[http://dx.doi.org/10.1002/smll.201101430] [PMID: 21953833]

[40] Jamkhande PG, Ghule NW, Bamer AH, Kalaskar MG. Metal nanoparticles synthesis: An overview on methods of preparation, advantages and disadvantages, and applications. J Drug Deliv Sci Technol 2019; 53: 101174.
[http://dx.doi.org/10.1016/j.jddst.2019.101174]

[41] Cachier H, Bremond MP, Buat-Ménard P. Determination of atmospheric soot carbon with a simple thermal method. Tellus B Chem Phys Meterol 1989; 41(3): 379-90.
[http://dx.doi.org/10.3402/tellusb.v41i3.15095]

[42] Williams P. The sputtering process and sputtered ion emission. Surf Sci 1979; 90(2): 588-634.
[http://dx.doi.org/10.1016/0039-6028(79)90363-7]

[43] Lyu SC, Zhang Y, Lee CJ, Ruh H, Lee HJ. Low-Temperature Growth of ZnO Nanowire Array by a Simple Physical Vapor-Deposition Method. Chem Mater 2003; 15(17): 3294-9.
[http://dx.doi.org/10.1021/cm020465j]

[44] Saporiti F, Juarez RE, Audebert F, Boudard M. Yttria and ceria doped zirconia thin films grown by pulsed laser deposition. Mater Res 2013; 16(3): 655-60.
[http://dx.doi.org/10.1590/S1516-14392013005000053]

[45] Chen SC, Hsieh DH, Jiang H, *et al.* Growth and characterization of Cu(In,Ga)Se2 thin films by nanosecond and femtosecond pulsed laser deposition. Nanoscale Res Lett 2014; 9(1): 280.
[http://dx.doi.org/10.1186/1556-276X-9-280] [PMID: 24423198]

[46] Willems VD. Roadmap Report on Nanoparticles. Barcelona, Spain: Espanasl WW 2005; p. 157.

[47] Kuppusamy P, Yusoff MM, Maniam GP, Govindan N. Biosynthesis of metallic nanoparticles using plant derivatives and their new avenues in pharmacological applications – An updated report. Saudi Pharm J 2016; 24(4): 473-84.
[http://dx.doi.org/10.1016/j.jsps.2014.11.013] [PMID: 27330378]

[48]   Keat CL, Aziz A, Eid AM, Elmarzugi NA. Biosynthesis of nanoparticles and silver nanoparticles. Bioresour Bioprocess 2015; 2(1): 47.
[http://dx.doi.org/10.1186/s40643-015-0076-2]

[49]   Li Y, Duan X, Qian Y, Yang L, Liao H. Nanocrystalline silver Particles: Synthesis, agglomeration, and sputtering induced by Electron beam. J Colloid Interface Sci 1999; 209(2): 347-9.
[http://dx.doi.org/10.1006/jcis.1998.5879] [PMID: 9885261]

[50]   Dhillon GS, Brar SK, Kaur S, Verma M. Green approach for nanoparticle biosynthesis by fungi: current trends and applications. Crit Rev Biotechnol 2012; 32(1): 49-73.
[http://dx.doi.org/10.3109/07388551.2010.550568] [PMID: 21696293]

[51]   Raveendran P, Fu J, Wallen SL, Sharma VK, Ria A. Silver nanoparticles: Green synthesis and their Antimicrobial activities. Chem Soc 2003; 125-139-140.

[52]   Shiv Shankar S, Rai A, Ahmad A, Sastry M. Rapid of Au, Ag, and bimetallic Au core–Ag shell using Neem (*Azadirachta indica*) leaf broth. J Colloid Interface Sci 2004; 275: 496-502.
[http://dx.doi.org/10.1016/j.jcis.2004.03.003] [PMID: 15178278]

[53]   Chandran SP, Chaudhary M, Pasricha R, Ahmad A, Sastry M. Synthesis of gold nanotriangles and silver nanoparticles using *Aloe vera* plant extract. Biotechnol Prog 2006; 22(2): 577-83.
[http://dx.doi.org/10.1021/bp0501423] [PMID: 16599579]

[54]   Kowshik M, Ashtaputre S, Kharrazi S, *et al.* Extracellular synthesis of silver Nanoparticles by a silver-tolerant yeast strain MKY3. Nanotechnol 2003; 14- 95–100.
[http://dx.doi.org/10.1088/0957-4484/14/1/321]

[55]   Dameron CT, Reese RN, Mehra RK, *et al.* Biosynthesis of cadmium sulphide quantum semiconductor crystallites. Nature 1989; 338(6216): 596-7.
[http://dx.doi.org/10.1038/338596a0]

[56]   Nair B, Pradeep T. Coalescence of nanoclusters and formation of submicron crystallites assisted by *Lactobacillus* Strains. Cryst Growth Des 2002; 2(4): 293-8.
[http://dx.doi.org/10.1021/cg0255164]

[57]   Sweeney RY, Mao C, Gao X, *et al.* Bacterial biosynthesis of cadmium sulfide nanocrystals. Chem Biol 2004; 11(11): 1553-9.
[http://dx.doi.org/10.1016/j.chembiol.2004.08.022] [PMID: 15556006]

[58]   Sánchez-López E, Gomes D, Esteruelas G, *et al.* Metal-Based Nanoparticles as Antimicrobial Agents: An Overview. Nanomaterials (Basel) 2020; 10(2): 292.
[http://dx.doi.org/10.3390/nano10020292] [PMID: 32050443]

[59]   Dibrov P, Dzioba J, Gosink KK, Häse CC. Chemiosmotic mechanism of antimicrobial activity of Ag($^+$) in *Vibrio cholerae.* Antimicrob Agents Chemother 2002; 46(8): 2668-70.
[http://dx.doi.org/10.1128/AAC.46.8.2668-2670.2002] [PMID: 12121953]

[60]   Sarkar S, Jana AD, Samanta SK, Mostafa G. Facile synthesis of silver nano particles with highly efficient anti-microbial property. Polyhedron 2007; 26(15): 4419-26.
[http://dx.doi.org/10.1016/j.poly.2007.05.056]

[61]   Stoimenov PK, Klinger RL, Marchin GL, Klabunde KJ. Metal Oxide Nanoparticles as Bactericidal Agents. Langmuir 2002; 18(17): 6679-86.
[http://dx.doi.org/10.1021/la0202374]

[62]   Huang L, Li DQ, Lin YJ, Wei M, Evans DG, Duan X. Controllable preparation of Nano-MgO and investigation of its bactericidal properties. J Inorg Biochem 2005; 99(5): 986-93.
[http://dx.doi.org/10.1016/j.jinorgbio.2004.12.022] [PMID: 15833320]

[63]   Klabunde KJ, Stark J, Koper O, *et al.* Nanocrystals as Stoichiometric Reagents with Unique Surface Chemistry. J Phys Chem 1996; 100(30): 12142-53.
[http://dx.doi.org/10.1021/jp960224x]

[64] Koper OB, Klabunde JS, Marchin GL, Klabunde KJ, Stoimenov P, Bohra L. Nanoscale powders and formulations with biocidal activity toward spores and vegetative cells of bacillus species, viruses, and toxins. Curr Microbiol 2002; 44(1): 49-55.
[http://dx.doi.org/10.1007/s00284-001-0073-x] [PMID: 11727041]

[65] Richards R, Li W, Decker S, *et al.* Consolidation of Metal Oxide Nanocrystals. Reactive Pellets with Controllable Pore Structure That Represent a New Family of Porous, Inorganic Materials. J Am Chem Soc 2000; 122(20): 4921-5.
[http://dx.doi.org/10.1021/ja994383g]

[66] Abbasi E, Aval SF, Akbarzadeh A, *et al.* Dendrimers: synthesis, applications, and properties. Nanoscale Res Lett 2014; 9(1): 247.
[http://dx.doi.org/10.1186/1556-276X-9-247]

[67] Dhermendra K, Tiwari J, Behari Sen P. Application of Nanoparticles in Waste Water Treatment. World Appl Sci J 2008; 3(3): 417-33.

[68] Ottaviani MF, Favuzza P, Bigazzi M, Turro NJ, Jockusch S, Tomalia DA. TEM and EPR investigation of the competitive Binding of uranyl ions to starburst dendrimers and liposomes: Potential use of dendrimers as uranyl ion sponges. Langmuir 2000; 16(19): 7368-72.
[http://dx.doi.org/10.1021/la000355w]

[69] Smith RJ, Gorman C, Menegatti S. Synthesis, structure, and function of internally functionalized dendrimers. J Polym Sci 2021; 59(1): 10-28.
[http://dx.doi.org/10.1002/pol.20200721]

[70] Tomalia DA. Birth of a new macromolecular architecture: Dendrimers as Quantized building blocks for nanoscale Synthetic organic chemistry. Aldrichim Acta 2004; 37: 39-57.

[71] Diallo MS, Christie S, Swaminathan P, Johnson JH, Goddard WA. Dendrimer enhanced ultra-filtration recovery of Cu (II) from aqueous solutions using Gx-NH2-PAMAM Dendrimers with ethylene diamine core. Environ Sci Technol 2005; 39: 1366-77.
[http://dx.doi.org/10.1021/es048961r] [PMID: 15787379]

[72] Stoimenov PK, Klinger RL, Marchin GL, Klabunde KJ. Metal oxide nanoparticles as Bactericidal agents. Langmuir 2002; 18(17): 6679-86.
[http://dx.doi.org/10.1021/la0202374]

[73] Jain P, Pradeep T. Potential of silver nanoparticle-coated polyurethane foam as an antibacterial water filter. Biotechnol Bioeng 2005; 90(1): 59-63.
[http://dx.doi.org/10.1002/bit.20368] [PMID: 15723325]

[74] Jain P, Pradeep T. Potential of silver nanoparticle-coated polyurethane foam as an antibacterial water filter. Biotechnol Bioeng 2005; 90(1): 59-63.
[http://dx.doi.org/10.1002/bit.20368] [PMID: 15723325]

[75] Son WK, Youk JH, Lee TS, Park WH. Preparation of antimicrobial ultrafine Cellulose acetate fibers with silver nanoparticles. Macromol Rapid Commun 2004; 25(18): 1632-7.
[http://dx.doi.org/10.1002/marc.200400323]

[76] Ouki SK, Kavannagh M. Treatment of metals-contaminated wastewaters by use of natural zeolites. Water Sci Technol 1999; 39(10-11): 115-22.
[http://dx.doi.org/10.2166/wst.1999.0638]

[77] Blanchard G, Maunaye M, Martin G. Removal of heavy metals from waters by means of natural zeolites. Water Res 1984; 18(12): 1501-7.
[http://dx.doi.org/10.1016/0043-1354(84)90124-6]

[78] Kallo D. DénesKalló Applications of natural zeolites in water and wastewater treatment. Rev Mineral Geochem 2001; 45(1): 519-50.
[http://dx.doi.org/10.2138/rmg.2001.45.15]

[79]   Wang S, Peng Y. Natural zeolites as effective adsorbents in water and wastewater treatment. Chem Eng J 2010; 156(1): 11-24.
[http://dx.doi.org/10.1016/j.cej.2009.10.029]

[80]   Di Natale F, Gargiulo V, Alfè M. Adsorption of heavy metals on silica-supported hydrophilic carbonaceous nanoparticles (SHNPs). J Hazard Mater 2020; 393: 122374.
[http://dx.doi.org/10.1016/j.jhazmat.2020.122374] [PMID: 32135363]

[81]   Li G, Shen J. A study of pyridinium-type functional polymers. IV. Behavioral features of the antibacterial activity of insoluble pyridinium-type polymers. J Appl Polym Sci 2000; 78(3): 676-84.
[http://dx.doi.org/10.1002/1097-4628(20001017)78:3<676::AID-APP240>3.0.CO;2-E]

[82]   Lakshmi S, Kumar SSP, Jayakrishnan A. Bacterial adhesion onto azidated poly(vinyl chloride) surfaces. J Biomed Mater Res 2002; 61(1): 26-32.
[http://dx.doi.org/10.1002/jbm.10046] [PMID: 12001242]

[83]   Wang H, Keller AA, Li F. Natural organic matter removal by adsorption ontocarbonaceous nanoparticles and coagulation. J Environ Eng 2010; 136(10): 1075-81.
[http://dx.doi.org/10.1061/(ASCE)EE.1943-7870.0000247]

[84]   Liang M, Guo LH. Application of nanomaterials in environmental analysis and monitoring. J Nanosci Nanotechnol 2009; 9(4): 2283-9.
[http://dx.doi.org/10.1166/jnn.2009.SE22] [PMID: 19437965]

[85]   Elshawy OE, Helmy EA, Rashed LA. Preparation, Characterization and *in Vitro* Evaluation of the Antitumor Activity of the Biologically Synthesized Silver Nanoparticles. Adv Nanopart 2016; 5(2): 149-66.
[http://dx.doi.org/10.4236/anp.2016.52017]

[86]   Chen M, Goodman DW. Catalytically active gold: from nanoparticles to ultrathin films. Acc Chem Res 2006; 39(10): 739-46.
[http://dx.doi.org/10.1021/ar040309d] [PMID: 17042474]

[87]   Kowalczyk P, Holyst R. Efficient adsorption of super greenhouse gas (tetrafluoromethane) in carbon nanotubes. Environ Sci Technol 2008; 42(8): 2931-6.
[http://dx.doi.org/10.1021/es071306+] [PMID: 18497146]

[88]   Zhang X, Liu W, Tang J, Xiao P. Study on PD detection in SF$_6$ using multi-wall carbon nanotube films sensor. IEEE Trans Dielectr Electr Insul 2010; 17(3): 833-8.
[http://dx.doi.org/10.1109/TDEI.2010.5492256]

[89]   Pitoniak E, Wu CY, Mazyck DW, Powers KW, Sigmund W. Adsorption enhancement mechanisms of silica-titania nanocomposites for elemental mercury vapor removal. Environ Sci Technol 2005; 39(5): 1269-74.
[http://dx.doi.org/10.1021/es049202b] [PMID: 15787366]

[90]   Lead JR, Wilkinson KJ. Aquatic Colloids and Nanoparticles: Current Knowledge and Future Trends. Environ Chem 2006; 3(3): 159-71.
[http://dx.doi.org/10.1071/EN06025]

[91]   Blaser SA, Scheringer M, MacLeod M, Hungerbühler K. Estimation of cumulative aquatic exposure and risk due to silver: Contribution of nano-functionalized plastics and textiles. Sci Total Environ 2008; 390(2-3): 396-409.
[http://dx.doi.org/10.1016/j.scitotenv.2007.10.010] [PMID: 18031795]

[92]   Khandel P, Shahi SK. Mycogenic nanoparticles and their bio-prospective applications: current status and future challenges. J Nanostructure Chem 2018; 8(4): 369-91.
[http://dx.doi.org/10.1007/s40097-018-0285-2]

[93]   Abdelghany TM, Al-Rajhi AMH, Al Abboud MA, *et al.* Recent Advances in Green Synthesis of Silver Nanoparticles and Their Applications: About Future Directions. A Review. Bionanoscience 2018; 8(1): 5-16.

[http://dx.doi.org/10.1007/s12668-017-0413-3]

[94] Zhang W. Nanoscale Iron Particles for Environmental Remediation: An Overview. J Nanopart Res 2003; 5(3/4): 323-32.
[http://dx.doi.org/10.1023/A:1025520116015]

[95] Sardar R, Funston AM, Mulvaney P, Murray RW. Gold nanoparticles: past, present, and future. Langmuir 2009; 25(24): 13840-51.
[http://dx.doi.org/10.1021/la9019475] [PMID: 19572538]

[96] Umamaheswari K, Baskar R, Chandru K, Rajendiran N, Chandirasekar S. Antibacterial activity of gold nanoparticles and their toxicity assessment. BMC Infect Dis 2014; 14(S3): P64.
[http://dx.doi.org/10.1186/1471-2334-14-S3-P64]

[97] Khare P, Sonane M, Pandey R, Ali S, Gupta K, Satish A. Adverse effects of TiO2 and ZnO nanoparticles in soil nematode, *Caenorhabditis elegans.* J Biomed Nanotechnol 2011; 7(1): 116-7.
[http://dx.doi.org/10.1166/jbn.2011.1229] [PMID: 21485831]

[98] Helland A, Wick P, Koehler A, Schmid K, Som C. Reviewing the environmental and human health knowledge base of carbon nanotubes. Cien Saude Colet 2008; 13(2): 441-52.
[http://dx.doi.org/10.1590/S1413-81232008000200019] [PMID: 18813560]

[99] Sajid M, Ilyas M, Basheer C, *et al.* Impact of nanoparticles on human and environment: review of toxicity factors, exposures, control strategies, and future prospects. Environ Sci Pollut Res Int 2015; 22(6): 4122-43.
[http://dx.doi.org/10.1007/s11356-014-3994-1] [PMID: 25548015]

[100] Ealias AM, Saravanakumar MP. A review on the classification, characterisation, synthesis of nanoparticles and their application'. Mater Sci Eng 2017; 263: 032019.

[101] Losert S, von Goetz N, Bekker C, *et al.* Human exposure to conventional and nanoparticle--containing sprays-a critical review. Environ Sci Technol 2014; 48(10): 5366-78.
[http://dx.doi.org/10.1021/es5001819] [PMID: 24821461]

[102] Vílchez-Maldonado S, Calderó G, Esquena J, Molina R. UV protective textiles by the deposition of functional ethylcellulose nanoparticles. Cellulose 2014; 21(3): 2133-45.
[http://dx.doi.org/10.1007/s10570-014-0217-3]

CHAPTER 8

Removal of Micropollutants and Pathogens from Water using Nanomaterials

Bhupinder Dhir[1,*] and **Raman Kumar**[2]

[1] *School of Life Sciences, Indira Gandhi National Open University, New Delhi-110078, India*

[2] *Department of Biotechnology, Maharishi Markandeshwar (Deemed to be University), Mullana, Ambala 133207, India*

Abstract: Presence of micro pollutants and pathogens in water has become a concern worldwide. Micropollutants such as pharmaceutically active compounds, personal care products, organic compounds and pathogens/microbes (viral, bacterial and protozoa) pose a threat to humans. Nanotechnology has proved effective in developing strategies for the treatment of contaminated water. Nanomaterials have found application in the removal of different categories of pollutants, from water. The properties such as high reactivity and effectiveness establish nanomaterials as ideal materials suitable for treatment of contaminated water/wastewater. Nanomaterials such as carbon nanotubes, graphene-based composites and metal oxides, have shown potential to remove dyes, pathogens from wastewater. Research efforts are required to develop an eco-friendly, economic and sustainable technology for the removal of micropollutants and biological agents such as microbes using nanomaterials.

Keywords: Carbon tubes, Composites, Graphene, Metal oxides, Microorganisms, Nanomaterials, Nanotechnology, Nanoadsorbents, Pharmaceuticals.

INTRODUCTION

Micropollutants are trace levels of synthetic organic substances/molecules that contaminate ground and surface water. The micropollutants released from household activities and industries mainly include pharmaceutical compounds and personal care products (PCPs) such as cosmetics, detergents, toxic chemicals, dyes, fertilizers, and endocrine disruptive compounds (EDCs). Other sources of micropollutants include runoff from agriculture and livestock areas, leakage from landfills, septic tanks and industrial sources. Micropollutants are usually present at trace levels (from ng/L to µg/L) in water, thus their detection and quantification

* **Corresponding author Bhupinder Dhir:** School of Life Sciences, Indira Gandhi National Open University, New Delhi-110078, India; E-mail: bhupdhir@gmail.com

are difficult. These pollutants need to be treated properly; otherwise, they enter the food chain and affect living beings in a negative way. The presence of micropollutants in water proves toxic to aquatic organisms and produces a detrimental effect on the health of humans *via* disruption of the endocrine-system [1 - 4].

Most of these micropollutants and their metabolites are non-biodegradable and do not get removed easily after treatment procedures. Various physico-chemical techniques used for the removal of organic pollutants from water include precipitation, membrane filtration, coagulation, ion exchange, adsorption and chemical oxidation [5]. Other conventional treatment techniques include ozonation and reverse osmosis [6]. The physical and chemical methods adopted for the removal of micropollutants involve operational setup and energy input, hence eco-friendly and cost-effective alternate techniques were explored. Nanotechnology has emerged as one of the promising technologies that play a role in areas of biomedical, health care, mechanics, environment and energy. Properties such as high specific surface area (fast dissolution, high reactivity, strong sorption), excellent mechanical properties, superparamagnetism, low cost and energy requirements, high chemical reactivity, recyclability, ease of fabrication, and functionalization establish nanomaterials such as ideal materials to be used in the removal of contaminants from wastewater. Nanotechnology benefits the existing environmental technologies with effective performance and less consumption of energy and materials. Nanobased characteristics allow the development of novel high-tech materials, namely membranes, adsorption materials, nanocatalysts, functionalized surfaces, coatings, and reagents for use in water treatment processes. Water and wastewater treatment using nanosized materials have been tried in the last few decades. The application of nanomaterials ranging from 1 nm to about 100 nm in environmental monitoring, remediation and efficiency for avoiding environmental contamination has been explored. Nanotechnologies offered proved useful in developing innovative methods to treat contaminated water [7].

Different nanoscale materials that have shown immense potential in environmental applications include nanoscale zeolites, metal oxides and titanium dioxide, carbon nanotubes and fibers and enzymes. Nanoengineered materials adsorb and/or degrade various pollutants such as metal ions, dyes, pesticides, pharmaceuticals, other organic pollutants and waterborne microbes (bacteria, viruses, protozoa, *etc*.). Some commercially available nanomaterials are carbon nanotubes, graphitic carbon nitride ($CNT/g-C_3N_4$) composites, graphene-based composites and metal oxides. Iron oxide nanoparticles (IONPs), magnetite (Fe_3O_4), maghemite ($\gamma-Fe_2O_3$) and hematite ($\alpha-Fe_2O_3$) particles are non-toxic, hence considered safe for their application in the area of removing contaminants from water. Nanotechnology can be used directly or indirectly in the treatment of surface water, groundwater and wastewater polluted with toxic metal ions, organic/inorganic compounds and microorganisms [8, 9].

NANOADSORBENTS

Nanoadsorbents are adsorbents having nanoscale pores and show high selectivity, surface area, permeability and good mechanical and thermal stability. Nanoadsorbents mainly include nanotubes, nanomesh, nano-filtration membranes, nanofibrous alumina filters, magnetic nanoparticles, nanoporous ceramics and clays, cyclodextrin nanoporous polymer, polypyrrole–carbon nanotube composite, *etc* [9, 10].

1. Nanomaterials have been classified into four types.
2. Carbon-based nanoadsorbents [carbon nanotubes (CNTs)]
3. Metal-based nanoadsorbents
4. Polymeric nanoadsorbents
5. Zeolites

Carbon-based nanoadsorbents are composed mostly of carbon and available in various shapes, such as the sphere, ellipsoid or tube. Spherical and ellipsoidal carbon nanomaterials are known as fullerenes, while cylindrical ones are called nanotubes. CNTs are categorized as single-walled nanotubes and multi-walled nanotubes. Carbon nanotubes have large specific surface area that makes them ideal adsorbents for removal of various organic pollutants and metal ions. Carbon nanotubes establish π-π electrostatic interactions and can easily be modified by chemical treatment to increase the adsorption capacity. The smooth interior of the nanotubes provides faster flow rates. This helps in saving the energy required to push the water through the tubes. They can be used for the adsorption of persistent contaminants as well as to preconcentrate and detect contaminants. Metal ions get adsorbed by CNTs through electrostatic attraction and chemical bonding. CNTs exhibit antimicrobial properties by causing oxidative stress in bacteria and destroying the cell membranes. As a result of chemical oxidation, no toxic byproducts are produced, hence can prove useful in disinfection processes like chlorination and ozonation [11, 12].

Metal-based nanomaterials include quantum dots, nanogold, nanosilver, and metal oxides, such as titanium dioxide. Nanoscale metal oxides act as effective adsorbents and help in the removal of heavy metals and radionuclides. They possess a high specific surface area. The nanoscale metal oxides such as nanomaghemite and nanomagnetite are superparamagnetic, and facilitate separation by a low-gradient magnetic field. They can be employed for adsorptive media filters and slurry reactors. Nano iron hydroxide [α-FeO(OH)] is an adsorbent with a huge specific surface area that enables the adsorption of arsenic from waste and drinking water. Nanosilver exhibits strong and broad-spectrum antimicrobial activity. It is applied to point-of-use water disinfection systems and antibiofouling surfaces. Nano-titanium dioxide (TiO_2), showing high chemical stability, is also used in disinfection and decontamination processes. TiO_2 acts as a catalyst and helps in the degradation of organic compounds and micro-organisms.

Polymeric nanoadsorbents mainly include dendrimers. Dendrimers are nanosized polymers built from branched units (repetitively branched molecules) and are used for removing organics and heavy metals. Organic compounds can be adsorbed by the interior hydrophobic shells, whereas heavy metals can be adsorbed by the tailored exterior branches. The bioadsorbent is biodegradable, biocompatible, and non-toxic. A high removal rate of dyes (~99%) has been achieved using these materials [13].

Zeolites have a porous structure in which nanoparticles such as silver ions can be embedded. There they are released from the zeolite matrix by exchange with other cations in solution [14].

Nano-zero valent iron can be used for the remediation of groundwater contaminated with chlorinated hydrocarbon fluids and perchlorates. *In situ* treatment of groundwater can be achieved by injecting a suspension of nano-zero valent iron into the groundwater. Zero-valent iron acts as a reducing agent and helps in removing contaminants such as polychlorinated biphenyls, pesticides, herbicides, aromatic hydrocarbons and metals [15].

Magnetic nanoparticles (magnetite Fe_3O_4) have been widely used for the separation of water pollutants, particularly arsenic. Magnetic nanoparticles can be injected directly into the contaminated ground, and loaded particles can be removed simply through a magnetic field. Magnetic nanoparticles act as an ideal compound to increase the osmotic pressure of draw solutions used in forward osmosis. Forward osmosis, as a contrary process, to reverse osmosis, draws water from a low osmotic pressure to one with a higher osmotic pressure (draw solution) using the osmotic gradient. Magnetic nanoparticles in the adsorption and degradation of organic pollutants have been reviewed [16]. The treatment technique was able to reduce the concentrations of these micropollutants by 80%.

The composite materials are porous solids and possess superior properties. Composites are also combined with other nanoparticles. Nanoparticles, such as nanosized clays, are added to enhance mechanical, thermal, barrier, and flame-retardant properties. Researchers proposed clay/Fe_3O_4 composites as adsorbents to remove heavy metals and dye molecules from modelled wastewater through magnetic separation. Composites show coordination bonds between metal ions and organic ligands.

Graphene is a kind of carbon nanomaterial that has attracted tremendous attention in water purification and various fields due to its unique physical and chemical properties [17].

Nanofiltration

Nanofiltration is one of the membrane filtration techniques in which molecules and particles less than 0.5 nm to 1 nm are not allowed to move across the membrane. Membranes used in nanofiltration show a unique charge-based repulsion mechanism allowing the separation of various ions. This technique helps in reducing hardness, color, odor, and heavy metal ions from groundwater. The technique also proves useful in removing dissolved salts and micropollutants, and softening of water softening. Nanomaterials used for this kind of treatment include chitosan, silver nanoparticles and carbon nanomaterials [18]. table

Nanocomposite membranes are a group of filtration materials with mixed matrix and surface-functionalized membranes. In mixed matrix membranes, nanofillers are added to the matrix material [19]. The nanofillers are inorganic and feature a larger surface area. Metal oxide nanoparticles (Al_2O_3, TiO_2) increase the mechanical and thermal stability as well as permeate flux of polymeric membranes. The addition of zeolites improves the hydrophilicity of membranes resulting in raised water permeability. Nanosilver, CNTs nanoparticles and (photo)catalytic nanomaterials (bimetallic nanoparticles, TiO_2) help in increasing resistance to fouling [20].

Nanofibers are in the range of nanometers used in separation and filtration processes. They ensure a high specific surface porosity and high surface-to-mass ratio.

Nanoclays

Various types of nanoclays have become an important part of water treatment technologies. Clay nanoparticles absorb various pollutants, including aliphatic/aromatic hydrocarbons, organic dyes, pesticides such as atrazine, trifluralin, parathion, malathion and inorganic compounds (such as metals) from soil and wastewater. Clay-based sorbents have unique properties such as high specific surface area, reusability and low cost. Removal of amoxicillin antibiotic from liquid suspensions using bentonite has been reported. Adsorption of trimethoprim using montmorillonite clay has also been reported.

Photocatalysis

Photocatalysis is an advanced oxidation process that is engaged in the treatment of water and wastewater, particularly the oxidative elimination of micropollutants and microbial pathogens. Nanomaterials-based composites (*e.g.,* with metal oxide/sulfides, noble metal NPs, GO, *etc.*) act as catalysts and help in the degradation of organic pollutants. TiO_2 is widely utilized as a photocatalyst. In catalyzed advanced oxidation processes, the principal mechanisms are the ligand-to-metal charge transfer and the excitation of metal-oxo clusters [21]. Rojas and Horcajada [22] found the application of MOFs in the elimination of emerging organic pollutants from water.

ROLE OF NANOTECHNOLOGY IN THE REMOVAL OF SPECIFIC POLLUTANTS

Dyes

Magnetic nanoparticles, carbonaceous nanomaterials, nano-sized TiO_2, and graphitic carbon nitride ($g-C_3N_4$) have been considered nanomaterials with high capacity for removing dyes [23, 24]. Adsorption, photocatalytic degradation, and biological treatment have been found to be the major mechanisms involved in the removal of dyes. Monometallic/bimetallic nanoparticles have shown an adsorption capacity of 875.0 mg/g for ethylene blue. NZVIs show high efficiency in the degradation of reactive dye present in wastewater. Removal of dye RB5 by NZVIs has been reported. The mechanism for removal of RB5 dye involved adsorption on the surface of NZVIs followed by scavenging of the azo bond ($-N = N$) by the strong reduction of NZVIs [25]. A nanofibrous membrane is used as an adsorbent to remove anionic dyes from aqueous solutions.

HEAVY METAL

Nanoadsorbents showed the capacity to chelate metal ions [26, 27]. Fe_3O_4 magnetic nanoparticles (MNPs) modified with 3- aminopropyltriethoxysilane (APS), copolymers of acrylic acid (AA) and crotonic acid (CA) have been reported to be strong adsorbents and remove heavy metal ions such as Cd (II), Zn (II), Pb (II) and Cu (II) from aqueous solutions. Adsorption increases gradually with an increase in alkalinity.

Amino and imino groups on the PEI macromolecular chains provide the main adsorption sites. Bifunctional properties help in adsorbing cationic and anionic target compounds at different pH values in aqueous solutions. This can lead to strong electrostatic repulsion of the cationic metal ions to be adsorbed. A metal complex can be formed when neutral nitrogen of amine and imino group with lone pair of electrons binds a metal ion.

PPCPs

Nanotechnology has provided a great alternative for removing the PPCPs/PhACs. Nano-enhanced techniques involved in the removal of PPCPs/PhACs from contaminated water include adsorption, nano-enhanced photocatalysis, and nanofiltration [28, 29].

Pesticides

Nanomaterials, namely CNTs, both single-walled nanotubes (SWNTs) and multi-walled nanotubes (MWNTs), have shown good potential to remove various pesticides. Adsorption of pollutants by CNTs is affected by the pore structure and broad spectrum of surface functional groups. These groups can be modified by chemical or thermal modifications to improve their optimal performance [30, 31]. Adsorption of organic chemicals on CNTs involves one or more mechanisms such as hydrophobic effect, covalent bonding, π-π interactions, hydrogen bonding, and electrostatic interactions. Adsorption may also take place through hydrogen bonding between functional groups such as - COOH, -OH, -NH$_2$ and organic molecules [32, 33]. Adsorption of atrazine by SWNTs and MWNTs has been demonstrated. Adsorption of diuron onto MWCNTs has been reported. It indicated that oxidation treatment of MWCNTs gave rise to a high surface area and pore volume and, subsequently, an increase in adsorption capacity [34]. SWCNTs have been demonstrated to have a higher adsorption capacity for 4-chloro-2- methylphenoxyacetic acid (MCPA), a phenoxy acid herbicide.

Graphene, a carbon nanomaterial, showed good adsorption capacity for the removal of pesticides (ranging from 600 to 2000 mg/g) [30]. Researchers reported dehalogenation and removal of persistent halocarbon pesticides from water using graphene.

Nanocrystalline metal oxides such as ferric oxides, manganese oxides, aluminum oxides, titanium oxides, magnesium oxides and cerium oxides proved to be effective in removing a broad range of pesticides. These materials showed high adsorption capacity, and faster kinetics because of the higher specific surface area, shorter intraparticle diffusion distance and larger number of surface reaction sites. Some researchers also reported the removal of organophosphorus pesticides by nanometal oxides. Nano metal oxides proved effective in removing organophosphorus pesticides. The removal of pesticides by magnetic nanoparticles has also been reported. High surface area and hydroxyl groups on the surface of nanocrystalline alumina proved useful in the removal of organophosphate pesticides.

Removal of a wide range of pesticides by photocatalytic degradation has been reported [35]. The materials involved in removal included ZnO, TiO_2, Fe_2O_3, CdS and WO_3. Among all, titanium dioxide has been used most widely because of its low toxicity, chemical stability, low cost, and abundance. Photodecomposition of pesticides by TiO_2 has been reported [36, 37]. Another study on the Photocatalytic degradation of dicofol with TiO_2 nanoparticles under UV light irradiation has been reported. The compound showed complete degradation.

INACTIVATION OF MICROORGANISMS USING NANOMATERIALS

Bacteria, viruses and protozoa are the major causal agents of waterborne diseases. They affect human beings after ingestion of contaminated water.

Bacterial Pathogens

Bacterial agents, including *E. coli, Pseudomonas, Shigella, Salmonella, Myco bacterium avium, Vibrio cholerae, Campylobacter, Helicobacter, Legion ella, etc.*, have been considered major agents causing waterborne diseases. Many nanomaterials have proved useful in removing pathogens from water, thus proving very useful in the treatment of water [38].

Zinc oxide nanoparticles have been shown to have antibacterial activity. Total (100%) inhibition of *S. aureus* and *P. aeruginosa* has been reported by ZnO NPs [39]. Inhibition of *S. aureus* by ZnO NPs has been reported [40]. Inactivation of *E. coli* and *S. aureus* by DOX-ZnO/PEG nanocomposites has been reported. The antimicrobial activity of silver nanoparticles has also been reported. Silver nanoparticles (AgNPs) inhibited biofilm formation by *S. aureus* . AgNPs also showed antimicrobial activity against *Methylobacterium* spp. *Gordonia* sp. and *E. coli.*

Photocatalysis with TiO_2 also proved useful in reducing bacterial disinfection (*E. coli, S. typhimurium* and *S. sonnei*). Nanomaterials that act as photocatalysts, namely TiO_2, Pt-TiO_2 and Ag-TiO_2 showed disinfection potential against *E. coli* [41, 42]. Ag-TiO_2 photocatalyst was found to be more effective for the reduction of microbes in water. The Ag–TiO_2 nanocomposite was shown to possess an enhanced antibacterial activity against *E. coli*; 1g/L of the nanocomposite resulted in 99.99% inhibition of the bacterial growth. Ti/TiO_2–Ag coupled into a photoelectrocatalytic process helped in achieving 100% inactivation of *Mycobacterium smegmati.*

Co- Mn- and binary Mn/Co-doped TiO_2 catalysts showed antibacterial activity against *K. pneumonia* and *E. coli* [43]. ZnO quantum dots decorated CuO nanosheets and TiO_2 quantum dots decorated WO_3 nanosheet composites showed inactivation of gram-positive and gram-negative bacterial strains, *Enterococcus faecalis* and *Micrococcus luteus* [44]. The antimicrobial activity could be attributed to the inhibition of bacterial growth through the electrostatic interaction of negative bacterial charges and the positive metal oxide nanostructures. TiO_2 nanowires showed antibacterial activity against *E. coli* and *S. epidermidis*. Nanoporous copper aluminosilicate (CAS) material showed antimicrobial action and inhibitory action against different target microorganisms *viz.Escherichia coli, Salmonella enterica, Pseudomonas aeruginosa, Listeria monocytogenes, Staphylococcus aureus, Enterococcus faecalis, Candida albicans* and *Aspergillus niger*) [45]. Inactivation of fecal coliform showed using nanoparticulated films of TiO_2 and TiO_2/Ag (4% w/w) [46]. Bacterial inactivation was the cell wall rapture. The nano-photocatalysts based on colloidal nanocarbon–metal composition (NCMC) were successfully used to destroy *E. coli* bacteria in 10–30 min in water [47]. SnO_2-doped nanocomposites (with SnO_2 being the dopant in sulphonated GO and CNT), showed significant dose-dependent bactericidal activity against *E. coli, P. graminis* and a variety of bacterial species *S. aureus, L. monocytogenes, B. subtilis, S. typhi* and *T. viride* isolated from a domestic wastewater treatment plant [48]. A simple method for the rapid removal of pathogenic microorganisms (both bacteria and virus) from water, using modifiedcore–shell Fe_3O_4–SiO_2–NH_2 nanoparticles, was developed by Zhan *et al.* [49]. $SnO_2/PSi/NH_2$ nanocomposite showed antibacterial potential toward *E. coli* and *S. aureus*.

Jin *et al.* modify Fe_3O_4 nanoparticles using an antibacterial agent, cetyltrimethylammonium bromide (CTAB), to produce bactericidal paramagnetic nanoparticles (Fe_3O_4@CTAB). The nanoparticles have great potential in water disinfection and inactivate more than 99% of *E. coli* and *B. subtilis* bacteria.

Hassouna *et al.* found that AgNPs-loaded clay (at 0.1 mg/L) showed the most pronounced antibacterial efficacy against *Salmonella* spp (90%), *E. coli, Klebsiella pneumonia* and *Shigella flexneri* (80%) and *Klebsiella aerogenes* (70%) after 2 h of exposure time. The antibacterial activity of CNTs-loaded clay was found effective against *Salmonella* spp. and *Klebsiella pneumonia* (70% inhibition), while it was 60% inhibition was recorded for *E. coli* strains. Hybrid polyaniline/graphene/carbon nanotube materials achieved 99.5% and 99.2% removal for *S. aureus* and *E. coli*, respectively.

The $-C_3N_4$-based photocatalysts showed antimicrobial effects for water disinfection. The GO/g-C_3N_4 showed an antibacterial effect against *E. coli* (97.9%).

Viruses

Pathogenic enteric viruses pose a significant risk to human health. The human pathogenic viruses belong to the families of *Caliciviridae, Adenoviridae,*

Hepeviridae, *Picornaviridae* and *Reoviridae*. Nano P25 TiO_2 was found to be more effective for MS2 phage removal [50]. A removal efficiency of up to 100% could be achieved. P25 has advanced adsorption and photocatalysis performance. $Cu–TiO_2$ nanofibers showed disinfection potential for bacteriophage f2 and its host *E. coli* [51]. Amine-functionalized magnetic $Fe_3O_4–SiO_2–NH_2$ nanoparticles showed good potential to remove bacteriophage f2 and Poliovirus-1 at 76.7% and 81.5% [49]. Silver doping of TiO_2 nanoparticles was an effective way to increase TiO_2 photocatalytic activity for virus inactivation of bacteriophage MS2 in aqueous media [52]. $g-C_3N_4$ showed enhanced virucidal performance against MS2. The $g–C_3N_4$/EP composite (carrier of expanded perlite-EP) resulted in complete inactivation of 8-log *E. coli* and MS2 within 3 and 4 h of visible light irradiation. Photocatalytic composites showed antimicrobial potential for source water disinfection.

Protozoa

Protozoan parasites are among the most important waterborne pathogens. These include *Cryptosporidium*, *Giardia*, *Cyclospora*, *Acanthamoeba*, *Isospora*, *etc.* silver salt and nanoparticles showed negative effects against and the removal of protozoan pathogen *Cryptosporidium parvu* [53]. The removal efficiencies ranged from 96.4% to 99.2%. Silver nanoparticle disinfection contributed to the treatment of *C. parvum* using silver-impregnated ceramic water filters. A composite fabricated by embedding silver nanospheres onto aragonitic cuttlefish bone (CB)-stabilized samarium-doped zinc oxide (Sm-doped ZnO) nanorods [54]. Ag@Sm-doped ZnO/CB showed significant biocidal efficiency against pathogenic bacteria and parasites. Nnanocomposite exhibited enhanced disinfection efficiencies for *Staphylococcus aureus* (80%), *Pseudomonas aeruginosa* (60%) and *Schistosoma mansoni* cercariae (100%). Nanocomposite also showed a negative effect on *Schistosoma mansoni* adult worms showing 100% mortality accompanied by significant disintegration of the worm body. MgO NPs showed a significant effect against *Cyclospora* oocysts [55, 56]. Sunnotel *et al.* [57] showed disinfection of surface water contaminated with *Cryptosporidium* oocysts, by TiO_2 photo catalysis.

Apart from all these contaminants, nanomaterials have shown immense potential in removing various organic compounds, antibiotics from wastewater [58 - 64], thus ensuring the availability of safe water [65, 66].

CONCLUSION

Nanotechnology has emerged as an effective tool for environmental clean-up. The use of nanomaterials thus can be a sustainable approach for the remediation of the environment. Nanomaterials (NMs) have proved to be efficient, cost-effective, and environmental friendly alternatives to existing water and wastewater treatment processes because of their properties. Nanomaterials have proved effective in removing heavy metals, dyes, pesticides, pharmaceuticals, and many other organic compounds. Removal of pathogens such as bacteria and viruses has also been achieved using nanomaterials. More studies need to be conducted to achieve the maximal removal of contaminants. Further research based on an assessment of nanobased material for their potential risks needs to be analyzed.

REFERENCES

[1]   Tixier C, Singer HP, Oellers S, Müller SR. Occurrence and fate of carbamazepine, clofibric acid, diclofenac, ibuprofen, ketoprofen, and naproxen in surface waters. Environ Sci Technol 2003; 37(6): 1061-8.
        [http://dx.doi.org/10.1021/es025834r] [PMID: 12680655]

[2]   Kummerer K. Pharmaceuticals in the environment: Sources, fate, effects and risks; with 77 tables. Berlin, Heidelberg: Springer 2004.
        [http://dx.doi.org/10.1007/978-3-662-09259-0]

[3]   Sanderson H, Brain RA, Johnson DJ, Wilson CJ, Solomon KR. Toxicity classification and evaluation of four pharmaceuticals classes: antibiotics, antineoplastics, cardiovascular, and sex hormones. Toxicology 2004; 203(1-3): 27-40.
        [http://dx.doi.org/10.1016/j.tox.2004.05.015] [PMID: 15363579]

[4]   Monteiro SC, Boxall AB. Occurrence and fate of human pharmaceuticals in the environment. Rev Environ Contam Toxicol 2010; 202: 53-154.
        [PMID: 19898761]

[5]   Singhal N, Perez-Garcia O. Degrading Organic Micropollutants: The Next Challenge in the Evolution of Biological Wastewater Treatment Processes. Front Environ Sci 2016; 4: 36.
        [http://dx.doi.org/10.3389/fenvs.2016.00036]

[6]   Derco J, Gotvajn AŽ, Čižmárová O, Dudáš J, Sumegová L, Šimovičová K. Removal of Micropollutants by Ozone-Based Processes. Processes (Basel) 2021; 9(6): 1013.
        [http://dx.doi.org/10.3390/pr9061013]

[7]   Kokkinos P, Mantzavinos D, Venieri D. Current Trends in the Application of Nanomaterials for the Removal of Emerging Micropollutants and Pathogens from Water. Molecules 2020; 25(9): 2016.
        [http://dx.doi.org/10.3390/molecules25092016] [PMID: 32357416]

[8]   Bagal MV, Raut-Jadhav S. The process for the removal of micropollutants using nanomaterialsHandbook of Nanomaterials for Wastewater Treatment Fundamentals and Scale Up Issues. Micro and Nano Technologies 2021; pp. 957-1007.

[9]   Mudhoo A, Sillanpää M. Magnetic nanoadsorbents for micropollutant removal in real water treatment: a review. Environ Chem Lett 2021; 19(6): 4393-413.
        [http://dx.doi.org/10.1007/s10311-021-01289-6] [PMID: 34341658]

[10]   Khajeh M, Laurent S, Dastafkan K. Nanoadsorbents: classification, preparation, and applications (with emphasis on aqueous media). Chem Rev 2013; 113(10): 7728-68.
        [http://dx.doi.org/10.1021/cr400086v] [PMID: 23869773]

[11]   Ali ME, Hoque ME, Safdar Hossain SK, Biswas MC. Nanoadsorbents for wastewater treatment: next generation biotechnological solution. Int J Environ Sci Technol 2020; 17(9): 4095-132.
        [http://dx.doi.org/10.1007/s13762-020-02755-4]

[12]   Piaskowski K, Zarzycki PK. Carbon-Based Nanomaterials as Promising Material for Wastewater Treatment Processes. Int J Environ Res Public Health 2020; 17(16): 5862.
        [http://dx.doi.org/10.3390/ijerph17165862] [PMID: 32823500]

[13] Gomez-Maldonado D, Erramuspe IBV, Peresin MS. Natural polymers as alternative adsorbents and treatment agents for water remediation. BioResources 2019; 14(4): 10093-160.
[http://dx.doi.org/10.15376/biores.14.4.Gomez-Maldonado]

[14] Jiang N, Shang R, Heijman SGJ, Rietveld LC. High-silica zeolites for adsorption of organic micro-pollutants in water treatment: A review. Water Res 2018; 144: 145-61.
[http://dx.doi.org/10.1016/j.watres.2018.07.017] [PMID: 30025266]

[15] Kim S, Park CM, Jang M, *et al.* Aqueous removal of inorganic and organic contaminants by graphene-based nanoadsorbents: A review. Chemosphere 2018; 212: 1104-24.
[http://dx.doi.org/10.1016/j.chemosphere.2018.09.033] [PMID: 30286540]

[16] Khan MF, Jamal A, Rosy PJ, Alguno AC, Ismail M, Khan I, Ismail A, Zahid M. Eco-friendly elimination of organic pollutants from water using graphene oxide assimilated magnetic nanoparticles adsorbent. Inorganic Chem Comm 2022; 139: 109422.

[17] Khan FSA, Mubarak NM, Khalid M, *et al.* A comprehensive review on micropollutants removal using carbon nanotubes-based adsorbents and membranes. J Environ Chem Eng 2021; 9(6): 106647.
[http://dx.doi.org/10.1016/j.jece.2021.106647]

[18] Bodzek M, Konieczny K. Membranes In Organic Micropollutants Removal. Curr Org Chem 2018; 22(11): 1070-102.
[http://dx.doi.org/10.2174/1385272822666180419160920]

[19] Niedergall K, Bach M, Hirth T, Tovar GEM, Schiestel T. Removal of micropollutants from water by nanocomposite membrane adsorbers. Separ Purif Tech 2014; 131: 60-8.
[http://dx.doi.org/10.1016/j.seppur.2014.04.032]

[20] Zusman OB, Perez A, Mishael YG. Multi-site nanocomposite sorbent for simultaneous removal of diverse micropollutants from treated wastewater 2021. Applied Clay Sci
[http://dx.doi.org/10.1016/j.clay.2021.106300]

[21] Russo V, Hmoudah M, Broccoli F, Iesce MR, Jung O, Serio MD. Applications of Metal Organic Frameworks in Wastewater Treatment: A Review on Adsorption and Photodegradation. Front Chem Eng 2020.
[http://dx.doi.org/10.3389/fceng.2020.581487]

[22] Rojas S, Horcajada P. Metal–Organic Frameworks for the Removal of Emerging Organic Contaminants in Water. Chem Rev 2020; 120(16): 8378-415.
[http://dx.doi.org/10.1021/acs.chemrev.9b00797] [PMID: 32023043]

[23] Cai Z, Sun Y, Liu W, Pan F, Sun P, Fu J. An overview of nanomaterials applied for removing dyes from wastewater. Environ Sci Pollut Res Int 2017; 24(19): 15882-904.
[http://dx.doi.org/10.1007/s11356-017-9003-8] [PMID: 28477250]

[24] Batool S, Akib S, Ahmad M, Balkhair KS, Ashraf MA. Study of Modern Nano Enhanced Techniques for Removal of Dyes and Metals. 2014.
[http://dx.doi.org/10.1155/2014/864914]

[25] Khashij M, Dalvand A, Mehralian M, Ebrahimi AA, Khosravi R. Removal of reactive black 5 dye using zero valent iron nanoparticles produced by a novel green synthesis method. Pigm Resin Technol 2020; 49(3): 215-21.
[http://dx.doi.org/10.1108/PRT-10-2019-0092]

[26] Yang J, Hou B, Wang J, *et al.* Nanomaterials for the Removal of Heavy Metals from Wastewater. Nanomaterials (Basel) 2019; 9(3): 424.
[http://dx.doi.org/10.3390/nano9030424] [PMID: 30871096]

[27] Liu L, Luo X, Ding L, Luo S. Application of Nanotechnology in the Removal of Heavy Metal From WaterNanomaterials for the Removal of Pollutants and Resource Reutilization. Micro Nano Technol 83-147.2019;

[28] Kokkinos P, Mantzavinos D, Venieri D. Current Trends in the Application of Nanomaterials for the Removal of Emerging Micropollutants and Pathogens from Water. Molecules 2020; 25(9): 2016.
[http://dx.doi.org/10.3390/molecules25092016] [PMID: 32357416]

[29] Wang Y, Fan L, Khan SJ, Roddick FA. Fugacity modelling of the fate of micropollutants in aqueous systems — Uncertainty and sensitivity issues. Sci Total Environ 2020; 699: 134249.
[http://dx.doi.org/10.1016/j.scitotenv.2019.134249] [PMID: 31522051]

[30] Maliyekkal SM, Sreenivasan ST, Krishnan D, *et al.* Graphene: A Reusable Substrate for Unprecedented Adsorption of Pesticides 9: 273-83.2013;

[31] Firozjaee TT, Mehrdadi N, Baghdadi M, Nabi Bidhendi GR. Application of Nanotechnology in Pesticides Removal from Aqueous Solutions - A review. Int J Nanosci Nanotechnol 2018; 14: 43-56.

[32] Peng J, Zhang Y, Zhang C, *et al.* Removal of triclosan in a Fenton-like system mediated by graphene oxide: Reaction kinetics and ecotoxicity evaluation. Sci Total Environ 2019; 673: 726-33.
[http://dx.doi.org/10.1016/j.scitotenv.2019.03.354] [PMID: 31003100]

[33] Li T, Wang T, Qu G, Liang D, Hu S. Synthesis and photocatalytic performance of reduced graphene oxide–TiO_2 nanocomposites for orange II degradation under UV light irradiation. Environ Sci Pollut Res Int 2017; 24(13): 12416-25.
[http://dx.doi.org/10.1007/s11356-017-8927-3] [PMID: 28361396]

[34] Pourzamani H, Mengelizadeh N, Hajizadeh Y, Mohammadi H. Electrochemical degradation of diclofenac using three-dimensional electrode reactor with multi-walled carbon nanotubes. Environ Sci Pollut Res Int 2018; 25(25): 24746-63.
[http://dx.doi.org/10.1007/s11356-018-2527-8] [PMID: 29923052]

[35] Petsas AS, Vagi MC. Photocatalytic Degradation of Selected Organophosphorus Pesticides Using Titanium Dioxide and UV Light, Titanium Dioxide - Material for a Sustainable Environment. Dongfang Yang, IntechOpen 2017.
[http://dx.doi.org/10.5772/intechopen.72193]

[36] Yu B, Zeng J, Gong L, Zhang M, Zhang L, Chen X. Investigation of the photocatalytic degradation of organochlorine pesticides on a nano-TiO_2 coated film. Talanta 2007; 72(5): 1667-74.
[http://dx.doi.org/10.1016/j.talanta.2007.03.013] [PMID: 19071814]

[37] Kumar K, Chowdhury A. Use of novel nanostructured photocatalysts for the environmental sustainability of wastewater treatments. Encycl Renew Sustain Mater 2020; 1: 949-64.
[http://dx.doi.org/10.1016/B978-0-12-803581-8.11149-X]

[38] Kassem A, Ayoub GM, Malaeb L. Antibacterial activity of chitosan nano-composites and carbon nanotubes: A review. Sci Total Environ 2019; 668: 566-76.
[http://dx.doi.org/10.1016/j.scitotenv.2019.02.446] [PMID: 30856567]

[39] Pinto RM, Lopes-de-Campos D, Martins MCL, *et al.* Impact of nanosystems in *Staphylococcus aureus* biofilms treatment. FEMS Microbiol Rev 2019; 43(6): 622-41.
[http://dx.doi.org/10.1093/femsre/fuz021] [PMID: 31420962]

[40] Jasim NA, Al-Gasha'a FA, Al-Marjani MF, *et al.* ZnO nanoparticles inhibit growth and biofilm formation of vancomycin-resistant S. aureus (VRSA). Biocatal Agric Biotechnol 2020; 29: 101745.
[http://dx.doi.org/10.1016/j.bcab.2020.101745]

[41] Suri RPS, Thornton HM, Muruganandham M. Disinfection of water using Pt- and Ag-doped TiO_2 photocatalysts. Environ Technol 2012; 33(14): 1651-9.
[http://dx.doi.org/10.1080/09593330.2011.641590] [PMID: 22988625]

[42] Foster HA, Ditta IB, Varghese S, Steele A. Photocatalytic disinfection using titanium dioxide: spectrum and mechanism of antimicrobial activity. Appl Microbiol Biotechnol 2011; 90(6): 1847-68.
[http://dx.doi.org/10.1007/s00253-011-3213-7] [PMID: 21523480]

[43] Venieri D, Fraggedaki A, Kostadima M, *et al.* Solar light and metal-doped TiO_2 to eliminate water-

transmitted bacterial pathogens: Photocatalyst characterization and disinfection performance. Appl Catal B 2014; 154-155: 93-101.
[http://dx.doi.org/10.1016/j.apcatb.2014.02.007]

[44]   Fakhri A, Azad M, Fatolahi L, Tahami S. Microwave-assisted photocatalysis of neurotoxin compounds using metal oxides quantum dots/nanosheets composites: Photocorrosion inhibition, reusability and antibacterial activity studies. J Photochem Photobiol B 2018; 178: 108-14.
[http://dx.doi.org/10.1016/j.jphotobiol.2017.10.038] [PMID: 29131989]

[45]   Hemdan BA, El Nahrawy AM, Mansour AFM, Hammad ABA. Green sol–gel synthesis of novel nanoporous copper aluminosilicate for the eradication of pathogenic microbes in drinking water and wastewater treatment. Environ Sci Pollut Res Int 2019; 26(10): 9508-23.
[http://dx.doi.org/10.1007/s11356-019-04431-8] [PMID: 30729438]

[46]   Domínguez-Espíndola RB, Varia JC, Álvarez-Gallegos A, Ortiz-Hernández ML, Peña-Camacho JL, Silva-Martínez S. Photoelectrocatalytic inactivation of fecal coliform bacteria in urban wastewater using nanoparticulated films of TiO$_2$ and TiO$_2$/Ag. Environ Technol 2017; 38(5): 606-14.
[http://dx.doi.org/10.1080/09593330.2016.1205148] [PMID: 27384128]

[47]   Khaydarov RA, Khaydarov RR, Gapurova O. Nano-photocatalysts for the destruction of chloro-organic compounds and bacteria in water. J Colloid Interface Sci 2013; 406: 105-10.
[http://dx.doi.org/10.1016/j.jcis.2013.05.067] [PMID: 23800371]

[48]   Pandiyan R, Mahalingam S, Ahn YH. Antibacterial and photocatalytic activity of hydrothermally synthesized SnO$_2$ doped GO and CNT under visible light irradiation. J Photochem Photobiol B 2019; 191: 18-25.
[http://dx.doi.org/10.1016/j.jphotobiol.2018.12.007] [PMID: 30557789]

[49]   Zhan S, Yang Y, Shen Z, *et al.* Efficient removal of pathogenic bacteria and viruses by multifunctional amine-modified magnetic nanoparticles. J Hazard Mater 2014; 274: 115-23.
[http://dx.doi.org/10.1016/j.jhazmat.2014.03.067] [PMID: 24769848]

[50]   Cheng R, Xue XY, Liu L, *et al.* Removal of waterborne pathogen by nanomaterial-membrane coupling system. J Nanosci Nanotechnol 2018; 18(2): 1027-33.
[http://dx.doi.org/10.1166/jnn.2018.13963] [PMID: 29448528]

[51]   Zheng X, Shen Z, Cheng C, Shi L, Cheng R, Yuan D. Photocatalytic disinfection performance in virus and virus/bacteria system by Cu-TiO$_2$ nanofibers under visible light. Environ Pollut 2018; 237: 452-9.
[http://dx.doi.org/10.1016/j.envpol.2018.02.074] [PMID: 29510364]

[52]   Liga MV, Bryant EL, Colvin VL, Li Q. Virus inactivation by silver doped titanium dioxide nanoparticles for drinking water treatment. Water Res 2011; 45(2): 535-44.
[http://dx.doi.org/10.1016/j.watres.2010.09.012] [PMID: 20926111]

[53]   Abebe LS, Su YH, Guerrant RL, Swami NS, Smith JA. Point-of-use removal of *Cryptosporidium parvum* from water: Independent effects of disinfection by silver nanoparticles and silver ions and by physical filtration in ceramic porous media. Environ Sci Technol 2015; 49(21): 12958-67.
[http://dx.doi.org/10.1021/acs.est.5b02183] [PMID: 26398590]

[54]   Darwish AS, Bayaumy FEA, Ismail HM. Photoactivated water-disinfecting, and biological properties of Ag NPs@Sm-doped ZnO nanorods/cuttlefish bone composite: In-vitro bactericidal, cercaricidal and schistosomicidal studies. Mater Sci Eng C 2018; 93: 996-1011.
[http://dx.doi.org/10.1016/j.msec.2018.09.007] [PMID: 30274138]

[55]   Hussein EM, Ahmed SA, Mokhtar AB, *et al.* Antiprotozoal activity of magnesium oxide (MgO) nanoparticles against *Cyclospora cayetanensis* oocysts. Parasitol Int 2018; 67(6): 666-74.
[http://dx.doi.org/10.1016/j.parint.2018.06.009] [PMID: 29933042]

[56]   Hossain F, Perales-Perez OJ, Hwang S, Román F. Antimicrobial nanomaterials as water disinfectant: Applications, limitations and future perspectives. Sci Total Environ 2014; 466-467: 1047-59.
[http://dx.doi.org/10.1016/j.scitotenv.2013.08.009] [PMID: 23994736]

[57] Sunnotel O, Verdoold R, Dunlop PSM, *et al.* Photocatalytic inactivation of *Cryptosporidium parvum* on nanostructured titanium dioxide films. J Water Health 2010; 8(1): 83-91.
[http://dx.doi.org/10.2166/wh.2009.204] [PMID: 20009250]

[58] Ahmadi M, Ramezani Motlagh H, Jaafarzadeh N, *et al.* Enhanced photocatalytic degradation of tetracycline and real pharmaceutical wastewater using MWCNT/TiO$_2$ nano-composite. J Environ Manage 2017; 186(Pt 1): 55-63.
[http://dx.doi.org/10.1016/j.jenvman.2016.09.088] [PMID: 27852522]

[59] Abazari R, Mahjoub AR, Sanati S, Rezvani Z, Hou Z, Dai H. Ni-Ti layered double hydroxide@graphitic carbon nitride nanosheet: A novel nanocomposite with high and ultrafast sonophotocatalytic performance for degradation of antibiotics. Inorg Chem 2019; 58(3): 1834-49.
[http://dx.doi.org/10.1021/acs.inorgchem.8b02575] [PMID: 30648385]

[60] Zhong H, Wang Y, Cui C, Zhou F, Hu S, Wang R. Facile fabrication of Cu-based alloy nanoparticles encapsulated within hollow octahedral N-doped porous carbon for selective oxidation of hydrocarbons. Chem Sci (Camb) 2018; 9(46): 8703-10.
[http://dx.doi.org/10.1039/C8SC03531H] [PMID: 30595835]

[61] Hennebel T, De Corte S, Verstraete W, Boon N. Microbial production and environmental applications of Pd nanoparticles for treatment of halogenated compounds. Curr Opin Biotechnol 2012; 23(4): 555-61.
[http://dx.doi.org/10.1016/j.copbio.2012.01.007] [PMID: 22321940]

[62] Wang X, Tan F, Wang W, Qiao X, Qiu X, Chen J. Anchoring of silver nanoparticles on graphitic carbon nitride sheets for the synergistic catalytic reduction of 4-nitrophenol. Chemosphere 2017; 172: 147-54.
[http://dx.doi.org/10.1016/j.chemosphere.2016.12.103] [PMID: 28068566]

[63] Saleem H, Zaidi SJ. Developments in the Application of Nanomaterials for Water Treatment and Their Impact on the Environment. Nanomaterials (Basel) 2020; 10: 1764.

[64] Saad SM, Abdel-Shafy HI, Mona SM. Mansour. Removal of pyrene and benzo(a)pyrene micropollutant from water *via* adsorption by green synthesized iron oxide nanoparticles. Adv Nat Sci: Nanosci Nanotechnol 2018; 9: 015006.

[65] Chauhan A, Sillu D, Agnihotri S. Removal of Pharmaceutical Contaminants in Wastewater Using Nanomaterials: A Comprehensive Review. Curr Drug Metab 2019; 20(6): 483-505.
[http://dx.doi.org/10.2174/1389200220666181127104812] [PMID: 30479212]

[66] Qu X, Brame J, Li Q, Alvarez PJJ. Nanotechnology for a safe and sustainable water supply: enabling integrated water treatment and reuse. Acc Chem Res 2013; 46(3): 834-43.
[http://dx.doi.org/10.1021/ar300029v] [PMID: 22738389]

CHAPTER 9

The Potential of Magnetic Nanoparticles in Environmental Remediation

Bhupinder Dhir[1,*]

[1] *School of Sciences, Indira Gandhi National Open University, New Delhi, India*

Abstract: Magnetic nanoparticles (MNPs) possess inherent properties that help them in improving the quality of the environment *via* the detection, remediation, and removal of pollutants and contaminants. The properties such as high reactivity, high surface-to-volume ratios, superparamagnetism, large surface area and biocompatibility are responsible for the extensive use of magnetic nanoparticles in environmental remediation. MNPs act as adsorbents or catalysts and help in the removal of contaminants from environmental matrices. High pollutant removal efficiency of magnetic nanoparticles can be exploited in framing low-cost-effective technologies for environmental remediation.

Keywords: Chlorinated solvents, Environment, Magnetic nanoparticles, Organic compounds.

INTRODUCTION

Nanotechnology has been used in various areas such as the environment, industrial, medical, material science, and engineering [1, 2]. It has emerged as a promising alternative to high-cost environmental remediation technologies [3, 4]. Nanoscale materials are being developed for potential use to adsorb or remove contaminants under *in situ* or *ex-situ* conditions [5, 6]. Different kinds of nanomaterials, including magnetic nanomaterials (MNMs), are being used in various areas of the environment, such as sensing and monitoring environmental contaminants [7]. Magnetic nanomaterials such as nano zero-valent iron (nZVI), magnetite and maghemite nanoparticles, Fe_3O_4 and $g\text{-}Fe_2O_3$ have found their use in the treatment of contaminated water. Nanomaterials possess good potentials in remediating the environment, such as the removal of pollutants and reduction of toxicity.

[*] **Corresponding author Bhupinder Dhir:** School of Sciences, Indira Gandhi National Open University, New Delhi, India; E-mail: bhupdhir@gmail.com

Magnetic nanoparticles possess properties such as high contaminant removal capacity and fast reaction rate. High surface-area-to-volume ratio and availability of a large number of active sites for the reaction are some of the properties of magnetic nanoparticles that support the removal/remediation of contaminants from the environment. Small particle size, high surface-area-to-volume ratio and property of magnetism support the larger rate of contaminant removal, fast kinetics and high reactivity of magnetic nanoparticles. The capacity for removal of contaminants and reactivity of magnetic nanoparticles depends on the surface area and surface properties. The capacity for the removal of contaminants increases with a decrease in particle size [8, 9]. When the particle size was reduced from 500 to 100nm, the reactivity of ZVI ranged between 50 to 90 times higher [10]. Iron oxide magnetic nanoparticles (IONPs), in particular, show a high surface area-to-volume ratio, fast kinetics, strong adsorption capacities, high reactivity and magnetism [9].

ZVI is classified into two types: (1) Nanoscale ZVI (nZVI) and (2) Reactive nanoscale iron product (RNIP). nZVI particles possess a diameter of 100–200 nm and are formed of iron (Fe) with zero valency, whereas RNIP particles are made up of 50% Fe and 50% Fe_3O_4. The contaminant removal capacity of Fe_3O_4 nanoparticles (8 nm) is much higher (about seven times) than coarse-grained counterparts (50 mm) [11].

ENVIRONMENTAL APPLICATIONS OF MAGNETIC NANO MATERIALS

Nanoscale materials have been used for applications in the field environment for the last few years [3, 12]. Remediation of contaminated soil and groundwater at hazardous sites has been achieved using nanoscale materials [13, 14]. The small size and high surface area of magnetite nanoparticles contribute to their high adsorption capacity. Iron oxide nanoparticles separate adsorbents from the system *via* magnetic property. Magnetic nanoparticles remove contaminants mainly *via* electrostatic interaction [15 - 19]. Ion exchange is another way by which cont aminants are removed.

Iron is a strong reducing agent, hence this property of nZVI is used in the degradation of a wide range of organic and inorganic pollutants.

A charged nanoparticle surface is preferred for the removal of charged contaminants. Environmental conditions that can affect the performance of magnetic nanoparticles include background ions, humic substances and pH. The chemistry and composition of groundwater vary from site to site. When magnetic nanoparticles are used in wastewater treatment, high ionic strength and extreme pH are usually a concern.

The reduction of nZVI and adsorption performance of Fe_3O_4 and Fe_2O_3 nanoparticles was affected by factors such as pH value. The surface property of magnetic nanoparticles gets changed by coating with organic surfactants or polymers [20 - 23]. Hu *et al.* [24] suggested a way to use the surface property of magnetic nanoparticles to remove contaminants from the environment. Heavy metals can be separated and recovered with the help of magnetic nanoparticles by changing pH values.

Adsorption is the process by which organic and inorganic contaminants get removed from water and wastewater. Small size, higher surface area, more sorption sites and the surface chemistry of iron oxide nanoparticles (IONPs) contribute to enhanced selectivity.

REMOVAL OF HEAVY METALS

Anions like arsenic, fluoride and chromium, and cations like copper, nickel and mercury get removed from water and wastewater with the help of MNPs. Magnetic nanoparticles act as good sorbents and help in the removal of metals. Iron oxide-based nanomaterials are very effective in removing heavy metals from water [25, 26]. Small size, high surface area and magnetic properties contribute to heavy metal removal. Nano zero-valent iron showed a capacity to remove arsenic, chromium and organic pollutants like chlorinated solvents. Functionalized sorbents remove heavy metals *via* affinity or selective removal from complicated matrices. Iron oxide/hydroxide shell is formed when nZVI comes in contact with air or water. Heavy metal removal occurs *via* surface sorption and co-precipitation [27, 28]. Coating the surface with a thin layer of Fe_3O_4, silica or polymers reduces contact with oxygen, thereby retaining the reactivity of nZVI [29 - 31]. Heavy metals diffuse on the active surface of Fe_3O_4 nanoparticles.

Nanoadsorbents prepared from iron oxide nanoparticles such as magnetite (Fe_3O_4), maghemite (γ-Fe_2O_3), and hematite (α-Fe_2O_3 remove heavy metals from water/wastewater. Contaminants get adsorbed on the surface of magnetic nanoparticles. Efficiency of nanoparticles for adsorption decreases after aggregation. Magnetic nanoparticles, therefore, play a role in the sustainable water treatment process [26].

Improvement in the surface of iron oxide nanoparticles can help in retaining their activity. Carboxylic acids, phosphoric acid, silanol, thio and amine are functional groups besides other small organic molecules, biomolecules, polymers and other metal nanoparticles that modify the surface of iron oxide nanoparticles. The removal of arsenic [As(III) and As(V)] from an aqueous solution occurs *via* adsorption. Flower-like magnetic adsorbent showed a maximum adsorption capacity of 51 and 30 mg/g for As(V) and Cr(V), respectively. Hollow nest-like magnetic adsorbent showed a high adsorption capacity of 75.3 and 58.5 mg/g for As(V) and Cr(V). The Amine group present on the surface of magnetic nanoparticles showed 98% removal of copper from the polluted river and tap water. It effectively removed metals such as Cu(II), Zn(II), and Pb(II). The maximum adsorption capacity of 25.77 mg/g was noted at pH 6. Adsorption capacity increases with an increase in pH. The high contaminant removal efficiency of magnetic adsorbent is due to the presence of multiple hydroxy and carboxyl groups.

The removal efficiency of adsorbents depends upon strong complexation that takes place between the surface of adsorbents and metal ions, and weak electrostatic interaction with the surface. The adsorption capacities of adsorbent increased with increasing the concentration of adsorbent. The metals such as Pb, Ag, and Hg showed removal efficiency of more than 95%, while copper showed not more than 80%. The low efficiency of copper is supposed to be due to weak binding to the thiol group.

Studies showed that Cr(VI) can be reduced by nZVI, forming precipitates. The reactivity of nZVI was high because of its smaller particle size and larger surface area. High-reactive ZVI reacts with the oxygen present in water or dissolved air. In the presence of oxygen, nZVI gets oxidized to Fe^{+2} and/or Fe^{+3} ions. nZVI is produced through reductive precipitation of $FeCl_3$ with $NaBH_4$, or reduction of goethite and hematite particles with H_2 at high temperatures (200-600°C).

The removal of contaminants by Fe_3O_4 nanoparticles occurs by both physical and chemical adsorption [16, 32 - 34]. A study on the removal of Cr(VI) with Fe_3O_4 nanoparticles showed that a reduction of Cr(VI) to Cr(III) occurs, and this is followed by surface precipitation of Cr(III) onto the magnetic nanoparticles [16, 23]. Physical adsorption is the major contaminant removal mechanism of Fe_2O_3 nanoparticles. The pollutant removal by g-Fe_2O_3 nanoparticles occurs because of electrostatic interaction.

Removal of contaminants by g-Fe_2O_3 and Fe_3O_4 nanoparticles, TiO_2, SAMMS™, nanotubes, ferritin, dendrimers, metalloporphyrinogens, and SOMS occurs *via* physical adsorption, hence desorption of contaminants and regeneration of nanoparticles is feasible.

The crystallite structure of g-Fe_2O_3 nanoparticles did not change after the removal of pollutants. This indicated that the removal mechanism did not involve chemical reactions [17, 18]. The pollutant removal by g-Fe_2O_3 nanoparticles occurs due to electrostatic interactions. The performance of Fe_3O_4 and g-Fe_2O_3 nanoparticles depends on pH [16, 19, 35, 36]. Oxygen atoms present on the surface of iron oxides (Fe_3O_4 and g-Fe_2O_3) get polarized under various pH values. In conditions where the pH value is below the zero point of charge (pHZPC), the positive charge on the surface of iron oxides attracts negatively charged pollutants (*e.g.,* Cr(VI) and As(V)) [8, 15, 35, 37].

REMOVAL OF ORGANIC COMPOUNDS

Magnetic nanoparticles show a high capacity for the removal of high concentrations of organic compounds [38]. They have found their use as a low-efficient adsorbent for the treatment of textile effluents. Nanomaterials have also found their application in the field of environment as they show the capacity to remove chlorinated solvents. Magnetic nanoparticles also show the capacity to treat chlorinated solvents or oil spills from contaminated sites [39]. These NPs show a high capacity to remove high concentrations of organic compounds. Highly reactive nZVI showed an inherent capacity to degrade persistent contaminants such as polycyclic aromatic hydrocarbons (PAHs) and pesticides [40 - 42].

Removal of organic pollutants like phenol and dyes from wastewater using MNPs has also been reported [43]. MNPs act as a catalyst and carry out photocatalytic degradation of dyes and pollutants [44, 45]. Bench- and pilot-scale research studies demonstrated the role of particles such as TiO_2, self-assembled monolayers on mesoporous supports (SAMMS™), dendrimers, carbon nanotubes, metalloporphyrinogens, and swellable organically modified silica (SOMS) in remediation of the environment [46, 47].

Chlorinated solvents get degraded to less harmful substances through dechlorination. Highly reactive nZVI effectively degraded persistent contaminants, such as polycyclic aromatic hydrocarbons (PAHs) and pesticides [41, 42].

ROLE OF NANOPARTICLES IN COMBATING AIR POLLUTION

Filtration techniques can be used to purify indoor air. The entry of contaminants can be prevented by using nano-filters in automobile tailpipes and factory smokestacks. Nanosensors that help in the detection of toxic gases at extremely low concentrations have been developed.

Materials such as titanium dioxide (TiO_2), zinc oxide (ZnO), iron (III) oxide (Fe_2O_3) and tungsten oxide (WO_3) act as photocatalysts. They are able to remove contaminants from groundwater containing 1,1-dichloroethane, cis- 1,2-dichloroethane, 1,1,1-trichloroethane, xylene and toluene. They oxidize organic pollutants into nontoxic materials. TiO_2 is used for the photochemical oxidation of contaminants present in water. It shows low levels of toxicity, high photoconductivity and high photostability, and it is an easily available and inexpensive material. A pilot scale study showed that TiO_2 was able to eliminate benzene, toluene, ethylbenzene and xylene (BTEX) from groundwater. Another photocatalyst, ZnO helps in the detection and remediation of contaminants. Laboratory studies showed that ZnO successfully detected and eliminated 4-chlorocatechol.

CONCLUSION

Nanoscience plays an important role in cleaning the environment by removing contaminants. Superparamagnetism, large surface area, stronger reactivity, fast kinetics and biocompatibility are some of the properties that establish nanoparticles as ideal materials as adsorbents or catalysts for removing contaminants from environmental matrices. Ultrahigh sensitivity of nanoparticles helps in the detection, remediation and removal of micropollutants. Magnetic nanoparticles (MNPs), in particular, possess great potential for use in the field of environment. MNPs possess high extraction efficiency because of a high surface area-to-volume ratio, faster separation, magnetic characteristics and high stability. Iron nanoparticles form an important component of water and wastewater treatment. These nanoparticles, in particular, act as efficient and cost-effective nanoadsorbents that help in removing heavy metals from the aqueous systems. They show potential for pilot testing, up-scaling and commercialization. More research needs to be conducted to increase the application of magnetic nanoparticles in environmental remediation to achieve sustainability in the environment.

REFERENCES

[1]   Gil PR, Parak WJ. Composite Nanoparticles Take Aim at Cancer 2(11)2008; ACS Nanotechnol

[2]   Klaine SJ, Alvarez PJJ, Batley GE, *et al.* Nanomaterials in the environment: behavior, fate, bioavailability, and effects. Environ Toxicol Chem 2008; 27(9): 1825-51.
        [http://dx.doi.org/10.1897/08-090.1] [PMID: 19086204]

[3]   Tang SCN, Lo IMC. Magnetic nanoparticles: Essential factors for sustainable environmental applications. Water Res 2013; 47(8): 2613-32.
        [http://dx.doi.org/10.1016/j.watres.2013.02.039] [PMID: 23515106]

[4]   Kim I, Yang H, Park CW, Yoon I, Sihn Y. Environmental applications of magnetic nanoparticles. Magnetic Nanoparticle-Based Hybrid Materials Fundamentals and Applications, Woodhead Publishing Series in Electronic and Optical Materials 2021; pp. 529-45.

[5]   Pratt A. Environmental applications of magnetic nanoparticles. Frontiers of Nanoscience 2014; 6: 259-307.
        [http://dx.doi.org/10.1016/B978-0-08-098353-0.00007-5]

[6]   Gutierrez AM, Dziubla TD, Hilt JZ. Recent advances on iron oxide magnetic nanoparticles as sorbents of organic pollutants in water and wastewater treatment. Rev Environ Health 2017; 32(1-2): 111-7.
        [http://dx.doi.org/10.1515/reveh-2016-0063] [PMID: 28231068]

[7]   U.S. Environmental Protection Agency (EPA). 2007.Nanotechnology White Paper. Senior Policy Council. EPA 100/B-07/001. www.epa.gov/sites/production/files/2015-01/documents/nanotechnology

_whitepaper.pdf

[8]   Yean S, Cong L, Yavuz CT, *et al.* Effect of magnetite particle size on adsorption and desorption of arsenite and arsenate. J Mater Res 2005; 20(12): 3255-64.
        [http://dx.doi.org/10.1557/jmr.2005.0403]

[9]   Tratnyek PG, Johnson RL. Nanotechnologies for environmental cleanup. Nano Today 2006; 1(2): 44-8.
        [http://dx.doi.org/10.1016/S1748-0132(06)70048-2]

[10]   Lin Y, Weng C, Chen F. Effective removal of AB24 dye by nano/micro-size zero-valent iron. Separ Purif Tech 2008; 64(1): 26-30.
        [http://dx.doi.org/10.1016/j.seppur.2008.08.012]

[11]   Shen YF, Tang J, Nie ZH, Wang YD, Ren Y, Zuo L. Preparation and application of magnetic Fe_3O_4 nanoparticles for wastewater purification. Separ Purif Tech 2009; 68(3): 312-9.
        [http://dx.doi.org/10.1016/j.seppur.2009.05.020]

[12]   Gallo-Cordova A, Streitwieser DA, Morales M del P, Ovejero JG. 2021.https://www .intechopen.com/online-first/magnetic-iron-oxide-colloids-for-environmental-applications
        [http://dx.doi.org/10.5772/intechopen.95351]

[13]   Lien Hl, Elliott DW, San YP, Zhang WX. Recent Progress in Zero-Valent Iron Nanoparticles for Groundwater Remediation. J Environ Econ Manage 2006; 16(6): 371-80.

[14]   Alazaiza MYD, Albahnasawi A, Ali GAM, *et al.* Recent Advances of Nanoremediation Technologies for Soil and Groundwater Remediation: A Review. Water 2021; 13(16): 2186.
        [http://dx.doi.org/10.3390/w13162186]

[15]   Chowdhury SR, Yanful EK. Arsenic and chromium removal by mixed magnetite–maghemite nanoparticles and the effect of phosphate on removal. J Environ Manage 2010; 91(11): 2238-47.
        [http://dx.doi.org/10.1016/j.jenvman.2010.06.003] [PMID: 20598797]

[16]   Hu J, Lo IMC, Chen G. Removal of Cr(VI) by magnetite. Water Sci Technol 2004; 50(12): 139-46.
        [http://dx.doi.org/10.2166/wst.2004.0706] [PMID: 15686014]

[17]   Hu J, Chen G, Lo IMC. Removal and recovery of Cr(VI) from wastewater by maghemite nanoparticles. Water Res 2005; 39(18): 4528-36. a
        [http://dx.doi.org/10.1016/j.watres.2005.05.051] [PMID: 16146639]

[18]   Hu J, Lo IMC, Chen G. Fast removal and recovery of Cr(VI) using surface-modified jacobsite ($MnFe_2O_4$) nanoparticles. Langmuir 2005; 21(24): 11173-9. b
        [http://dx.doi.org/10.1021/la051076h] [PMID: 16285787]

[19]   Yang W, Kan AT, Chen W, Tomson MB. pH-dependent effect of zinc on arsenic adsorption to magnetite nanoparticles. Water Res 2010; 44(19): 5693-701.
        [http://dx.doi.org/10.1016/j.watres.2010.06.023] [PMID: 20598730]

[20]   Huang SH, Chen DH. Rapid removal of heavy metal cations and anions from aqueous solutions by an amino-functionalized magnetic nano-adsorbent. J Hazard Mater 2009; 163(1): 174-9.
        [http://dx.doi.org/10.1016/j.jhazmat.2008.06.075] [PMID: 18657903]

[21]   Hao YM, Man C, Hu ZB. Effective removal of Cu (II) ions from aqueous solution by amino-functionalized magnetic nanoparticles. J Hazard Mater 2010; 184(1-3): 392-9.
        [http://dx.doi.org/10.1016/j.jhazmat.2010.08.048] [PMID: 20837378]

[22]   He J, Huang M, Wang D, Zhang Z, Li G. Magnetic separation techniques in sample preparation for biological analysis: A review. J Pharm Biomed Anal 2014; 101: 84-101.
        [http://dx.doi.org/10.1016/j.jpba.2014.04.017] [PMID: 24809747]

[23]   Ge F, Li MM, Ye H, Zhao BX. Effective removal of heavy metal ions Cd2+, Zn2+, Pb2+, Cu2+ from aqueous solution by polymer-modified magnetic nanoparticles. J Hazard Mater 2012; 211-212: 366-72.

[http://dx.doi.org/10.1016/j.jhazmat.2011.12.013] [PMID: 22209322]

[24]   Hu J, Chen G, Lo IMC. Selective removal of heavy metals from industrial wastewater using maghemite nanoparticle: performance and mechanisms. J Environ Eng 2006; 132(7): 709-15.
[http://dx.doi.org/10.1061/(ASCE)0733-9372(2006)132:7(709)]

[25]   Dave PN, Chopda LV. Application of iron oxide nanomaterials for the removal of heavy metals. J Nanotechnol 2014; 2014: 1-14.
[http://dx.doi.org/10.1155/2014/398569]

[26]   Huang Y, Keller AA. EDTA functionalized magnetic nanoparticle sorbents for cadmium and lead contaminated water treatment. Water Res 2015; 80: 159-68.
[http://dx.doi.org/10.1016/j.watres.2015.05.011] [PMID: 26001282]

[27]   Fang Z, Qiu X, Huang R, Qiu X, Li M. Removal of chromium in electroplating wastewater by nanoscale zero-valent metal with synergistic effect of reduction and immobilization. Desalination 2011; 280(1-3): 224-31.
[http://dx.doi.org/10.1016/j.desal.2011.07.011]

[28]   Scott TB, Popescu IC, Crane RA, Noubactep C. Nano-scale metallic iron for the treatment of solutions containing multiple inorganic contaminants. J Hazard Mater 2011; 186(1): 280-7.
[http://dx.doi.org/10.1016/j.jhazmat.2010.10.113] [PMID: 21115222]

[29]   Liu Y, Majetich SA, Tilton RD, Sholl DS, Lowry GV. TCE dechlorination rates, pathways, and efficiency of nanoscale iron particles with different properties. Environ Sci Technol 2005; 39(5): 1338-45.
[http://dx.doi.org/10.1021/es049195r] [PMID: 15787375]

[30]   Tang NJ, Chen W, Zhong W, Jiang HY, Huang SL, Du YW. Highly stable carbon-coated Fe/SiO2 composites: Synthesis, structure and magnetic properties. Carbon 2006; 44(3): 423-7.
[http://dx.doi.org/10.1016/j.carbon.2005.09.001]

[31]   Wilson JL, Poddar P, Frey NA, *et al.* Synthesis and magnetic properties of polymer nanocomposites with embedded iron nanoparticles. J Appl Phys 2004; 95(3): 1439-43.
[http://dx.doi.org/10.1063/1.1637705]

[32]   Shen YF, Tang J, Nie ZH, Wang YD, Ren Y, Zuo L. Tailoring size and structural distortion of Fe_3O_4 nanoparticles for the purification of contaminated water. Bioresour Technol 2009; 100(18): 4139-46.
[http://dx.doi.org/10.1016/j.biortech.2009.04.004] [PMID: 19414249]

[33]   Wang L, Li J, Jiang Q, Zhao L. Water-soluble Fe_3O_4 nanoparticles with high solubility for removal of heavy-metal ions from waste water. Dalton Trans 2012; 41(15): 4544-51.
[http://dx.doi.org/10.1039/c2dt11827k] [PMID: 22358186]

[34]   Xu P, Zeng GM, Huang DL, *et al.* Use of iron oxide nanomaterials in wastewater treatment: A review. Sci Total Environ 2012; 424: 1-10.
[http://dx.doi.org/10.1016/j.scitotenv.2012.02.023] [PMID: 22391097]

[35]   Hu J, Lo I, Chen G. Performance and mechanism of chromate (VI) adsorption by δ-FeOOH-coated maghemite (γ-Fe2O3) nanoparticles. Separ Purif Tech 2007; 58(1): 76-82.
[http://dx.doi.org/10.1016/j.seppur.2007.07.023]

[36]   Yantasee W, Warner CL, Sangvanich T, *et al.* Removal of heavy metals from aqueous systems with thiol functionalized superparamagnetic nanoparticles. Environ Sci Technol 2007; 41(14): 5114-9.
[http://dx.doi.org/10.1021/es0705238] [PMID: 17711232]

[37]   Miehr R, Tratnyek PG, Bandstra JZ, Scherer MM, Alowitz MJ, Bylaska EJ. Diversity of Contaminant Reduction Reactions by Zerovalent Iron: Role of the Reductate 38: 139-47.2004; Environ Sci Technol

[38]   Huang Y, Keller AA. Magnetic nanoparticle adsorbents for emerging organic contaminants. ACS Sustain Chem& Eng 2013; 1(7): 731-6.
[http://dx.doi.org/10.1021/sc400047q]

[39]   Hussain CM. CHAPTER 19. Magnetic Nanomaterials for Environmental Analysis. Detection Science 2016; 2: 1-13.
[http://dx.doi.org/10.1039/9781782629139-00001]

[40]   Chang MC, Kang HY. Remediation of pyrenecontaminated soil by synthesized nanoscale zero-valent iron particles J Environ Sci Health e Part A Toxic/Hazard Subs Environ Eng 44: 576-82.2009;

[41]   Li L, Fan M, Brown RC, *et al.* Synthesis, properties, and environmental applications of nanoscale iron-based materials: a review. Crit Rev Environ Sci Technol 2006; 36(5): 405-31. a
[http://dx.doi.org/10.1080/10643380600620387]

[42]   Li X, Elliott DW, Zhang W. Zero-valent iron nanoparticles for abatement of environmental pollutants: materials and engineering aspects. Crit Rev Solid State Mater Sci 2006; 31(4): 111-22. b
[http://dx.doi.org/10.1080/10408430601057611]

[43]   Ul-Islam M, Ullah MW, Khan S, *et al.* Current advancements of magnetic nanoparticles in adsorption and degradation of organic pollutants. Environ Sci Pollut Res Int 2017; 24(14): 12713-22.
[http://dx.doi.org/10.1007/s11356-017-8765-3] [PMID: 28378308]

[44]   Qadri S, Ganoe A, Haik Y. Removal and recovery of acridine orange from solutions by use of magnetic nanoparticles. J Hazard Mater 2009; 169(1-3): 318-23.
[http://dx.doi.org/10.1016/j.jhazmat.2009.03.103] [PMID: 19406571]

[45]   Afkhami A, Moosavi R. Adsorptive removal of Congo red, a carcinogenic textile dye, from aqueous solutions by maghemite nanoparticles. J Hazard Mater 2010; 174(1-3): 398-403.
[http://dx.doi.org/10.1016/j.jhazmat.2009.09.066] [PMID: 19819070]

[46]   Jiang B, Lian L, Xing Y, *et al.* Advances of magnetic nanoparticles in environmental application: environmental remediation and (bio)sensors as case studies. Environ Sci Pollut Res Int 2018; 25(31): 30863-79.
[http://dx.doi.org/10.1007/s11356-018-3095-7] [PMID: 30196461]

[47]   Mishra S, Sundaram B. Efficacy and challenges of carbon nanotube in wastewater and water treatment. Environ Nanotechnol Monit Manage 2023; 19: 100764.

CHAPTER 10

Role of Nanotechnology in Water Treatment

Rashmi Verma[1,*]

[1] *Department of Genetics, University of Delhi South Campus, New Delhi, India*

Abstract: Nanotechnology has emerged as an alternative to conventional water treatment methods that involve high costs and processes. Nanomaterials offer great potential for cleaning wastewater. Various nanomaterials have shown the potential to remove pollutants such as organic and inorganic content, and toxic heavy metal ions from wastewater. Nanoparticles with nanofibers and carbon nanotubes form an important part of ultrafiltration membrane, osmosis, sorption, advanced oxidation process, water remediation as well as disinfection processes. The rate of removal of contaminants from wastewater depends upon the physical and chemical characteristics of the nanomaterial, the contaminant, and wastewater.

Keywords: Nanoparticles, Nanotechnology, Pollution, Purification.

INTRODUCTION

Industrialization and urbanization have increased tremendously in recent years. This involve activities such as manufacturing, transportation, construction, petroleum refining, mining, depletion of natural resources and production of hazardous wastes. These activities cause pollution and threaten human health and the environment [1]. The wastes include atmospheric pollutants such as toxic gases (ozone, nitrogen oxides, sulfur oxides, carbon oxides) and suspended airborne particles, while soil and water pollutants comprise organic substances (pesticides, insecticides, phenols, hydrocarbons, volatile organic compounds), heavy metals (lead, cadmium, arsenic, mercury) and microbial pathogens. These environmental pollutants adversely influence human health [2, 3]. They enter the human body either *via* inhalation, ingestion or absorption. It is necessary to save the environment to build better ecosystems.

The entry of pollutants into water contaminates it. Thermal treatment, chemical oxidation and surface solvent flushing are some techniques used for cleaning water. The conventional methods of water treatment are lengthy, expensive and not very effective. The development of sustainable, efficient, and low-cost techno-

* **Corresponding author Rashmi Verma:** Department of Genetics, University of Delhi South Campus, New Delhi, India; E-mail:verma15rashmi@gmail.com

logies that assist in the treatment of toxic environmental pollutants is required. These technologies help in reducing or optimizing the use of natural resources. Nanotechnology has emerged as an alternative to conventional technologies.

ROLE OF NANOTECHNOLOGY FOR THE SUSTAINABLE ENVIRONMENT

Nanotechnology has found its applications at a wide scale in developing innovative methods to form new products, substitute existing ones, reform new materials with less energy consumption, reduce harm caused to the environment, and assist in environmental remediation. Nanotechnology is one of the most promising approaches that have proved useful in environmental remediation. Nanotechnology can be applied in many fields and is considered a sustainable technology. Nanotechnology involves the use of nanoparticles (particle size of less than 100 nm range) to produce materials, devices, and systems that can find their use in various applications [4, 5] Fig (**1**). Nanotechnology has found its application in fields like the cosmetics industry, agricultural industry, medicine, food industry, energy, and control/treatment of pollution [6 - 14]. Nanotechnology represents the ''key technology" of the present century and is expected to bring improvement in the environmental zone considerably. The chemical and physical properties of nanomaterials make them suitable for numerous applications in the field of the environment [15]. The research conducted by various researchers proved that nanotechnology can be used in environmental sectors such as saving raw materials, treatment of wastewater and contaminated soil, and hazardous waste management [16].

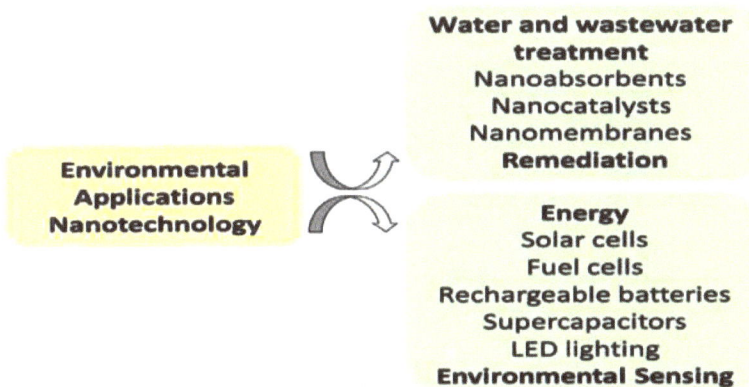

Water and wastewater treatment
Nanoabsorbents
Nanocatalysts
Nanomembranes
Remediation

Environmental Applications Nanotechnology

Energy
Solar cells
Fuel cells
Rechargeable batteries
Supercapacitors
LED lighting
Environmental Sensing

Fig. (1). Nanotechnology used for the environmental application (adopted from Pathakoti *et al.* [18]).

Nanotechnology can contribute to resolving different environmental issues like the cleaning of drinking water and the transformation/detoxification of contaminants (PCBs, heavy metals, organochlorine pesticides, and solvents) [17]. Environmental applications of nanotechnology can address the issues related to existing environmental problems and can provide preventive measures for future problems.

Nanotechnology has been used in increasing the yield and quality of agricultural products, improving cosmetic products and direct delivery of medicines and sensor applications. Innovations have led to the development of nanosensors that can easily recognize disease-causing elements, toxins and elements in environmental samples, and food nutrients. The main focus of nanotechnology in the past few decades has been on the use of nanoparticles in various sectors, and this resulted in its unrestricted development [9].

For the benefit of the environment, nanotechnology has been involved in the development of products, *i.e.*, nanomaterials or products that can clean hazardous waste sites directly, desalinate water, treat pollutants, and monitor environmental pollutants. Lightweight nanocomposites developed for automobiles and transportation help in saving fuel. Nanotechnology-based fuel cells and LED (light-emitting diodes) lighting aim to reduce environmental pollution, and help in energy generation. The conservation of fossil fuels reduces the use of materials for production. A nanoscale self-cleaning (surface coatings) material reduces/eliminates the use of many chemicals that may harm the environment. Nanotechnology also improved battery life, resulting in less use of material and less waste production (Fig. **1**) [18].

Nanotechnology for Environmental Remediation

Abatement of pollution and environmental protection are the two major concerns that need special attention throughout the world. The numbers of landfills, military setups, oil fields, manufacturing units and industrial operations are mainly responsible for the contamination of the environment. Techniques that use biological systems such as bioremediation, phytoremediation and rhizoremediation have been successfully tried to eliminate environmental pollution and restore contaminated sites. Recently nanotechnology has been applied to remediate environmental contamination, and it was termed nanoremediation, which provides an effective solution for environmental clean-up. Nanoremediation plays a significant role detection, prevention, monitoring and remediation of pollution [19]. Nanoremediation that involves the use of nanoscale particles or nanomaterials has helped in the remediation of affected sites and improved the efficiency of the remediation processes in a cost-effective manner. Contaminated soil and groundwater at hazardous waste sites have been successfully remediated using nanomaterials. Various kinds of nanoscale materials, including nanoscale zeolites, carbon nanotubes, nanofibers, metal oxides and bimetallic nanoparticles, have been used in the processes such as the transformation and detoxification of pollutants. Nanoscale zero-valent iron (nZVI) is the most widely used nanomaterial [20, 21]. High surface area and high reactivity are some properties that make nanomaterials an ideal agent that is capable of transforming or degrading contaminants in soils and water.

NANOTECHNOLOGY FOR WATER TREATMENT

Nanotechnology's has become a promising technique as far as the treatment of water is concerned. Nanotechnology has been successfully applied in water purification [22, 23]. Research studies have now been focused on water management and desalination. Nanotechnology involves the use of various nanomaterials for the treatment of surface water, groundwater, wastewater and the removal of contaminants, such as toxic metal ions, inorganic solutes, organic compounds and microorganisms. Research studies have been conducted to find out the way by which recalcitrant contaminants can be treated using nanomaterials [24]. Nanomaterials have shown the capacity to efficiently help in seawater desalination, water recycling and water remediation. Many developments in the nanomaterials that are used for wastewater treatment have been noted recently. Nanofiltration has been tried for the remediation of biologically treated sewage. Removal of organic colorants from the paper industry has been done using manganese-doped ZnO particles. Hence, nanotechnology has shown promising results and the potential to improve the quality, availability and *via*bility of water. Water purification involves the use of different kinds of processes, such as nanofiltration (NF), reverse osmosis (RO), microfiltration (MF), and ultrafiltration (UF) [25].

Nanotechnology for Water Cleaning

Nanoscopic materials (carbon nanotubes) and nanofiltration (alumina fibers) have been used n the process of water purification by nanotechnology. Nanoscopic pores in zeolite filtration membranes, nanocatalysts and magnetic nanoparticles form an important component of nanotechnology used in the cleaning of water. Titanium oxide nanowires or palladium nanoparticles are nanosensors used in the analytical detection of contaminants present in water samples.

Nanotechnology can tackle a different type of impurities present in water. This depends on the stage of the purification of water. Removal of sediments, charged particles, chemical effluents, and microorganism (bacteria and other pathogens) has been done using the technique of nanotechnology. Toxic trace elements like arsenic and viscous liquid impurities such as oil can also be removed using nanotechnology. Nanofilters act as efficient systems whose large surface area can easily be cleaned by back-flushing in comparison with the conventional methods. Carbon nanotube membranes remove various types of water contaminants such as turbidity, oil, bacteria, viruses and organic contaminants. Negatively charged contaminants such as viruses, bacteria, and organic and inorganic colloids get easily removed by nanofibrous alumina filters and other nanofiber products at a faster rate in comparison to conventional filters [26].

Nanotechnology for Water Filtration

Nanofiltration is a membrane filtration process used to treat water containing a low quantity of total dissolved solids, such as surface water and fresh groundwater. It is also used to soften (remove polyvalent cation) water and disinfection it [27, 28]. The technique of nanofiltration is widely used in dairy and food processing. Nanometer-sized cylindrical pores (size 1-10 Angstrom) present in the nanofilter membranes are larger than those used in reverse osmosis but smaller than that used in microfiltration and ultrafiltration. The polymer thin films of materials such as polyethylene terephthalate or metals like aluminum are used in the formation of these membranes. Membranes made up of polyethylene terephthalate, and similar materials are referred to as "track-etch" membranes. The name is based on the pore made on the membranes. Pore densities range from 1 to 106 pores per cm^2, while the dimensions are regulated by temperature, pH, and time taken during the development of pores. Membranes having metal-like alumina are made by electrochemical processes by growing a thin layer of aluminum oxide in an acidic medium.

Nanotechnology to Disinfect Water

The increasing demand for clean water due to the continuous increase in population is a matter of serious concern all over the globe. Nanotechnology offers an alternate solution to clean water by removing germs and other pollutants. Antimicrobial nanotechnology is an approach that has been followed. It is based on the fact that several nanomaterials possess strong antimicrobial properties over mechanisms such as photocatalytic production of reactive oxygen species (damaged cell components and viruses). Synthetically-fabricated nanometallic particles are also produced that show antimicrobial actionoligodynamic disinfection and can inactivate microorganisms at low concentrations. In commercial systems, titanium oxide is used for photocatalysis, and technology showed complete inactivation of fecal coliforms in 15 minutes after activation by sunlight [25]. Dendrimers, zeolites, carbonaceous, and metal-containing nanomaterials are mainly used in water treatment.

CONCLUSION

Advanced nanotechnology has emerged as a technique with great potential for environmental protection. Features such as cost-effectiveness and multifunctional dimensions make nanotechnology a technique that can help in monitoring of pollution, and treatment of pollutants and can be applied in many other aspects related to improvement in the quality of water. Nanotechnology has shown the potential to act as a good, sustainable option for environmental remediation by preventing the formation of by-products and decomposition of toxic pollutants by producing no waste.

REFERENCES

[1]   Ibrahim RK, Hayyan M, AlSaadi MA, Hayyan A, Ibrahim S. Environmental application of nanotechnology: air, soil, and water. Environ Sci Pollut Res Int 2016; 23(14): 13754-88.
[http://dx.doi.org/10.1007/s11356-016-6457-z] [PMID: 27074929]

[2]   Kampa M, Castanas E. Human health effects of air pollution. Environ Pollut 2008; 151(2): 362-7.
[http://dx.doi.org/10.1016/j.envpol.2007.06.012] [PMID: 17646040]

[3]   Balali-Mood M, Ghorani-Azam A, Riahi-Zanjani B. Effects of air pollution on human health and practical measures for prevention in Iran. J Res Med Sci 2016; 21(1): 65.
[http://dx.doi.org/10.4103/1735-1995.189646] [PMID: 27904610]

[4]   Mansoori GA, Soelaiman TF. Nanotechnology—an introduction for the standards community. J ASTM Int 2005; 2: 1-21.

[5]   Ramsden J. Essentials of nanotechnology. Ventus Publishing ApS 2014.

[6]   Kumar A, Gupta K, Dixit S, Mishra K, Srivastava S. A review on positive and negative impacts of nanotechnology in agriculture. Int J Environ Sci Technol 2019; 16(4): 2175-84.
[http://dx.doi.org/10.1007/s13762-018-2119-7]

[7]   Müller B, Zumbuehl A, Walter MA, *et al.* Translational medicine: nanoscience and nanotechnology to improve patient care The nano-micro interface: bridging the micro and nano worlds. 289-310.2015;

[8]   Kiparissides C, Kammona O. Nanotechnology Advances in Diagnostics, Drug Delivery, and Regenerative Medicine In book The Nano-Micro Interface. 311-40. 2015; pp.

[9]   Duncan TV. Applications of nanotechnology in food packaging and food safety: Barrier materials, antimicrobials and sensors. J Colloid Interface Sci 2011; 363(1): 1-24.
[http://dx.doi.org/10.1016/j.jcis.2011.07.017] [PMID: 21824625]

[10]   Shanthilal J, Bhattacharya S. Nanoparticles and nanotechnology in food.Conventional and advanced food processing technologies. Wiley 2014; pp. 567-94.
[http://dx.doi.org/10.1002/9781118406281.ch23]

[11]   Serrano E, Rus G, García-Martínez J. Nanotechnology for sustainable energy. Renew Sustain Energy Rev 2009; 13(9): 2373-84.
[http://dx.doi.org/10.1016/j.rser.2009.06.003]

[12]   Hussein AK. Applications of nanotechnology in renewable energies—A comprehensive overview and understanding. Renew Sustain Energy Rev 2015; 42: 460-76.
[http://dx.doi.org/10.1016/j.rser.2014.10.027]

[13]   Karn B, Kuiken T, Otto M. Nanotechnology and in situ remediation: a review of the benefits and potential risks. Environ Health Perspect 2009; 117(12): 1813-31.
[http://dx.doi.org/10.1289/ehp.0900793] [PMID: 20049198]

[14] Brame J, Li Q, Alvarez PJJ. Nanotechnology-enabled water treatment and reuse: emerging opportunities and challenges for developing countries. Trends Food Sci Technol 2011; 22(11): 618-24.
[http://dx.doi.org/10.1016/j.tifs.2011.01.004]

[15] Satapanajaru T, Anurakpongsatorn P, Pengthamkeerati P, Boparai H. Remediation of atrazine-contaminated soil and water by nano zerovalent iron. Water Air Soil Pollut 2008; 192(1-4): 349-59.
[http://dx.doi.org/10.1007/s11270-008-9661-8]

[16] Wang Z, Dai Z. Carbon nanomaterial-based electrochemical biosensors: an overview. Nanoscale 2015; 7(15): 6420-31.
[http://dx.doi.org/10.1039/C5NR00585J] [PMID: 25805626]

[17] A Framework for Biotechnology Statistics 2005.http://www.oecd.org/sti/sci-tech/34935605.pdf

[18] Pathakoti K, Manubolu M, Hwang H. Nanotechnology Applications for Environmental Industry. 2018.
[http://dx.doi.org/10.1016/B978-0-12-813351-4.00050-X]

[19] Rajan CSR. Nanotechnology in Groundwater Remediation. Int J Environ Sci Dev 2011; 2: 182-7.
[http://dx.doi.org/10.7763/IJESD.2011.V2.121]

[20] Tratnyek PG, Johnson RL. Nanotechnologies for environmental cleanup. Nano Today 2006; 1(2): 44-8.
[http://dx.doi.org/10.1016/S1748-0132(06)70048-2]

[21] Garner KL, Keller AA. Emerging patterns for engineered nanomaterials in the environment: a review of fate and toxicity studies. J Nanopart Res 2014; 16(8): 2503.
[http://dx.doi.org/10.1007/s11051-014-2503-2]

[22] Saed A, Tlili I. Numerical investigation of working fluid effect on Stirling engine performance Int J Therm Environ Eng 10: 31-6.2015;

[23] Khan MA, Khan T, Riaz MS, Ullah N, Ali H, Nadhman A. Plant cell nanomaterials interaction: Growth, physiology and secondary metabolism, Comprehensive Analytical Chemistry. Elsevier 2019; pp. 23-5.

[24] Tian F, Chen G, Yi P, *et al.* Fates of Fe_3O_4 and $Fe_3O_4@SiO_2$ nanoparticles in human mesenchymal stem cells assessed by synchrotron radiation-based techniques. Biomaterials 2014; 35(24): 6412-21.
[http://dx.doi.org/10.1016/j.biomaterials.2014.04.052] [PMID: 24814428]

[25] Khan S, Ul-Islam M, Ikram M, *et al.* Preparation and structural characterization of surface modified microporous bacterial cellulose scaffolds: A potential material for skin regeneration applications in vitro and in vivo. Int J Biol Macromol 2018; 117: 1200-10.
[http://dx.doi.org/10.1016/j.ijbiomac.2018.06.044] [PMID: 29894790]

[26] Gehrke I, Geiser A, Somborn-Schulz A. Innovations in nanotechnology for water treatment. Nanotechnol Sci Appl 2015; 8: 1-17.
[http://dx.doi.org/10.2147/NSA.S43773] [PMID: 25609931]

[27] Chong MN, Jin B, Chow CWK, Saint C. Recent developments in photocatalytic water treatment technology: A review. Water Res 2010; 44(10): 2997-3027.
[http://dx.doi.org/10.1016/j.watres.2010.02.039] [PMID: 20378145]

[28] Gomes SIL, Soares AMVM, Scott-Fordsmand JJ, Amorim MJB. Mechanisms of response to silver nanoparticles on *Enchytraeus albidus* (Oligochaeta): Survival, reproduction and gene expression profile. J Hazard Mater 2013; 254-255: 336-44.
[http://dx.doi.org/10.1016/j.jhazmat.2013.04.005] [PMID: 23644687]

CHAPTER 11

Use of Biodegradable Polymers and Plastics- A Suitable Alternate to Prevent Environmental Contamination

Chandrika Ghoshal[1,*], **Shashi Pandey**[1] and **Avinash Tomer**[2]

[1] *Division of Genetics, ICAR- Indian Agricultural Research Institute, New Delhi-110012, India*

[2] *Division of Vegetable Science, ICAR- Indian Agricultural Research Institute, New Delhi-110012 India*

Abstract: Bioplastics are plastics that are manufactured from biomass. These polymers have become increasingly popular as a means of conserving fossil fuels, lowering CO_2 emissions and minimising plastic waste. The biodegradability of bioplastics has been highly promoted, and the demand for packaging among merchants and the food industry is fast rising. It also has a lot of potential applications in the biological and automobile industries. The plastic on the market is extremely dangerous because it is non-biodegradable and harmful to the environment. As a result, the production and usage of biodegradable polymers are becoming increasingly popular. Some of the more recent formulations, partially as a result of third-party certifications, are more compliant than the initial generation of degradable plastics, which failed to achieve marketing claims. Many "degradable" plastics, on the other hand, do not degrade quickly, and it is unclear whether their use will lead to significant reductions in a litter. Biodegradable polymers, such as poly(lactic acid), are seen as *via*ble replacements for commodity plastics. In seawater, however, poly(lactic acid) is practically non-degradable. Other biodegradable polymers' degradation rates are further influenced by the habitats they wind up in, such as soil or marine water, or when utilised in healthcare equipment. All of these aspects are discussed in detail in this chapter, including bioplastic types, applications, production, degradation, problems in landfills and sea water, fermentation, synthesis, and sustainability. This chapter, taken as a whole, is intended to help evaluate the possibilities of biodegradable polymers as alternative materials to commercial plastics.

Keywords: Applications, Biopolymers, Poly hydroxybutyrate, Poly lactic acid.

* **Corresponding author Chandrika Ghoshal:** Division of Genetics, ICAR- Indian Agricultural Research Institute, New Delhi-110012, India; E-mail: chandrika.ghoshal@gmail.com

Bhupinder Dhir (Ed.)

INTRODUCTION

Plastics have substantially replaced traditional packaging materials because of their superior physical properties, particularly strength and toughness, lightness, and barrier properties. Their capacity to keep perishable goods from rotting at a low cost has resulted in a revolution in food delivery, to the point where they are now considered necessary in modern retailing. Compared to conventional materials, plastics are also more energy efficient [1]. Paper requires double the weight of polyethylene to successfully protect goods, and if all polymers currently used in packaging were to be replaced with paper, the environmental consequences would be disastrous in terms of forest loss, increased energy consumption, and environmental harm [2, 3].

The volume of plastics, synthetic fibres, and rubber that end up in landfills is becoming a severe concern for trash management. Domestic and industrial garbage was once inexpensively disposed of in holes in the ground on the outskirts of towns and cities. The decrease in the number of such sites, along with the huge volumes of the waste, has resulted in an unacceptably high cost of transporting packaging wastes to available dump sites. There is also a growing awareness that waste should be treated as a resource that may be re-used by recycling it into usable items rather than burying it. Plastic trash is also a direct threat to wildlife, with many different species having been identified as being harmed by it. For most animals, the main risks connected with plastic materials are entanglement in and ingestion of those items [4]. Plastic waste entangles juvenile organisms in particular, causing serious harm as the animal matures [5, 6]. It restricts their movement, preventing animals from adequately feeding and breathing [7].

Biodegradable polymers provide a possible solution to this problem. As biodegradable polymers are generally obtained from natural sources, they are usually referred to as "biopolymers" (Fig **1**) [8]. Biodegradable biopolymers are few and far between on the list of biopolymers. Plastics like PLA, PHA, and starch are the most often utilised biopolymers that have little to no impact on the environment's rising carbon footprint.

TYPES OF BIOPOLYMERS

The most common and widely available biopolymers can be divided into four groups Fig. (**2**).

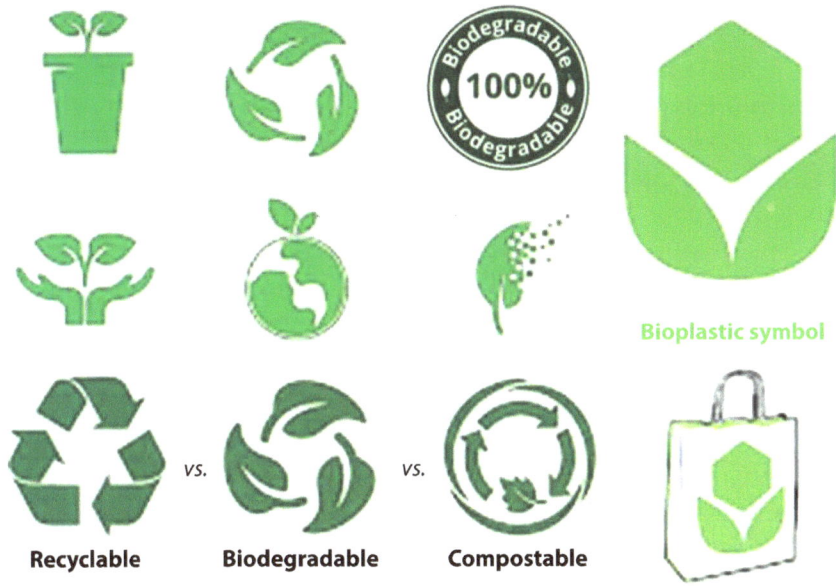

Fig. (1). Bioplastic symbol.

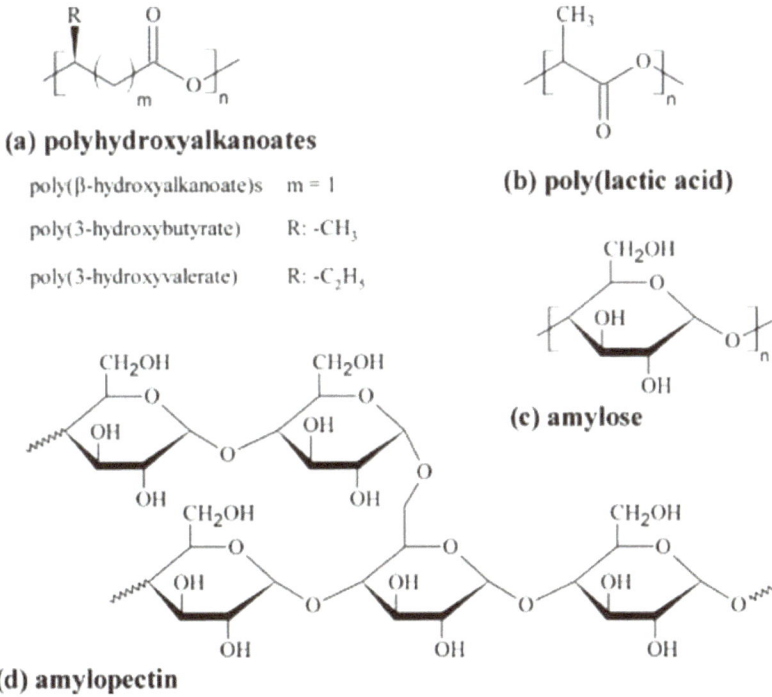

(a) polyhydroxyalkanoates

poly(β-hydroxyalkanoate)s   m = 1

poly(3-hydroxybutyrate)   R: -CH₃

poly(3-hydroxyvalerate)   R: -C₂H₅

(b) poly(lactic acid)

(c) amylose

(d) amylopectin

Fig. (2). Biopolymers.

Sugar Based Biopolymers

Poly(lactic acid) (PLA) is the most ubiquitous, versatile, and widely explored sugar-based biopolymer with enormous potential in a variety of applications, including the automobile industry. Lactic acid is the basic ingredient for PLA's monomer, which is made from lactose, a sugar found in milk. It can be found in natural sources like potatoes, sugarcane, and wheat. Poly(hydroxybutyrate) (PHB) is another type of biopolymer generated by bacteria and used as an energy storage molecule when the microorganism's nutrition supply is restricted. PHB is a polymer produced by microorganisms as a result of carbon absorption from various types of glucose.

Starch-based Biopolymers

Starch is a biopolymer, and a type of carbohydrate found exclusively in plants, and it is stored in their tissues. The predominant component of starch, however, is glucose, which can be turned into glucose using either temperature or other chemical stimulants. Tapioca, potatoes, corn, and wheat are just a few of the foods that contain them.

Cellulose-based Biopolymers

The most abundant biopolymer found in plant cell walls is cellulose. Glucose is the most important component of cellulose. Cellulose can be used as a filler or a plasticizer to improve the characteristics of other biopolymers.

Biopolymers Based on Synthetic Materials

Synthetic biopolymers, such as aliphatic or aromatic polyesters or copolyesters, are created by a chemical reaction from monomers sourced from petroleum feedstock. These synthetic biopolymers, like nature-derived biopolymers, are compostable. Three significant aliphatic polyesters are poly(ε-caprolactone) (PCL), poly(glycolide) (PGA), and poly(3-hydroxybutyrate-co-3-hydroxyva lerate) (PHBV), and poly(butylene adipate-co-terephthalate) (PBAT) is an aromatic copolyester chemically produced from petroleum feedstock [9].

APPLICATION OF BIOPOLYMERS

Biomedical Application

The phrase "biopolymer" is a technical term for bio-based polymers that are either biodegradable or not in nature. In biomedical engineering, by mimicking body parts, normal bodily functions are maintained. Biopolymers are widely employed in tissue engineering, medical devices, and the pharmaceutical industry due to their biocompatible qualities [10]. Due to their mechanical properties, many biopolymers can be employed in regenerative medicine, tissue engineering, drug delivery, and other medicinal applications [11]. Polypeptides, like collagen and silk, are biocompatible materials being exploited in cutting-edge research because they are affordable and readily available. Gelatin polymer is frequently used as an adhesive in wound dressings. Scaffolds and films made of gelatin can retain medications and other nutrients that can be delivered to a wound to help it heal [12].

Application in the Automobile Industry

Apart from its application in biomedical engineering, automobile industries are also using biopolymers in a rapid way. Car body parts comprised of petroleum-based polymers are increasingly being replaced by biobased polymers. The use of biobased polymers in car body parts by the automotive industry enhances fuel efficiency, design variance, comfort, and insulation. In addition, it not only reduces our reliance on fossil fuels but also helps to lessen our carbon impact to some level. Aside from PLA, various additional biobased biopolymers, such as PBS, PHAs, and starch, have found significant use in packaging and other commodity applications. However, these biopolymers' mechanical and thermal qualities are insufficient to compete with commercially available polymers in the automotive industry. These biopolymers are either biodegradable or compostable, but they lack mechanical competency in automotive engineering applications. Blending these biodegradable biopolymers with PLA, on the other hand, can keep the biodegradability standards if they need to be maintained after the product's service life [13].

Application in the Food Industry

Packaging, edible encapsulation films, and food coating are all examples of how biopolymers are employed in the food business. Due to its clear clarity and water resilience, polylactic acid (PLA) is widely used in the food business. Most polymers, on the other hand, are hydrophilic and begin to deteriorate when exposed to moisture. Biopolymers are also being used to encapsulate meals in edible films. Antioxidants, enzymes, probiotics, minerals, and vitamins can all be carried by these films. These nutrients can be delivered to the body by food encapsulated in biopolymer film [10, 14].

Application in Packaging

Polyhydroxyalkanoate (PHA), polylactic acid (PLA), and starch are the most prevalent biopolymers used in packaging. As starch and PLA are readily available and biodegradable, they are popular choices for packaging. Their barrier and thermal qualities, however, are not optimal. Hydrophilic polymers are not water resistant, allowing water to pass through the container and potentially contaminate the contents [15]. Polyglycolic acid (PGA) is a biopolymer with excellent barrier properties that are now being employed to overcome the barrier challenges posed by PLA and starch.

Blending or copolymerizing PBS with PLA has improved a number of qualities that can be used in food packaging, food service ware, mulch films, and other long-lasting applications in car interior body parts [13]. Due to the manifestation of unique features at a concentration of a few weight percentages, biopolymers from natural sources such as cellulose, starch, chitin and chitosan, cellulose nanofibers, and cellulose nanocrystals have gotten a lot of attention in academic disciplines [16].

CATEGORY OF BIOPOLYMERS

Commercial biodegradable polymers can be classified into three groups based on the raw ingredients used and the processes used to make them [17, 18] Fig. (**3**).

Produced from microbial fermentation

Polyhydroxyalkanoates (PHAs), such as poly(3-hydroxybutyrate) (P3HB),Poly---hydro xyvalerate (PHV), and their copolymers poly (3-hydroxybutyrate-c--3-hydroxy valerate) [P(3HB-co-3HV)], poly (3-hydroxybutyrate-c--4-hydroxybutyrate) [P(3HB-co-4HB)], are biodegradable plastics made from microbial fermentation procedure.

Produced from Reusable Monomers

The second group includes biodegradable polymers made with reusable monomer origins in industrial processes, such as PLA, bio-based PBS, and poly(butylene succinate-co-butylene adipate) (PBSA). PBS, polycaprolactone (PCL), poly(butylene adipate terephthalate) (PBAT), polyglycolide (PGA), PBSA, poly(propylene carbonate) (PPC), poly(vinyl alcohol) (PVA), and others are among the biodegradable plastics made from petrochemical sources (Fig **3**). As people become more concerned about global pollution, the demand for biodegradable plastics has increased. The cost has progressively approached that of commodity plastics, causing some plastics to be phased out. Starch, PLA, PBS, and PBAT are the most prolific biodegradable materials, accounting for 38.4 percent, 25.0 percent, 7.7 percent, and 24.1 percent of total biodegradable plastic capacity, respectively [19]. Starch is a commonly used natural degradable polymer with unique biodegradability in a variety of natural conditions. It's frequently plasticized and combined with other polymers to boost the product's bio-based content or cut expenses [20 - 22]. PLA is utilised in a variety of industries, including disposable tableware, medicinal applications, packaging, and so on, with the largest production figs of 449000 tonnes per year. PLA has a promising future not just because of its high mechanical strength and remarkable transparency, but also because its feedstock, lactic acid, is renewable and PLA is considered a "green" material [23, 24]. At ambient temperature, their high Tg makes them brittle and difficult to blow into thin films; this, combined with their poor heat resistance, has been a key hurdle for their large-scale commercial usage [25 - 27]. The introduction of PBS and PBAT more than compensates for this deficit. The most significant benefit of PBS is that it combines mechanical strength and toughness with heat resistance. This makes it appropriate for a wide range of applications, with the exception of those requiring high barrier properties [28 - 30]. PBS can also be regarded as a biomaterial because the feedstock succinic acid can be produced by biological fermentation from biomass such as corn or soybeans [24, 31]. The most important property of PBAT is its ability to make good films while maintaining mechanical strength. It can be used to make a number of disposable film goods, such as bags, and it's particularly good for agricultural mulch films [32]. It's worth noting that PBAT has the lowest production cost out of all biodegradable polyesters due to the cheap cost of terephthalic acid. With the growing use of biodegradable materials in disposable packaging, PBAT's production capacity and market share among all biodegradable plastics have steadily increased, rising from 11.6 percent in 2017 to 16.7 percent in 2018, and then to 24.1 percent in 2019. In most natural situations, other polyesters, such as PCL, fossil-based semicrystalline polyester, can be biodegraded by both aerobic and anaerobic microbes. However, its application range is limited by its low Tm (58 °C), poor temperature resistance, and exorbitant costs.

Produced From other Sources

PHAs, on the other hand, are a distinct class of commercially applied bio-based biodegradable and/or biocompatible polyesters that exhibit a wide range of properties based on the length of the side aliphatic chain at the β-carbon. However, the PHAs market and applications are still modest, with a market worth of 98 million US$ in 2021, owing to high manufacturing costs and the difficulty of extraction techniques.

DEGRADATION MECHANISM AND LABORATORY SIMULATION

Biodegradable polymers are described as "polymers susceptible to degradation by biological activity, with the degradation accompanied by a decrease in mass," according to the International Union of Pure and Applied Chemistry (IUPAC).

Other definitions call for a biodegradable material to be mineralized during biodegradation into carbon dioxide, water, and biomass (standard CEN/TR 15351:2006). So, let's start with a definition of "biodegradability." Do we merely need to convert polymers into water-soluble substances to alle*via*te the waste problem of piling plastics in the environment, or do we require full mineralization of the polymers. Furthermore, the biodegradation time range is crucial.

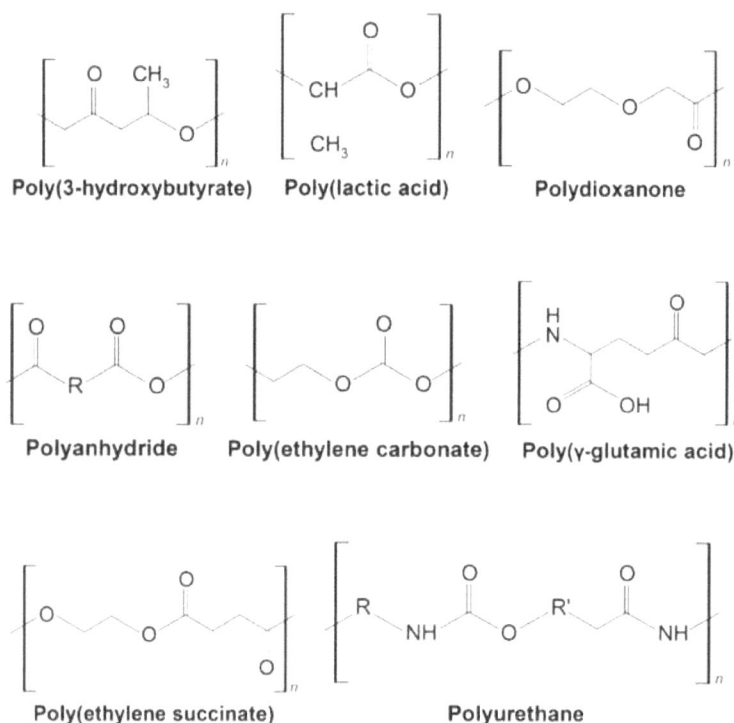

Poly(3-hydroxybutyrate)   Poly(lactic acid)   Polydioxanone

Polyanhydride   Poly(ethylene carbonate)   Poly(γ-glutamic acid)

Poly(ethylene succinate)   Polyurethane

Fig. (3). Most common biopolymers

Stages of Biodegradation

Biodegradation of polymers can be broken down into four stages: i) Biodeterioration, ii) Depolymerization, iii) Bioassimilation, and iv) Mineralization [33]. The establishment of a microbial biofilm leads to surface degradation, which fragments the polymeric material into smaller particles in the first stage. Extracellular enzymes are secreted by biofilm microorganisms, which accelerate the depolymerization of polymer chains into oligomers, dimers, or monomers. The process of assimilation occurs by uptaking the tiny molecules created in this way into the microbial cell, and thus the subsequent production of primary and secondary metabolites takes place. These metabolites are mineralized in the final stage, and end products such as CO_2, CH_4, H_2O, and N_2 are generated and discharged into the environment.

Abiotic and Biotic Factors Involved in the Biodegradation of Polymers

Polymer breakdown is regulated by both abiotic and biotic variables. Abiotic factors include mechanical stress, light, and temperature, while biotic aspects include the presence of naturally occurring microbes such as bacteria, fungi, and algae (Fig **4**) [34].

Abiotic Factors

When a polymer is exposed to environmental conditions, it can undergo mechanical, light, heat, and chemical transformations, all of which can affect the polymer's mechanical characteristics (Fig **4**). For example, UV irradiation can cause the polymer to become brittle [33]. Polymers containing heteroatoms such as esters, anhydrides, amides, or urethanes get chemically degraded *via* a hydrolysis mechanism [33]. Hydrolysis of a substance occurs either by mechanism of bulk or surface erosion. When degradation occurs uniformly across the thickness of a polymeric material, it is referred to as bulk erosion but the reduction in the thickness of the surface is known as surface erosion. Both the pathways and the parameters that influence hydrolysis were explained by Laycock *et al.* [35]. When the rate of hydrolysis exceeds the rate of water diffusion into the bulk polymer, surface erosion occurs. In other words, a catalyst (such as enzymes) is unable to pass through the bulk polymer. Surface erosion is commonly noted in hydrophobic and semi-crystalline polymers, as well as polymers with a high rate of hydrolysis. On the other hand, bulk erosion occurs when the rate of water diffusion is more than the rate of hydrolysis. Hydrolysis mechanism of a material can change from surface to bulk erosion when thickness of the sample reaches below a critical value [36]. The size of a material, in general, is significant since a bigger surface area promotes degradation [37]. The degradation of the polymer increases as the thickness of the foil decreases. The rate of hydrolysis is regulated by various external factors. Factors such as an increase in temperature and change in pH speed up the rate of hydrolysis. The rate of hydrolysis of poly (lactic-c--glycolic acid) (PLGA) increases under acidic and basic conditions [38]. Furthermore, pH levels affect erosion. At low pH, bulk erosion occurs, while at high pH, surface erosion is noted. The method of degradation changes depending on the conditions. Under basic conditions, PLA degrades to form intermediate dilactide *via* a dezipping mechanism, while under acidic conditions, lactic acid is directly formed. Pierre and Chiellini highlighted the importance of acid–base catalysis and noted that a one-unit rise in pH increases the rate of hydrolysis by ten times [39]. It was difficult to determine the rate of hydrolysis at a neutral pH. Despite the rarity of strongly acidic or basic circumstances in nature, high pH values are frequently employed to demonstrate the general (bio) degradability of specific polymers.

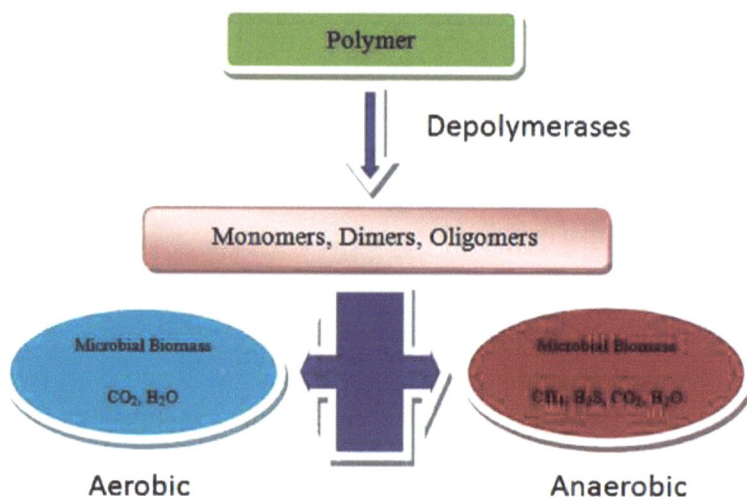

Fig. (4). Biodegradation of Polymers.

Biotic Factors and Enzymes

Enzymes play a significant function in catalysing hydrolysis in natural ecosystems. They function best at room temperature. At neutral pH, the speed of reactions increases by 10^8 to 10^{20} times [40]. High molecular weight (several kDa) prevents the enzymes from permeating the polymer matrix, therefore enzymatic hydrolysis occurs mainly *via* surface erosion [35]. The enzyme protease degrades a-ester bonds, whereas poly(hydroxybutyrate)(PHB)-depolymerase b-ester linkages and lipase g-w bonds are known to be degraded by specific enzymes [41]. Both lipase and protease, for example, aid in the breakdown of PLA [42]. Several reviews mention the enzymes that can degrade PLA or other biodegradable polymers, as well as their sources [35, 43, 44]. Nylon-6,6, a commodity plastic, can only be damaged by microorganisms with specific characteristics. The enzymes protease, lipase, and PHB depolymerise, PVA oxidase and lignin-degrading enzymes isolated from bacteria or fungi showed the ability to degrade Nylon-6,6 [45 - 49].

Methods to Assess Biodegradation

Different approaches were used to observe the degradation, which depicted different stages of biodegradation: The change in the crystal structure of the polymer was related to the direction of biodeterioration. This was measured by atomic force microscopy (AFM) [47]. Depolymerization, *i.e.*, decrease in molecular weight, is measured by size-exclusion chromatography (SEC) or lactic acid formation during PLA degradation (evaluated using an enzyme bioanalysis kit) [45, 49]. The decrease in weight of a polymer specimen indicates complete mineralization or formation of water-soluble byproducts. The quantity of mineralization is indicated by CO_2 detection [46]. A biodegradable material gets mineralized to form carbon dioxide, water, and biomass during biodegradation [50]. This ensures that the substance fades completely, which is not possible with a simple drop-in molecular weight. *In vivo* applications of the polymers were estimated using pH values and temperatures in the degradation tests with physiological values. It's important to note that enzyme activity changes with a change in temperature. Hoshino and Isono noted that lipase PL produced from the bacteria *Alcaligenes sp.* showed optimal reaction at a temperature of 50.8°C [51], and enzyme activity reduced by 60% at 30.8°C. The rate of degradation changes in a natural setting with temperatures below physiological levels needs to be explained. The diverse mixes of enzymes and microorganisms present in a natural environment are not always conducive to the degradation of the studied polymer. Thus, the conversion of laboratory biodegradation testing with separated bacteria into real-world circumstances seems to be difficult.

LANDFILL DISPOSAL OF PLASTICS

Millions of tonnes of plastic are disposed of away in landfills each year (60 percent of all plastic generated) [52]. The most common method of waste management is landfilling, though it is restricted due to space issues in developed countries and countries with high population densities [53]. Most of the plastic packaging and non-recyclable plastics are put in landfills. Landfills all over the world have been filled with plastic, which makes up around 18–20% of the trash [54, 55]. Toxic leachates released from plastic harm soil microbial flora resulting in soil infertility, and contamination of ground water. Packaging of organic waste, such as bio-inert and impervious plastic bags, leads to the persistence of the plastic [56, 57]. Therefore, plastic waste needs to be managed properly to overcome sustainability concerns. Biodegradable and non-biodegradable plastics need to be separated to get decomposition and composting. Alternative methods, such as gasification/pyrolysis and mechanical biological treatment, have gained pace in recent years for plastic disposal in addition to the widely used methods. Gasification/pyrolysis that converts a wide range of polymers into synthetic gas/petrochemical fuels at high temperatures and thermally breaks the plastics acquired popularity recently [58 - 61, 61, 62]. Exposure of plastics to high temperatures of 600–800 °C leads to gasification, which produces oil and condensable gases. These get transported through a gasifier along with air and steam at a temperature of 1200–1500 °C to produce synthetic gas (composed of hydrogen, carbon monoxide, CO_2, CH_4 and steam). During pyrolysis, plastic materials are exposed to a temperature of 500°C in the absence of oxygen. The plastics broke to produce a mixture of condensable vapours, which on cooling, form petrochemical fuels (oils, chars and waxes). Dioxins or other contaminants are not produced in the process of gasification/pyrolysis as in incineration [60]. The process is very costly. Purification is also required to get large-scale production of synthetic gas/petrochemicals [61]. Mechanical biological treatment involves both mechanical and biological processes [63]. In this method, recyclable wastes are separated, treated biologically and disposed of in landfill. Wastes left after mechanical treatment are subjected to biological treatments such as composting/anaerobic digestion [63]. This treatment is very effective for treating mixed garbage (which includes degradable and non-degradable plastics, glass, metal, rubber, *etc.*). The plastics cannot be treated *via* mechanical biological treatment. Alternative sources that are environmentally and economically *via*ble need to be searched so that the exploitation of fossil petroleum resources is reduced, and the rising demand for plastics can be fulfilled.

ALTERNATIVES TO CONVENTIONAL PLASTICS

Research studies have been conducted for the last 30–40 years to find an alternative to making polymers from renewable materials. The production of biologically generated plastics has gained attention as it can limit the disposal of non-degradable plastics in landfills along with reducing the use of fossil fuels and greenhouse gas emissions. Bio-plastics are polymers generated from renewable carbon sources that break to form CO_2 and water under aerobic settings or methane (CH_4,) CO_2 and water under anaerobic conditions [64, 65]. They are environmentally benign and sustainable (Fig **5**). The cradle-to-grave-to-cradle strategy is the theoretical concept behind this technique which restricts the landfilling of recyclable bio-plastics. The retention of the physico-chemical properties of petroleum-derived plastics, along with the maintenance of biocompatibility and biodegradability, is difficult for bio-polymers [66, 67]. Bio-plastics are made either from natural polymers (*e.g.,* starch-based plastics), fermentation or conventional chemical processes that polymerize bio-based monomers and oligomers (*e.g.,* polylactic acid), or microbial polymer (polyhydroxyalkanoates) synthesis. Polylactic acid (PLA) and polyhydroxyalkanoate (PHA) are two bio-plastics that are considered *via*ble alternatives to synthetic plastics.

Polylactic Acid

PLA is aliphatic thermo-softening polyester. It is made by chemically polymerizing D- and L-lactic acid produced from sugar beet, maize (corn), sugarcane, or wheat fermentation [68, 69]. PLA polymers show a glass transition temperature (Tg) of 55 °C, and are similar to petroleum-derived PET. They show great moisture resistance in packed items [70]. PLA has been successfully used in making loose-fill packaging, food packaging, beverage containers, and disposable food service tableware. PLA is also extensively used in biomedical applications such as the making of sutures, stents, dialysis devices, and drug capsules. It has also been evaluated as a matrix for tissue engineering [71 - 73]. PLA shows a faster degradation in composting facilities within 4–6 weeks, as seen in the case of the paper [74]. The production of PLA also requires 20–50% less fossil fuels compared to petro-plastics [65, 75], though energy consumption is higher. PLA is more expensive (25% more) than its petroleum-based counterpart, PET [55].

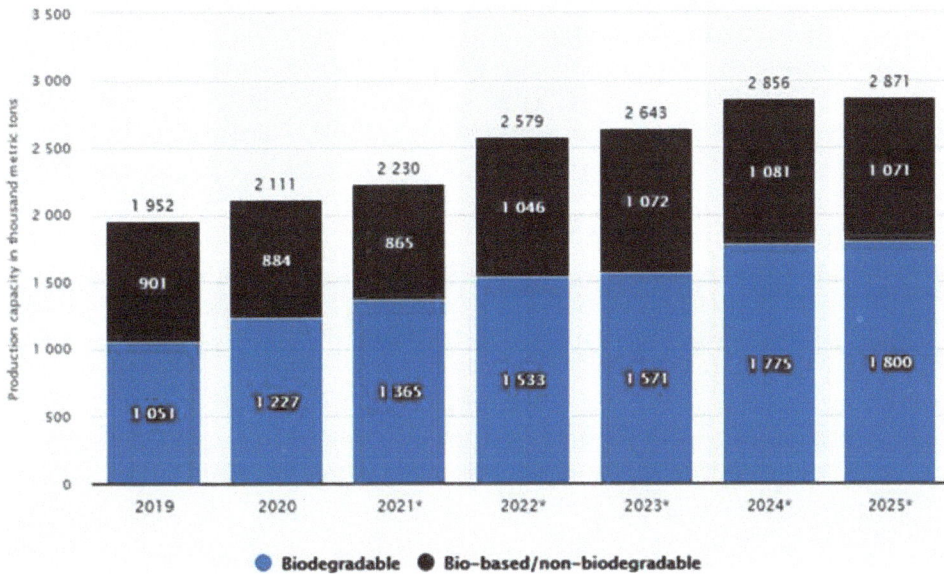

Fig. (5). Production of bioplastics worldwide from 2019 to 2025 by type (in 1000 metric tons).

Polyhydroxyalkanoates (PHA)

Polyhydroxyalkanoates (PHA) are produced by converting different carbon substrates into aliphatic polyesters *via* microbes [66, 67, 76]. PHAs are characterised by the number of carbon atoms and type of monomeric unit they possess. For example, polymers with 3–5 carbon atoms are referred to as short chain-length PHA whereas polymers with 6–14 carbon atoms are classified as medium chain-length PHA [67].

PHAs show similar material properties as traditional plastics [77], while their physical [78, 79] and mechanical properties can be adjusted by modifying the co-polymer content [76, 78]. PHA is used in various short-term packaging applications such as plastic films for bags and other kinds of packaging, containers, paper coatings, disposable goods (personal care products, surgical gowns) and upholstery [76, 79]. Their manufacturing cost is high, and they act as a promising material for use in medical applications as they are biodegradable and immunologically innocuous [66, 76, 79]. Poly-3-hydroxybutyrate (PHB) is a widely studied PHA. It is produced by a variety of Gram-positive and Gram-negative bacteria, including methanotrophs/methane-oxidizing bacteria [76, 79 - 83]. PHB polymers show properties such as resistance to moisture, UV, ageing, water insolubility, oxygen impermeability, indefinite stability in air, and higher optical purity [67, 84]. PHBs are good materials from the waste management approach.

Synthesis of Polyhydroxyalkanoates

PHA is produced in cell-free systems employing bacterial fermentation, genetically engineered plants, and enzymatic catalysts. Microbially produced PHA synthesis (*i.e.*, PHB) is the only technology available at an industrial scale. PHBs produced *via* microbial means use a variety of carbon-sources as feedstocks. These mainly include wheat bran, whey, molasses, cane starch, palm oil, cassava waste, sucrose, and glucose [75, 78]. These account for more than 30–50% of the total production cost [85]. The limiting step in the production of PHB by microorganisms is the feedstock. It is also the costliest part involved in microbial PHB production.

The cost of production of PHB ranges from USD 4–16 per kg of polymer. The substrate and bacteria utilised have a big impact on the price of PHB. PHB made from *Cupriavidus necator* (*Alcaligenes eutrophus*) costs USD 16 per kg of polymer. It costs more than polypropylene. PHB made from recombinant *Escherichia coli* costs USD 4 per kg of polymer. Its cost is comparable to other bio-degradable plastic materials like PLA and aliphatic polyesters. It was noted that at the global level, bio-plastic output increased from 890,000 tons to 2.5 million metric tonnes (Mt) between 2012 and 2017 [86] and is expected to reach 1.5–4.4 Mt by 2020 [85]. Feedstock *via*bility is likely to get affected by the requirements of production area [85, 86], as an increase in production might lead to competition for arable land between food and feedstock crops. According to the UN's Food and Agriculture Organization (FAO), food prices noted a record increase in 2010 with a food price index of 214.7 points [87]. CH_4, a potent greenhouse gas from landfills and other sources, can be used as a source of carbon for PHB production, and this can result in significant cost savings in terms of feedstock, land utilisation, and energy needs [80, 82, 88, 89]. Use of CH_4-fed methanotrophs/methane-oxidizing bacteria for the production of PHB results in many benefits, such as the sequestration of greenhouse gases CH_4 and CO_2. According to analysis, 60 percent of oxidised CO_2 from CH_4 is incorporated into biomass [90]. Higher PHB accumulation (maximal 80%) can be achieved at an industrial scale [80, 82].

DEGRADATION OF POLYESTERS IN SEAWATER

Polymers have become an important part of human life. They have also been recognized as ocean pollutants because of their extended lifespan and limited degradability. Several studies have demonstrated the negative effects of plastics on marine life, biodiversity and toxicity. Though the negative effects of plastics are debated, it is desirable to avoid plastic waste. Polymers that breakdown in seawater have been developed, and their prospects as replacements of commodity polymers in several applications are being explored. Biodegradable polymers that break or degrade *in vivo* or during composting often decay very slowly in saltwater. Most of these materials do not degrade or disintegrate easily in natural conditions, therefore techniques and novel routes for the creation of seawater-degradable polymers that show fast deterioration in natural environments have been explored. This is supposed to help in developing a better understanding of the development of seawater-degradable polymers.

Environmental Characteristics of Seawater

The marine habitat can be distinguished from terrestrial conditions of soil and compost in terms of their low temperature, high salinity, high pressure, currents, and low nutrition levels (*e.g.,* nitrate). The temperature of water changes with depth and also varies according to seasons and places [91].

An average surface temperature of 17.4 degrees Celsius has been noted in seawater. As the depth of water exceeds 2000 m, temperature decreases by 0–4°C [92]. The levels of inorganic salts present in saltwater get affected by evaporation, precipitation, river runoff, and ocean currents.

Salinity fluctuates at different depths in distinct locations. Seasonal fluctuations are also noted in the same sea area. The salinity of offshore and estuarine waters rarely exceeds 30%, and the ocean's surface salinity ranges from 32 to 37 percent, with an average of 35 percent. Seawater has an alkaline pH of 8.0 to 8.5.The pH of the surface seawater is normally constant at 8, but the pH of the intermediate and deep seawater ranges between 7.8 and 7.5.The dissolved oxygen level of saltwater is generally less than that of soil. Both pH and dissolved oxygen levels of saltwater get altered by temperature. The biological processes get affected by the presence and abundance of particular marine creatures. The biological habitat in the marine environment is structured vertically, similar to abiotic factors (such as salinity, temperature, pressure, UV radiation). The species composition of marine habitats shows variation depending on the zones. For example- the epipelagic zone (0– 200 m) is influenced by UV radiation, currents, depth and the abyssopelagic zone (4000 m down to the ground) is influenced by currents.

Microorganisms such as bacteria, archaea, unicellular algae, fungi, and protozoans are unicellular creatures that are commonly found in marine settings and play a crucial role in marine food webs. Autotrophic microorganisms (*e.g.,* algae and cyanobacteria) undergo photosynthesis to generate organic matter in the light-flooded strata (epipelagic zone). Heterotrophic organisms that breakdown organic compounds are found generally in deeper zones. They dissolve organic matter (DOM) and particulate organic matter (POM), as well as marine debris and organic materials such as broken plastics. The number of microorganisms present in marine habitats is around half that of those present in terrestrial soils. The organisms in the marine environment need to adapt well to environmental conditions such as low temperature, high pressure, high salt, and limited nutrient content. The density of heterotrophic bacteria in seawater is about 10^5–10^7 per mL. The number and composition of varies depending on the location and depth of the deep sea. Low temperatures limit microbial growth [93, 94]. Whitman *et al.* reported values of 5-10^5 prokaryotes (autotrophic and heterotrophic bacteria, as well as archaea) per mL in the upper ocean and 0.5-10^5 prokaryotic cells per mL in the lower ocean. The frequency of microbial species reduces at depths below 4000 m due to low temperatures and limited food supply. The deep sea bacteria can be found on buried organic debris [95, 96]. The composting is carried out at a temperature between 58 °C and 65 °C. The microbial species actively involved in biodegradation differ from those found in seawater in terms of their numbers. The numbers are large, exceeding 10^9 per mL [97]. Due to this pattern, polyester breakdown *via* biodegradation in seawater is different from that found in soil or composting settings.

Seawater Degradation of Biopolymers

The performance of some common biodegradable polymers (PLA, PBAT, PBS, PHAs, and PCL) found in seawater are summarized and compared. Their chemical characteristics are listed in Table (**1**), while Table (**2**) lists the experimental conditions and parameters such as time scale, characteristics of water and sample. Changes in morphology, mechanical strength, and molecular weight of the sample are also documented.

Table 1. Comparison of mechanical properties, crystallinity, degradability, processing properties and cost of commercial biodegradable resins with non-degradable resins.

POLYMER	Tg [°C]	Tm[°C]	Tensile strength[MPa]	Elastic modulus [MPa]	Elongation at break [%]	Market share	Product cost
LDPE (Low density polyethylene)	-100	98-115	8-20	300-500	100-1000	-	1150
Starch	-	-	-	-	-	38.4	400
PHB (Polyhydroxybutyrate)	5-10	177-182	40	700-1800	6-8	2.2	3800-6400

POLYMER	Tg [°C]	Tm[°C]	Tensile strength[MPa]	Elastic modulus [MPa]	Elongation at break [%]	Market share	Product cost
PHBV (Polyhydroxybutyrates-c--betahydroxyvalerate)	0-10	100-150	10-40	600-1000	10-500		
PLA (Polylactide)	40-70	130-18-	44-65	2800-3500	10-240	25	2200-2600
PCL (Polycaprolactone)	-60	59-65	4-28	390-470	700-1000	0-2.5	3000
PBS(Polybutylene succinate)	-32	114	40-60	500	170-500	7.7	2000
PBAT(polybutyleneadipateco terephthalate)	-30	110-115	25-40	65-90	500-800	24.1	1500
PGA (Polyglycolic acid)	30-40	225-230	89	7000-8400	30	0-2.5	-
PVA (Poly vinyl alcohol)	58-85	150-230	28-65	30-530	50-220	-	-

*Tg: Glass transition temperature; Tm: Melt temperature.

Table 2. Performance of biodegradable polymers in seawater.

Polymer	Shape/size[ab]	Time studied	Weight loss (%)	Mechanical stress decrease (%)	Ref
PLA (Poly l-lactide)					
PLA	PLLA Spline	180 days	-	unchanged	[100]
PLA	PLLA Spline	52 weeks	2	0	[116]
PLA	PLLA Film	1 year	Unchanged	-	[117]
PLA	PLLA film	10 weeks	Unchanged	unchanged	[98]
PLA	PLLA Film	5 weeks	25	100	[99]
PHB [Poly(3-hydroxybutyrate)]					
PHB	PHB film	1year	6	-	[117]
PHB	PHB film	6 weeks	40-100	-	[93]
PHB	PHB film	160 days	58	19 (0.1mm)	[155]
PHB	PHB film	10 weeks	9	40	[98]
PHB	PHB Spline	5 weeks	65	100	[99]
PHB	PHB Film	28 days	41(bay) 23 (ocean)	-	[104]
PHB	PHB film	49 days	30	-	[105]

(Table 2) cont.....

Polymer	Shape/size[ab]	Time studied	Weight loss (%)	Mechanical stress decrease (%)	Ref
P3HB [Poly(3-hydroxybutyrateco-3-hydroxyhexanoate)] copolymers					
P3HB	PH3B-co-HH Film	6 weeks	19	-	
P3HB	P(3HB-co-3HV) film	160 days	54 (0.005 mm) 13 (0.1 mm)	-	[155]
P3HB P(3HB-co-3HV) [Poly(3-hydroxybutyrateco-3-hydroxyvalerate)] PHB	P(3HB-co-4HB) fibers	8 weeks	65	100	[102]
P(3HB-co-3HV)	P(3HB-co-3HV) film	28 days	100 (bays and ocean)	-	[104]
PHB	P(3HB-co-4HB) film	28 days	70 (bay) 59 (Ocean)	-	[104]
P(3HB-co-3HV)	P(3HB-co-3HV) Spline	360 days	8	unchanged	[106]
P(3HB-co-3HV)	P(3HB-co-3HV) film	6 weeks	60	-	[156]
P(3HB-co-3HV)	P(3HB-co-3HV) fiber	12 months	-	80-100	[112]
P(3HB-co-3HV)	P(3HB-co-3HV) film	49 days	16	-	[105]
PCL Poly(□-caprolactone)					
PCL	PCL film	1 year	0.5	-	[117]
PCL	PCL film	6 weeks	98	-	[93]
PCL	PCL film	10 weeks	25	95	[98]
PCL	PCL film	5 weeks	34	100	[99]
PCL	PCL film	28 days	100 (bay), 67 (ocean)	-	[104]
PCL	PCL fiber	18 months	-	80-100	[112]
PCL	PCL fiber	2 months	100	100	[110]
PCL	PCL spline	52 weeks	30	20	[113, 116]
PBS (Polybutylene succinate)					
PBS	PBS spline	52 weeks	2	60	[116]
PBS	PBS film	6 weeks	2	-	[93]
PBS	PBS film	28 days	2	-	[104]
PBS	PBS fiber	12 months	-	0-10	[112]

(Table 2) cont.....

Polymer	Shape/size[ab]	Time studied	Weight loss (%)	Mechanical stress decrease (%)	Ref
PBAT [Poly(butylene adipateco-terephthalate)]					
PBAT	PBAT Spline	52 weeks	2	43	[115, 116]
PBAT	PBA film	6 weeks	7	-	[93]

*Film: the sample with thickness less than 1 mm; Spline: the sample with thickness equal or more than 1 mm.

Polylactide

PLA, a popular biodegradable plastic, degrades easily in compost and shows a significant reduction in degradability in seawater. Tsuji and Suzuyoshi studied the degrading properties of PLA films (0.05 mm thickness) in natural saltwater and collected seawater in the lab conditions *i.e.* static conditions. They found that general qualities of PLA sheets did not change appreciably after 10 weeks. The plasticization process increased tensile strength and Young's modulus of the experiment [98].

In natural ocean conditions, mechanical forces fracture the films within 5 weeks resulting in greater weight loss and mechanical reduction, though no significant changes in GPC molar mass has been noted [99]. Deroine *et al.* noted deterioration of PLA splines (4 mm thickness) in the Lorient harbour (France), loss in tensile strength but no significant change in molar mass or mechanical properties was noticed [100]. The absorption of water into the polyester is influenced by salinity of seawater as compared to distilled water. The rate of breakdown of polymers in seawater is slow than in pure water. Wang *et al.* and Agarwal *et al.* noted that degradation in saltwater took almost 1 year and this forecasts the lifetime of PLA splines (2 mm thickness).

Polyhydroxyalkanoates

PHAs get hydrolyzed quickly in seawater [101]. Research related to the breakdown of PHAs in seawater is published in 1992 [102 - 108]. The scientists discovered that the degradation rate is slower in seawater, and the mechanism for PHAs in seawater follows surface erosion, as seen in soil and compost. Rutkowska *et al.* reported that PHBV films (0.115 mm thickness) got fully deteriorated in compost within 6 weeks and 60% of their weight was lost when immersed in seawater for the same period. Volova *et al.* found that degradation of PHB and P(3HB-3HV) films by PHA-degrading microorganisms such as *Enterobacter* sp. (four strains), *Bacillus* sp., and *Gracilibacillus* sp. in the South

China Sea took 160 days. The shape and size of the material showed a substantial impact on the degradability of PHAs in seawater. Weight loss of 38 percent and 13 percent in PHB and P(3HB-co-3HV) films of 0.1mm thickness was noted after 160 days in saltwater. This suggested that degradation of PHB occurred faster than P(3HB-co-3HV) under same conditions. Furthermore, if the thickness of the film is reduced to 0.005 mm, weight loss increased to 58 percent and 54 percent, respectively. The rate of degradation is determined by the size of the sample (particularly thickness and surface area) in the surface corrosion process. Laycock *et al.* conducted research studies on the rate of biodegradation of PHA in the marine environment and used this information to estimate the lifetime of PHA products [109]. The rate of PHA breakdown in the ocean was found to be 0.04–0.09 mg day1 cm^2. The time taken for complete degradation depended on the thickness and size of the product. Time of about 25 to 2 months is required for the complete degradation of 0.035 mm thickness bags whereas time of 1.5 years is required for bottles of thickness 0.8 mm. several articles have reported breakdown of PCL in seawater [99, 110, 111].

Polycaprolactone

The degradation properties of PCL in river, lake, and seawater have been examined (Kasuya et al.) [104]. The studies demonstrated that PCL breakdown can occur in natural waterways as degrading bacteria are extensively found in various fluids. The deterioration of PCL fibres after soaking in deep seawaters for 12 months in Rausu, Toyama, and Kume, Japan has been reported by Enoki *et al.* [112]. Microbial action produced heterogeneous pinholes and fractures in the surface of PCL fibres with significant mechanical decline. Thus indicated that extensive biodegradation of the fibres occurs in the waters. Five PCL-degrading bacteria were screened and isolated from deep seawaters pumped up at three locations using a PCL granule-containing agar media. Isolates belonging to the genus *Pseudomonas* are distributed extensively in natural habitats. Multiple aliphatic polyester-degrading bacteria, *Alcanivorax* and *Tenacibaculum* have not been previously identified as aliphatic polyester-degrading bacteria. PCL films with 0.1 mm thickness degraded quickly in a laboratory when immersed in seawater from the bay [104]. The degrading qualities were found to be identical to those of PHBV in the same seawater from the bay, but reduction was noted in the ocean water. The As the thickness of the material increased, the rate of degradation decreased. The PCL films of thickness 0.05mm lost 100% of their mechanical strength and 30% of their original weight in natural saltwater after 5 weeks [98]. PCL splines of thickness 2mm when immersed in natural seawater in BoHai, China, showed a reduction of 30% in weight after 52 weeks compared to its original weight. Analysis of the size, molecular weight, and mechanical strength indicated that a surface erosion mechanism occurred and this is similar to that reported in the literature [107, 113].

Poly(butylene Adipate Terephthalate)

The PBAT class of polyesters forms the second-largest share of their market. PBAT has been developed recently and applied in the last few years. Not much research has been done on the degrading qualities of seawater. Alvarez-Zeferino *et al.* investigated the biodegradation of oxo-degradable LDPE and compostable PLA/PBAT mixes in marine environments using a respirometric lab test for 48 days with a marine inoculum [114]. The compostable plastic demonstrated a higher degree of mineralization (10%), whereas the polyolefins (2.06–2.78%) showed no change (2.06–2.78%) even with or without the presence of pro-oxidants or previous abiotic degradation. Exposure to UV radiation brought a great loss in elongation of the oxo-degradable plastic (>68%). Their findings highlight the modest biodegradation rates and also reveal a high rate of loss of physical integrity. The fragmentation was noted before significant biodegradation and development of microplastics.

The degradation of PBAT splines after 56 weeks in Tianjin BoHai Bay, China, is recently documented. Erroneous weight loss and mechanical properties due to film fragility is avoided after using a conventional spline with a thickness of 2 mm. No weight loss occurred despite the fact that the molecular weight and mechanics had fallen to roughly half of their initial levels. Microorganisms formed deteriorated pores on the surface of the spline. According to surface electron microscopy of the spline, the interior microstructure of the spline did not show many alterations [115].

Polybutylene Succinate

The degradation of PBS in seawater is not common. A study conducted in 1998 showed weight loss of 0.1 mm PBS film with 3,300 mol g^l Mn when saltwater obtained from bay and ocean showed less than 2% PBS degradation after 28 days [104]. A study conducted in Rausu, Toyama, and Kume, Japan demonstrated that PBS degrades more slowly than PCL and PHBV in deep seawaters. Besides mechanical damage, the surface of fibre of both PCL and PHBV showed the presence of biodegradation pinholes and fractures. The fibres of PBS showed a decrease in strength by less than 10% after 12 months. PBS fibre soaked in water of Rausu and Kume showed no modification, while Toyama water had rough surfaces with many spots [112].

Measurement of changes in biochemical oxygen demand (BOD) and weight loss of polyester film over time gives an idea of the degradability of polyesters in different types of waters. Polyesters poly(ethylene succinate) (PES), poly(ethylene adipate) (PEA), and poly(butylene adipate) (PBA) are not used at present [104]. PHB and its copolymers and PCL showed weight loss and a significant decrease in BOD in seawater (after 28 days), whereas PBS and poly(butylene adipate) showed biodegradation at the rate of 2% and 11% in seawater (ocean) after 28 days of incubation at 25 °C. Both PLA and PBAT were not studied.

The degradation potential of commonly used commercial polyesters such as PCL, PLA, PBS, and PBAT was studied. In BoHai Bay, China spline samples (thickness 2 mm) were immersed in natural seawater for 52 weeks. The breakdown rates for polyester noted a significant reduction. On the other hand, PCL, showed fast surface erosion and a weight loss of 32% after 52 weeks. The parameters such as surface morphology, molecular weight, and mechanical characteristics of PLA did not show any change even after 52 weeks. The spline surface of both PBS and PBAT showed extensive roughening indicating biodegradation, while sample weight and molar mass remained unaltered [116]. The degradability of biodegradable plastics in seawater showed the order PCL >> PBS > PBAT > PLA. This assessment was based on weight loss and changes in both molar mass and retention of the strength of biodegradable plastics in seawater [112, 117]. Only a rough estimate of the degradability of these materials in seawater in one or two years could be obtained through weight loss, molecular weight, or mechanical analysis. The degradation performance was supposed to be highly influenced by the environment and condition of the sample. This is because most of the materials breakdown slowly in seawater so the exact degradation rate of certain materials in seawater cannot be estimated.

Factors Influencing Seawater-Degradation of Polyesters

The reason for a decrease in the rate of biodegradation in saltwater is still to be elucidated. Ocean waves in the epipelagic zone induce mechanical that speed the change of large items to small fragments in the first stage, though strong UV radiation gets reflected from the sea surface. The pH and buffer capacity of the ocean at various ocean depths is affected by temperature, pressure, atmospheric CO_2 equilibrium, and diffusion. High ionic strength of seawater makes it similar to a weak alkaline buffer that enhances the force of degradation as compared to neutral freshwater [118]. Polyester degradation rates depend upon low temperatures and microbial diversity. These factors, though, produce an opposite effect. The materials that sink to deeper zones (temperature of about 2°C) show a low rate of hydrolysis. Low temperature and high salt in the marine environment result in increased microbial diversity (increase in number of communities), a decrease in numbers of microorganisms, and few microbial species that decompose biodegradable compounds. Since the number of active microorganisms gets reduced, the second vital stage of biotic hydrolysis gets hindered. Abiotic hydrolysis occurs at a slow rate due to low temperature. Slow degradation of PLA in seawater occurs due to a lack of efficient bacteria and hence slow abiotic hydrolysis. In marine habitats, microbes that degrade PBAT and PBS are scarce, and this results in slow degradation despite corrosion occurring at the spline's surface. Fast deterioration of PHAs and PCL in saltwater through surface erosion could be due to the presence of a large number of microorganisms capable of degrading these polyesters. The rate of breakdown of the polymers gets enhanced with an increase in a number of microorganisms [104, 119]. The deterioration of PCL splines in water supports this hypothesis [113]. Studies have indicated that only 3% of PCL weight was lost in clean water after 52 weeks. Immersion of PCL in sterilised artificial saltwater also resulted in only 3% of loss in molecular weight. Microbe-exposed water experience significant deterioration. The rate of degradation varies considerably depending on the type and number of microorganisms. A14% weight loss was noted in a static river after 52 weeks, whereas a 12% weight loss was noted in static saltwater. Immersion of PCL samples in natural saltwater for 52 weeks resulted in a weight loss of about 32%.

Microorganisms influence the performance of polyester in different ocean zones. Variation in environmental variables such as temperature, the presence of microbes in saltwater and different depths throughout the year affects the performance of material degradation performance. In the deep-water regions, bioplastic breakdown showed an initial lag phase which was due to the low quantity of microbial communities, but after that, degradation rates were equal to those at the other stations. Significant biodegradation occurred when microorganisms colonised the plastic, and this step is influenced by the presence of microbial populations. Therefore, it can be assumed that delay in degradation occurs in open saltwater when few microorganisms are present. The 28-day deterioration performance of polyester films at the seaside, ocean was compared to that of inland river and lake water [104]. The films that decomposed in lake water and river water exhibited a considerable reduction in degradation ability when entering seawater due to differences in a number of species and microorganisms. The rate of breakdown of polymers in deep sea waters was low in bay waters due to differences in the quantity and type of microorganisms. The degradation of PCL, PHAs, and PBS in deep seawaters was studied at three distinct sites (Rausu, Toyama, and Kume, Japan) by Sekiguchi and coworkers. The demonstrated that same polymer showed different degradation patterns at different locations [112]. The degradation ability of biodegradable plastics gets affected in deepwater due to the presence of less number of microbes as the density here is higher than saltwater.

Hence, a plastic bag with fast biodegradation kinetics takes more time to disintegrate on the shoreline then after entering the ocean or deep sea.

Biodegradation of Polymers in Seawater

The breakdown of polymeric material in seawater is assured *via* various techniques: i) Novel polymers with a selective saltwater breakdown profile need to be developed, or ii) Rate of deterioration of degradable polymers needs to be increased by blending or making chemical changes. Low temperatures and lack of appropriate microorganisms in seawater delay the hydrolysis process hence inhibiting the absorption by microbes.

Degradation of Polymers in Seawater without any Blending or Modification

A fast rate of hydrolysis converts high molecular polymer into small soluble oligomers and hence plays a key role in increasing polyester biodegradation in saltwater. In marine habitats, microbes are extensively dispersed for seawater-degradable polymers. The polymers based on starch or cellulose rapidly get degraded in saltwater because of the presence of microbial degraders. Mostly these polymers are used as fillers than the primary component of the material [120, 121]. Some polyesters such as PCL and PHAs get easily degraded in seawater, but at a slow rate than in soil.

The chemical routes to PHB will be explored. A perfectly stereoregular pure isotactic crystalline thermoplastic material [122], ROP of cyclic lactones (4-membered –butyrolactone [123], eight-membered cyclic diolide [124]) along bacterial poly(R)-3- hydroxybutyrate, P(R)-3HB can be conducted to obtain P3 [124]. The production of two monomeric hydrolytic products *viz*. 3HB and CA formed during the abiotic hydrolysis of poly[(R)-3HB] under acidic or basic conditions of chemical recycling has been reported by Yu and coworkers [125]. Hydrolytic products such as 3HB and CA were produced by alkaline hydrolysis but poly[(R)-3HB] was resistant to hydrolysis at mild acidic conditions. Under acidic conditions, poly[(R)-3HB] got fully decomposed to form 2% 3HB and 90% CA and 3HB was dehydrated to form CA. The loss of P3HB films was directly linked with the accumulation of monomeric products, 3HB and CA. These showed sequential degradation of P3HB polymers and oligomers. Research studies have shown that P3HB is a *via*ble seawater-degradable polymer and its breakdown product 3HB can be digested.

PGA and PLGA are polyesters that show a fast rate of abiotic hydrolysis and can be used as seawater-degradable materials. PGA is simple aliphatic polyesters produced by the ROP of glycolide (GA). It has low solubility, high melting temperature (Tm) (225–230 °C), and high mechanical strength (Young's modulus

E = 12.8 GPa) [126]. PGA hydrolyzes and loses its mass in 5 months and tensile strength in 1–2 months. Random copolymerization of GA units inside LA segments leads to destabilization of polymer's regularity and reduces the crystallinity of PLGA copolymers in pure water. Increased availability of ester linkages results in enhanced toughness and solubility of amorphous copolymer PLGA and rapid, tunable biodegradability [127, 128]. The degradation of PLGA in artificial saltwater and freshwater at 25 °C under fluorescence light (16 hours of light and 8 hours of darkness) for a year was investigated [117]. PLGA showed 100 percent breakdown in seawater (SW) after 270 days. This was measured by time-dependent weight loss and change in molar mass. The results were comparable, as noted for freshwater. The first PLGA eluted with a single-mode molar mass curve in GPC. First PLGA was eluted with a single-mode molar mass curve in GPC. Ring-opening polymerization from high-cost LA and GA intermediates leads to the development of commercial PGA and PLGA. This was done to achieve high molecular weight, narrow molecular weight distribution, high purity, and low by-product for use in medical applications. Very high prices prevent them from being used for other purposes [129]. PET, one of the most widely used synthetic polymers, is a potential seawater-degradable polymer. PETase was created chemically, A few marine bacteria and fungi that are capable of partly decomposing PET to oligomers or monomers have been characterised. Known PET hydrolases possess good turnover rates [130]. Polyesters that degrade in seawater might also act as a substitute for slowly deteriorating PET. Researchers created copolymers based on butyrolactone and transhexa hydrophthalide that showed barrier and mechanical characteristics equal to petroleum-based PET but superior to biobased PLLA [131]. These copolymers can be used in packaging applications because of their degradability or chemical recyclability, though no saltwater degradation has been reported for them.

Poly (vinyl alcohol) (PVA) and poly(ethylene glycol) (PEG) are water-soluble polymers and act as family of potential seawater-degradable polymers (PEG). Large scale production of PVA has been noted after hydrolysis (methanolysis) of polyvinyl acetate. The solubility of PVA's is determined by their degree of alcoholysis and molecular weight. PVA occurs in various grades due to varying mechanical characteristics and water solubility. It has found its use in various sectors such as food packaging, coatings, textiles, cosmetics, and paper [132 - 136]. PVA can be biodegraded by various microorganisms, including *Pseudomonas, Sphingomonas,* and *Sphingopyxis.* This can happen in both aerobic and anaerobic conditions. The variation and number of microorganisms that can degrade PVA in seawater are less than those that can degrade aliphatic polyesters [137]. Microorganisms that decompose PVA are not widely distributed. They exist in particular settings, such as wastewater released by PVA-containing textile and paper mills [132, 138, 139].

Nogi *et al.* discovered Thalassospira strains when searching for PVA-degrading microbes in saltwater [140]. *Thalassospira,* a Gram-negative halophilic bacteria, belongs to the family Rhodospirillaceae of the class Alphaproteobacteria. The bacteria degrade polycyclic aromatic hydrocarbons and are found in seas, waste oil pools, and petroleum-contaminated saltwater. Though microorganisms that can absorb PVA are not common in marine settings, possibility of PVA breakdown under these conditions cannot be ruled out.

Respirometric experiments were carried out to explore weight loss and CO_2 evolution studies for ethylene vinyl alcohol copolymers in the presence of marine sediments and marine microorganisms [141]. No considerable degradation was noted. A marine actinomycete that could survive in the presence of EVOH as a sole carbon source was discovered. PEG has to find its use in various products, including surfactants, creams, food additives, and biomedical studies [142]. The biodegradation of PEG in several aquatic environments has been investigated [143]. It was found that biodegradation of PEG in water depends on molar mass for example, PEG with a molar mass of 920 g mol^{-1} decomposed completely in one month, while PEGs with larger molar mass deteriorated slowly. A variety of polyphosphoesters, which are water-soluble polymers (PPEs), were investigated in Wurm's lab. The rate of their disintegration depended upon their chemical structure. The degradation of polyphosphonates was significantly faster (days to weeks) under basic circumstances (pH = 8, equivalent to saltwater), while polyphosphates showed a very slow degradation. The high rate of degradation in polyphosphonates was supposed to be linked to the backbiting degradation process and is likely to be caused by electron density of the central phosphorus [144]. Studies reported that polyphosphonates were used as a kinetic hydrate inhibitor polymer and degraded in saltwater (biodegradation of about 31%) [145].

The potential of water-soluble polymers to degrade in seawater has also been investigated. The degradation depends upon the molecular structure of the polymer and the presence of microorganisms that can use the polymer as a carbon source. The total breakdown of biodegradable polymers results in total mineralization or partial mineralization. The production of biogas or other organic breakdown products depends on the respiratory conditions (aerobic/ anaerobic) and the microorganisms involved [146, 147]. The environmental impact assessment of polymers is not governed by regulations such as European chemical legislation. There is a scarcity of ecotoxicological data on biodegradable polymers. Though studies related to the biodegradability of polymers in the marine environment are there, the influence of degradation products on the marine ecosystem is unclear. The biodegradable polymers are used in medicinal applications, but research needs to be conducted on human toxicity. Existing tests and biodegradability standards developed for aquatic ecosystems cannot indicate the final products and byproducts of degradation in natural seawater, and their impacts on multispecies communities and biogeochemical processes such as elemental cycling. Specific regulations and guidelines related to the ecotoxicity criteria of compostable plastics are specified.

The toxicity of breakdown products has been studied in several creatures in the lab. An increase in microbial activity (drop in pH value, increase in oxygen demand) has a significant effect on soil organisms throughout the breakdown process. Studies related to comparable effects on various marine species are required [149]. The breakdown of PBAT in aqueous media produced no harmful effects on luminescent bacteria or the crustacean *Daphnia magna* [150]. The breakdown of starch-cellulose fibre composites after sieving through 0.25mm and 0.75mm membrane and incubation in an aqueous medium (100 mg L^{-1}) for 48 days did not produced any harmful impact on luminous bacteria [151].

Research studies showed that commodity polymers produced negative impact on specific organisms. Polyethylene and poly(vinyl chloride) polymer microbeads, and poly(lactic acid) polymer microbeads affected lugworm *Arenicola marina* [152]. Polystyrene nanoparticles altered the body size and reproductive behaviour of *Daphnia magna.* They also reduced population growth in the algal species *Scenedesmus obliquus* [153]. Exposure to polystyrene micro plastics adversely affected the eating behaviour and reproductive production of copepod *Calanushelgolandicus* [154]. Research studies to determine the influence of degradation products on the marine environment are required [155, 156].

ADVANTAGES

Bioplastics provide a non-toxic, biodegradable, and recyclable alternative to traditional plastics. They reduce dependence on petroleum-based polymers and reduce plastic waste. Hence actively contribute to the reduction of pollution in the environment. It decreases carbon footprint and preserves non-renewable energy during manufacture. It does not produce any harmful effect, because it does not contain any health-harming additives.

FUTURE PROSPECT

The increase in the price of oil has increased our concern and forced us to reconsider the use of petrochemical plastics. Biopolymers as alternative materials replace synthetic plastics because of their renewable nature and biodegradability. The synthesis and processing of these polymers is a costly affair, therefore, polymers are still in the early stages of commercial development. The manufacturing costs of these plastics can be minimized if low-cost substrates, materials from recombinant microbial strains, mixed cultures and processes such as efficient fermentation and recovery purification are used. The future of bioplastics depends upon the need to meet price and performance requirements. Microbial biopolymer synthesis is a never-ending process. Homopolymers can be made using a variety of monomers, copolymers, or block copolymers in various combinations. Owing to their unique properties, biopolymers can prove to be very useful in the future and act as materials that can contribute to the sustainability of the environment.

CONCLUSION

Biodegradable polymers do not mean that they are totally "biodegradable". Humidity, temperature and microbe populations lead to variations in the rate of biodegradation. Polylactide is a well-known biodegradable plastic but is not biodegradable in water. Collaboration among scientists from other domains can help in developing novel polymers with improved biodegradation properties in natural surroundings. A better understanding of the microorganisms and their modes of biodegradation can prove useful in developing novel macromolecules that are biodegradable by microbes. Biodegradation parameters need to be considered while selecting biodegradable polymers for specific applications. The polymers need to be biodegradable. A single polymer cannot rapidly biodegrade in all types of ecosystems because biodegradation varies in other habitats, and requirements also vary. Therefore new biodegradable polymers based on risk assessment need to be developed. The term biodegradable polymers can be used for polymers that fully mineralize into CO_2, H_2O, biomass, and inorganic salts (without producing toxic byproducts) in natural conditions in the minimum possible time. These plastics ensure less environmental impact. Biodegradable polymers can successfully develop future plastics, provided that their degradability potential is high. Examples of these include biodegradable mulch films used in agriculture or drug delivery agents used in biomedicine. For a sustainable future, all commercial plastics need to be replaced with biodegradable alternatives.

REFERENCES

[1]   Scott G. Polymers and the Environment. New York: Springer 2003.

[2]   Mosthaf H. Plastics and the environment. How to draw up an ecological balance sheet. Plastverarbeiter 1990; 41: 50-2.

[3]   Scott G. Degradable polymers. Dordrecht: Springer 2011.

[4]   Pemberton D, Brothers NP, Kirkwood R. Entanglement of Australian fur seals in man-made debris in Tasmanian waters. Wildl Res 1992; 19(2): 151.
        [http://dx.doi.org/10.1071/WR9920151]

[5]   Sazima I, Gadig OBF, Namora RC, Motta FS. Plastic debris collars on juvenile carcharhinid sharks (Rhizoprionodon lalandii) in southwest Atlantic. Mar Pollut Bull 2002; 44(10): 1149-51.
        [http://dx.doi.org/10.1016/S0025-326X(02)00141-8] [PMID: 12474977]

[6] Gregory MR. Environmental implications of plastic debris in marine settings-entanglement, ingestion, smothering, hangers-on, hitch-hiking and alien invasions. Philos Trans R Soc Lond B Biol Sci 2009; 364(1526): 2013-25.
[http://dx.doi.org/10.1098/rstb.2008.0265] [PMID: 19528053]

[7] Azzarello MY, Van Vleet ES. Marine birds and plastic pollution. Mar Ecol Prog Ser 1987; 37: 295-303.
[http://dx.doi.org/10.3354/meps037295]

[8] Buggy M. Natural fibers, biopolymers, and biocomposites.Polymer Int. 2006; 55: pp. (12)1462-2.

[9] Gunatillake PA, Adhikari R. Biodegradable synthetic polymers for tissue engineering. Eur Cell Mater 2003; 5: 1-16.
[http://dx.doi.org/10.22203/eCM.v005a01] [PMID: 14562275]

[10] Babu RP, O'Connor K, Seeram R. Current progress on bio-based polymers and their future trends. Prog Biomater 2013; 2(1): 8.
[http://dx.doi.org/10.1186/2194-0517-2-8] [PMID: 29470779]

[11] Salerno A, Pascual CD. Bio-based polymers, supercritical fluids and tissue engineering. Process Biochem 2015; 50(5): 826-38.
[http://dx.doi.org/10.1016/j.procbio.2015.02.009]

[12] Kim BS, Baez CE, Atala A. Biomaterials for tissue engineering. World J Urol 2000; 18(1): 2-9.
[http://dx.doi.org/10.1007/s003450050002] [PMID: 10766037]

[13] Park JW, Im SS. Phase behavior and morphology in blends of poly(L-lactic acid) and poly(butylene succinate). J Appl Polym Sci 2002; 86(3): 647-55.
[http://dx.doi.org/10.1002/app.10923]

[14] Márquez Costa JP, Legrand V, Fréour S. Durability of Composite Materials under Severe Temperature Conditions: Influence of Moisture Content and Prediction of Thermo-Mechanical Properties During a Fire. Journal of Composites Science 2019; 3(2): 55.
[http://dx.doi.org/10.3390/jcs3020055]

[15] Acevedo-Fani A, Salvia-Trujillo L, Rojas-Graü MA, Martín-Belloso O. Edible films from essential-oil-loaded nanoemulsions: Physicochemical characterization and antimicrobial properties. Food Hydrocoll 2015; 47: 168-77.
[http://dx.doi.org/10.1016/j.foodhyd.2015.01.032]

[16] Elsabee MZ, Abdou ES. Chitosan based edible films and coatings: A review. Mater Sci Eng C 2013; 33(4): 1819-41.
[http://dx.doi.org/10.1016/j.msec.2013.01.010] [PMID: 23498203]

[17] Chiellini E, Solaro R. Biodegradable Polymeric Materials. Adv Mater 1996; 8(4): 305-13.
[http://dx.doi.org/10.1002/adma.19960080406]

[18] Vroman I, Tighzert L. Biodegradable Polymers. Materials (Basel) 2009; 2(2): 307-44.
[http://dx.doi.org/10.3390/ma2020307]

[19] Bioplastics market to touch $4.4 bn by 2026. Focus on Catalysts 2020; 2020(6): 2.
[http://dx.doi.org/10.1016/j.focat.2020.05.008]

[20] Reddy MM, Vivekanandhan S, Misra M, Bhatia SK, Mohanty AK. Biobased plastics and bionanocomposites: Current status and future opportunities. Prog Polym Sci 2013; 38(10-11): 1653-89.
[http://dx.doi.org/10.1016/j.progpolymsci.2013.05.006]

[21] Mekonnen T, Mussone P, Khalil H, Bressler D. Progress in bio-based plastics and plasticizing modifications. J Mater Chem A Mater Energy Sustain 2013; 1(43): 13379.
[http://dx.doi.org/10.1039/c3ta12555f]

[22] Polymer durability and radiation effects. Choice Rev Online 2008; 45(12): 45-6793.

[23]   Mehta R, Kumar V, Bhunia H, Upadhyay SN. Synthesis of Poly(Lactic Acid): A Review. J Macromol Sci Part C Polym Rev 2005; 45(4): 325-49.
[http://dx.doi.org/10.1080/15321790500304148]

[24]   Zhu Y, Romain C, Williams CK. Sustainable polymers from renewable resources. Nature 2016; 540(7633): 354-62.
[http://dx.doi.org/10.1038/nature21001] [PMID: 27974763]

[25]   Farah S, Anderson DG, Langer R. Physical and mechanical properties of PLA, and their functions in widespread applications - A comprehensive review. Adv Drug Deliv Rev 2016; 107: 367-92.
[http://dx.doi.org/10.1016/j.addr.2016.06.012] [PMID: 27356150]

[26]   Madhavan Nampoothiri K, Nair NR, John RP. An overview of the recent developments in polylactide (PLA) research. Bioresour Technol 2010; 101(22): 8493-501.
[http://dx.doi.org/10.1016/j.biortech.2010.05.092] [PMID: 20630747]

[27]   Jamshidian M, Tehrany EA, Imran M, Jacquot M, Desobry S. Poly-Lactic Acid: Production, Applications, Nanocomposites, and Release Studies. Compr Rev Food Sci Food Saf 2010; 9(5): 552-71.
[http://dx.doi.org/10.1111/j.1541-4337.2010.00126.x] [PMID: 33467829]

[28]   Supthanyakul R, Kaabbuathong N, Chirachanchai S. Random poly(butylene succinate-co-lactic acid) as a multi-functional additive for miscibility, toughness, and clarity of PLA/PBS blends. Polymer (Guildf) 2016; 105: 1-9.
[http://dx.doi.org/10.1016/j.polymer.2016.10.006]

[29]   Koitabashi M, Noguchi MT, Sameshima-Yamashita Y, *et al.* Degradation of biodegradable plastic mulch films in soil environment by phylloplane fungi isolated from gramineous plants. AMB Express 2012; 2(1): 40.
[http://dx.doi.org/10.1186/2191-0855-2-40] [PMID: 22856640]

[30]   Xu J, Guo BH. Poly(butylene succinate) and its copolymers: Research, development and industrialization. Biotechnol J 2010; 5(11): 1149-63.
[http://dx.doi.org/10.1002/biot.201000136] [PMID: 21058317]

[31]   Thakker C, San KY, Bennett GN. Production of succinic acid by engineered E. coli strains using soybean carbohydrates as feedstock under aerobic fermentation conditions. Bioresour Technol 2013; 130: 398-405.
[http://dx.doi.org/10.1016/j.biortech.2012.10.154] [PMID: 23313685]

[32]   Gan Z, Kuwabara K, Yamamoto M, Abe H, Doi Y. Solid-state structures and thermal properties of aliphatic–aromatic poly(butylene adipate-co-butylene terephthalate) copolyesters. Polym Degrad Stabil 2004; 83(2): 289-300.
[http://dx.doi.org/10.1016/S0141-3910(03)00274-X]

[33]   Lucas N, Bienaime C, Belloy C, Queneudec M, Silvestre F, Nava-Saucedo JE. Polymer biodegradation: Mechanisms and estimation techniques – A review. Chemosphere 2008; 73(4): 429-42.
[http://dx.doi.org/10.1016/j.chemosphere.2008.06.064] [PMID: 18723204]

[34]   Kyrikou I, Briassoulis D. Biodegradation of Agricultural Plastic Films: A Critical Review. J Polym Environ 2007; 15(2): 125-50.
[http://dx.doi.org/10.1007/s10924-007-0053-8]

[35]   Laycock B, Nikolić M, Colwell JM, *et al.* Lifetime prediction of biodegradable polymers. Prog Polym Sci 2017; 71: 144-89.
[http://dx.doi.org/10.1016/j.progpolymsci.2017.02.004]

[36]   Lyu S, Untereker D. Degradability of polymers for implantable biomedical devices. Int J Mol Sci 2009; 10(9): 4033-65.
[http://dx.doi.org/10.3390/ijms10094033] [PMID: 19865531]

[37] Andrady AL. Assessment of Environmental Biodegradation of Synthetic Polymers. J Macromol Sci Part C Polym Rev 1994; 34(1): 25-76.
[http://dx.doi.org/10.1080/15321799408009632]

[38] Shen J, Burgess DJ. Accelerated in-vitro release testing methods for extended-release parenteral dosage forms. J Pharm Pharmacol 2012; 64(7): 986-96.
[http://dx.doi.org/10.1111/j.2042-7158.2012.01482.x] [PMID: 22686344]

[39] Pierre TS, Chiellini E. Review : Biodegradability of Synthetic Polymers Used for Medical and Pharmaceutical Applications: Part 1- Principles of Hydrolysis Mechanisms. J Bioact Compat Polym 1986; 1(4): 467-97.
[http://dx.doi.org/10.1177/088391158600100405]

[40] Rittié L, Perbal B. Enzymes used in molecular biology: a useful guide. J Cell Commun Signal 2008; 2(1-2): 25-45.
[http://dx.doi.org/10.1007/s12079-008-0026-2] [PMID: 18766469]

[41] Eubeler JP, Bernhard M, Knepper TP. Environmental biodegradation of synthetic polymers II. Biodegradation of different polymer groups. Trends Analyt Chem 2010; 29(1): 84-100.
[http://dx.doi.org/10.1016/j.trac.2009.09.005]

[42] Sakai K, Kawano H, Iwami A, Nakamura M, Moriguchi M. Isolation of a thermophilic poly-l-lactide degrading bacterium from compost and its enzymatic characterization. J Biosci Bioeng 2001; 92(3): 298-300.
[http://dx.doi.org/10.1016/S1389-1723(01)80266-8] [PMID: 16233100]

[43] Tokiwa Y, Calabia BP. Biodegradability and biodegradation of poly(lactide). Appl Microbiol Biotechnol 2006; 72(2): 244-51.
[http://dx.doi.org/10.1007/s00253-006-0488-1] [PMID: 16823551]

[44] Qi X, Ren Y, Wang X. New advances in the biodegradation of Poly(lactic) acid. Int Biodeterior Biodegradation 2017; 117: 215-23.
[http://dx.doi.org/10.1016/j.ibiod.2017.01.010]

[45] Jarerat A, Tokiwa Y. Degradation of Poly(L-lactide) by a Fungus. Macromol Biosci 2001; 1(4): 136-40.
[http://dx.doi.org/10.1002/1616-5195(20010601)1:4<136::AID-MABI136>3.0.CO;2-3]

[46] Lee SH, Kim MN. Isolation of bacteria degrading poly(butylene succinate-co-butylene adipate) and their lip A gene. Int Biodeterior Biodegradation 2010; 64(3): 184-90.
[http://dx.doi.org/10.1016/j.ibiod.2010.01.002]

[47] Numata K, Hirota T, Kikkawa Y, *et al.* Enzymatic degradation processes of lamellar crystals in thin films for poly[(R)-3-hydroxybutyric acid] and its copolymers revealed by real-time atomic force microscopy. Biomacromolecules 2004; 5(6): 2186-94.
[http://dx.doi.org/10.1021/bm0497670] [PMID: 15530032]

[48] Vaclavkova T, Ruzicka J, Julinova M, Vicha R, Koutny M. Novel aspects of symbiotic (polyvinyl alcohol) biodegradation. Appl Microbiol Biotechnol 2007; 76(4): 911-7.
[http://dx.doi.org/10.1007/s00253-007-1062-1] [PMID: 17594087]

[49] Deguchi T, Kakezawa M, Nishida T. Nylon biodegradation by lignin-degrading fungi. Appl Environ Microbiol 1997; 63(1): 3 29-31.
[http://dx.doi.org/10.1128/aem.63.1.329-331.1997] [PMID: 8979361]

[50] BS PD CEN/TR 15351:2006 [Internet]. Techstreet.com. 2021 [cited 22 October 2021]. Available from: https://www.techstreet.com/standards/bs-pd-cen-tr-15351-2006?product_id=1310436

[51] Hoshino A, Isono Y. Degradation of aliphatic polyester films by commercially available lipases with special reference to rapid and complete degradation of poly(L-lactide) film by lipase PL derived from Alcaligenes sp. Biodegradation 2002; 13(2): 141-7.
[http://dx.doi.org/10.1023/A:1020450326301] [PMID: 12449316]

[52]   Hopewell J, Dvorak R, Kosior E. Plastics recycling: challenges and opportunities. Philos Trans R Soc Lond B Biol Sci 2009; 364(1526): 2115-26.
[http://dx.doi.org/10.1098/rstb.2008.0311] [PMID: 19528059]

[53]   Al-Salem SM, Lettieri P, Baeyens J. Recycling and recovery routes of plastic solid waste (PSW): A review. Waste Manag 2009; 29(10): 2625-43.
[http://dx.doi.org/10.1016/j.wasman.2009.06.004] [PMID: 19577459]

[54]   Ishigaki T, Sugano W, Nakanishi A, Tateda M, Ike M, Fujita M. The degradability of biodegradable plastics in aerobic and anaerobic waste landfill model reactors. Chemosphere 2004; 54(3): 225-33.
[http://dx.doi.org/10.1016/S0045-6535(03)00750-1] [PMID: 14575734]

[55]   Criddle CS, Billington SL, Frank CW. Renewable bioplastics and biocompositesfrom biogas methane and waste-derived feedstock: development of enabling technology, life Cycle Assessment and analysis of costs. California Dept of Resour Recycling Recovery 2014; p. 169.

[56]   Oehlmann J, Schulte-Oehlmann U, Kloas W, *et al.* A critical analysis of the biological impacts of plasticizers on wildlife. Philos Trans R Soc Lond B Biol Sci 2009; 364(1526): 2047-62.
[http://dx.doi.org/10.1098/rstb.2008.0242] [PMID: 19528055]

[57]   Teuten EL, Rowland SJ, Galloway TS, Thompson RC. Potential for plastics to transport hydrophobic contaminants. Environ Sci Technol 2007; 41(22): 7759-64.
[http://dx.doi.org/10.1021/es071737s] [PMID: 18075085]

[58]   Kunwar B, Cheng HN, Chandrashekaran SR, Sharma BK. Plastics to fuel: a review. Renew Sustain Energy Rev 2016; 54: 421-8.
[http://dx.doi.org/10.1016/j.rser.2015.10.015]

[59]   Kuppusamy S, Thavamani P, Megharaj M, Naidu R. Biodegradation of polycyclic aromatic hydrocarbons (PAHs) by novel bacterial consortia tolerant to diverse physical settings – Assessments in liquid- and slurry-phase systems. Int Biodeterior Biodegradation 2016; 108: 149-57.
[http://dx.doi.org/10.1016/j.ibiod.2015.12.013]

[60]   Andrady A. Plastics and Environmental Sustainability. New York: John Wiley & Sons 2015.
[http://dx.doi.org/10.1002/9781119009405]

[61]   Arena U. Process and technological aspects of municipal solid waste gasification. A review. Waste Manag 2012; 32(4): 625-39.
[http://dx.doi.org/10.1016/j.wasman.2011.09.025] [PMID: 22035903]

[62]   Anuar Sharuddin SD, Abnisa F, Wan Daud WMA, Aroua MK. A review on pyrolysis of plastic wastes. Energy Convers Manage 2016; 115: 308-26.
[http://dx.doi.org/10.1016/j.enconman.2016.02.037]

[63]   Bilitewski B, Oros C, Christensen T. Mechanical biological treatment 628-38.2010;

[64]   Pei L, Schmidt M, Wei W. Conversion of Biomass into Bioplastics and Their Potential Environmental Impacts. Biotechnol Biopolymers 2011; pp. 57-75.
[http://dx.doi.org/10.5772/18042]

[65]   Lackner M. Bioplastics. Kirk-Othmer Encyclopedia of Chemical Technology 2015; pp. 1-41.

[66]   Keshavarz T, Roy I. Polyhydroxyalkanoates: bioplastics with a green agenda. Curr Opin Microbiol 2010; 13(3): 321-6.
[http://dx.doi.org/10.1016/j.mib.2010.02.006] [PMID: 20227907]

[67]   Reddy CSK, Ghai R, Rashmi , Kalia VC. Polyhydroxyalkanoates: an overview. Bioresour Technol 2003; 87(2): 137-46.
[http://dx.doi.org/10.1016/S0960-8524(02)00212-2] [PMID: 12765352]

[68]   Madhavan Nampoothiri K, Nair NR, John RP. An overview of the recent developments in polylactide (PLA) research. Bioresour Technol 2010; 101(22): 8493-501.
[http://dx.doi.org/10.1016/j.biortech.2010.05.092] [PMID: 20630747]

[69] Tsuji H, Tashiro K, Bouapao L, Hanesaka M. Corrigendum to "Synchronous and separate homo-crystallization of enantiomeric poly(L-lactic acid)/poly(d-lactic acid) blends" [Polymer 53 (3) (2012) 747–754 Polymer (Guildf) 2013; 54(13): 3426. [Polymer 53 (3) (2012) 747–754
[http://dx.doi.org/10.1016/j.polymer.2013.04.033]

[70] Pang X, Zhuang X, Tang Z, Chen X. Polylactic acid (PLA): Research, development and industrialization. Biotechnol J 2010; 5(11): 1125-36.
[http://dx.doi.org/10.1002/biot.201000135] [PMID: 21058315]

[71] Garrison T, Murawski A, Quirino R. Bio-Based Polymers with Potential for Biodegradability. Polymers (Basel) 2016; 8(7): 262.
[http://dx.doi.org/10.3390/polym8070262] [PMID: 30974537]

[72] Gentile P, Chiono V, Carmagnola I, Hatton P. An overview of poly(lactic-co-glycolic) acid (PLGA)-based biomaterials for bone tissue engineering. Int J Mol Sci 2014; 15(3): 3640-59.
[http://dx.doi.org/10.3390/ijms15033640] [PMID: 24590126]

[73] Lasprilla AJR, Martinez GAR, Lunelli BH, Jardini AL, Filho RM. Poly-lactic acid synthesis for application in biomedical devices - A review. Biotechnol Adv 2012; 30(1): 321-8.
[http://dx.doi.org/10.1016/j.biotechadv.2011.06.019] [PMID: 21756992]

[74] Kimura T, Ihara N, Ishida Y, Saito Y, Shimizu N. Hydrolysis Characteristics of Biodegradable Plastic(Poly Lactic Acid). Nippon Shokuhin Kagaku Kogaku Kaishi 2002; 49(9): 598-604.
[http://dx.doi.org/10.3136/nskkk.49.598]

[75] Gerngross TU, Slater SC. How green are green plastics? Sci Am 2000; 283(2): 36-41.
[http://dx.doi.org/10.1038/scientificamerican0800-36] [PMID: 10914397]

[76] Lee SY. Bacterial polyhydroxyalkanoates. Biotechnol Bioeng 1996; 49(1): 1-14.
[http://dx.doi.org/10.1002/(SICI)1097-0290(19960105)49:1<1::AID-BIT1>3.0.CO;2-P]   [PMID: 18623547]

[77] Steinbüchel A, Füchtenbusch B. Bacterial and other biological systems for polyester production. Trends Biotechnol 1998; 16(10): 419-27.
[http://dx.doi.org/10.1016/S0167-7799(98)01194-9] [PMID: 9807839]

[78] Harding K, Dennis J, Vonblottnitz H, Harrison S. Environmental analysis of plastic production processes: Comparing petroleum-based polypropylene and polyethylene with biologically-based poly-β-hydroxybutyric acid using life cycle analysis. J Biotechnol 2007; 130(1): 57-66.
[http://dx.doi.org/10.1016/j.jbiotec.2007.02.012] [PMID: 17400318]

[79] Ojumu TV, Yu J, Solomon BO. Production of Polyhydroxyalkanoates, a bacterial biodegradable polymer. Afr J Biotechnol 2004; 3(1): 18-24.
[http://dx.doi.org/10.5897/AJB2004.000-2004]

[80] Wendlandt KD, Stottmeister U, Helm J, Soltmann B, Jechorek M, Beck M. The potential of methane-oxidizing bacteria for applications in environmental biotechnology. Eng Life Sci 2010; 10: NA.
[http://dx.doi.org/10.1002/elsc.200900093]

[81] Lee SY. Plastic bacteria? Progress and prospects for polyhydroxyalkanoate production in bacteria. Trends Biotechnol 1996; 14(11): 431-8.
[http://dx.doi.org/10.1016/0167-7799(96)10061-5]

[82] Karthikeyan OP, Chidambarampadmavathy K, Cirés S, Heimann K. Review of Sustainable Methane Mitigation and Biopolymer Production. Crit Rev Environ Sci Technol 2015; 45(15): 1579-610.
[http://dx.doi.org/10.1080/10643389.2014.966422]

[83] Pieja AJ, Rostkowski KH, Criddle CS. Distribution and selection of poly-3-hydroxybutyrate production capacity in methanotrophic proteobacteria. Microb Ecol 2011; 62(3): 564-73.
[http://dx.doi.org/10.1007/s00248-011-9873-0] [PMID: 21594594]

[84] Chanprateep S. Current trends in biodegradable polyhydroxyalkanoates. J Biosci Bioeng 2010; 110(6):

621-32.
[http://dx.doi.org/10.1016/j.jbiosc.2010.07.014] [PMID: 20719562]

[85]   Shen L, Haufe J, Patel M. Product overview and market projection of emerging biobasedplastics. Netherlands: Utrech University 2009; pp. 1-227.

[86]   Holst D, Triolo R, Neal S, Bayrami B. Bioplastics in California Economic assessment of market conditions for PHA/PHB bioplastics produced from waste methane California. Berkeley: University of California 2013; pp. 1-76.

[87]   News BBC. World Food Prices at Fresh High. Says UN 2008.

[88]   Rostkowski KH, Criddle CS, Lepech MD. Cradle-to-gate life cycle assessment for a cradle-to-cradle cycle: biogas-to-bioplastic (and back). Environ Sci Technol 2012; 46(18): 9822-9.
[http://dx.doi.org/10.1021/es204541w] [PMID: 22775327]

[89]   Listewnik HF, Wendlandt KD, Jechorek M, Mirschel G. Process Design for the Microbial Synthesis of Poly-β-hydroxybutyrate (PHB) from Natural Gas. Eng Life Sci 2007; 7(3): 278-82.
[http://dx.doi.org/10.1002/elsc.200620193]

[90]   Yang S, Matsen JB, Konopka M, *et al.* Global Molecular Analyses of Methane Metabolism in Methanotrophic Alphaproteobacterium, *Methylosinus trichosporium* OB3b. Part II. Metabolomics and 13C-Labeling Study. Front Microbiol 2013; 4: 70.
[http://dx.doi.org/10.3389/fmicb.2013.00070] [PMID: 23565113]

[91]   Fricke AH, Thum AB. TEMPERATURE RECORDING IN SHALLOW MARINE ENVIRONMENTS. Trans R Soc S Afr 1975; 41(4): 351-7.
[http://dx.doi.org/10.1080/00359197509519449]

[92]   Russell NJ. Cold adaptation of microorganisms. Philos Trans R Soc Lond B Biol Sci 1990; 326(1237): 595-611.
[http://dx.doi.org/10.1098/rstb.1990.0034] [PMID: 1969649]

[93]   Nakayama A, Yamano N, Kawasaki N. Biodegradation in seawater of aliphatic polyesters. Polym Degrad Stabil 2019; 166: 290-9.
[http://dx.doi.org/10.1016/j.polymdegradstab.2019.06.006]

[94]   Schut F, de Vries EJ, Gottschal JC, *et al.* Isolation of Typical Marine Bacteria by Dilution Culture: Growth, Maintenance, and Characteristics of Isolates under Laboratory Conditions. Appl Environ Microbiol 1993; 59(7): 2150-60.
[http://dx.doi.org/10.1128/aem.59.7.2150-2160.1993] [PMID: 16348992]

[95]   Whitman WB, Coleman DC, Wiebe WJ. Prokaryotes: The unseen majority. Proc Natl Acad Sci USA 1998; 95(12): 6578-83.
[http://dx.doi.org/10.1073/pnas.95.12.6578] [PMID: 9618454]

[96]   a) Klopfer K. Book Review: Das Mittelmeer: Fauna, Flora,Ökologie. By R. Hofrichter. Feddes Repert 2003; 114(56): 387-9.

[97]   Nazareth M, Marques MRC, Leite MCA, Castro ÍB. Commercial plastics claiming biodegradable status: Is this also accurate for marine environments? J Hazard Mater 2019; 366: 714-22.
[http://dx.doi.org/10.1016/j.jhazmat.2018.12.052] [PMID: 30583241]

[98]   Tsuji H, Suzuyoshi K. Environmental degradation of biodegradable polyesters 1. Poly(ε-caprolactone), poly[(R)-3-hydroxybutyrate], and poly(L-lactide) films in controlled static seawater. Polym Degrad Stabil 2002; 75(2): 347-55.
[http://dx.doi.org/10.1016/S0141-3910(01)00240-3]

[99]   Tsuji H, Suzuyoshi K. Environmental degradation of biodegradable polyesters 2. Poly(ε-caprolactone), poly[(R)-3-hydroxybutyrate], and poly(L-lactide) films in natural dynamic seawater. Polym Degrad Stabil 2002; 75(2): 357-65.
[http://dx.doi.org/10.1016/S0141-3910(01)00239-7]

[100] Deroiné M, Le Duigou A, Corre YM, *et al.* Accelerated ageing of polylactide in aqueous environments: Comparative study between distilled water and seawater. Polym Degrad Stabil 2014; 108: 319-29.
[http://dx.doi.org/10.1016/j.polymdegradstab.2014.01.020]

[101] Gabirondo E, Sangroniz A, Etxeberria A, Torres-Giner S, Sardon H. Poly(hydroxy acids) derived from the self-condensation of hydroxy acids: from polymerization to end-of-life options. Polym Chem 2020; 11(30): 4861-74.
[http://dx.doi.org/10.1039/D0PY00088D]

[102] Doi Y, Kanesawa Y, Tanahashi N, Kumagai Y. Biodegradation of microbial polyesters in the marine environment. Polym Degrad Stabil 1992; 36(2): 173-7.
[http://dx.doi.org/10.1016/0141-3910(92)90154-W]

[103] Mergaert J, Wouters A, Swings J, Anderson C. In situ biodegradation of poly(3-hydroxybutyrate) and poly(3-hydroxybutyrate- *co* -3-hydroxyvalerate) in natural waters. Can J Microbiol 1995; 41(13) (Suppl. 1): 154-9.
[http://dx.doi.org/10.1139/m95-182] [PMID: 7606659]

[104] Kasuya K, Takagi K, Ishiwatari S, Yoshida Y, Doi Y. Biodegradabilities of various aliphatic polyesters in natural waters. Polym Degrad Stabil 1998; 59(1-3): 327-32.
[http://dx.doi.org/10.1016/S0141-3910(97)00155-9]

[105] Thellen C, Coyne M, Froio D, Auerbach M, Wirsen C, Ratto JA. A Processing, Characterization and Marine Biodegradation Study of Melt-Extruded Polyhydroxyalkanoate (PHA) Films. J Polym Environ 2008; 16(1): 1-11.
[http://dx.doi.org/10.1007/s10924-008-0079-6]

[106] Deroiné M, Le Duigou A, Corre YM, *et al.* Seawater accelerated ageing of poly(3-hydroxybutyrat--co-3-hydroxyvalerate). Polym Degrad Stabil 2014; 105: 237-47.
[http://dx.doi.org/10.1016/j.polymdegradstab.2014.04.026]

[107] Mergaert J, Anderson C, Wouters A, Swings J, Kersters K. Biodegradation of polyhydroxyalkanoates. FEMS Microbiol Lett 1992; 103(2-4): 317-21.
[http://dx.doi.org/10.1111/j.1574-6968.1992.tb05853.x] [PMID: 1476776]

[108] Imam SH, Gordon SH, Shogren RL, Tosteson TR, Govind NS, Greene RV. Degradation of starch-poly(β-hydroxybutyrate-co-β-hydroxyvalerate) bioplastic in tropical coastal waters. Appl Environ Microbiol 1999; 65(2): 431-7.
[http://dx.doi.org/10.1128/AEM.65.2.431-437.1999] [PMID: 9925564]

[109] Dilkes-Hoffman LS, Lant PA, Laycock B, Pratt S. The rate of biodegradation of PHA bioplastics in the marine environment: A meta-study. Mar Pollut Bull 2019; 142: 15-24.
[http://dx.doi.org/10.1016/j.marpolbul.2019.03.020] [PMID: 31232288]

[110] Rutkowska M, Jastrzębska M, Janik H. Biodegradation of polycaprolactone in sea water. React Funct Polym 1998; 38(1): 27-30.
[http://dx.doi.org/10.1016/S1381-5148(98)00029-7]

[111] Rutkowska M, Krasowska K, Heimowska A, Steinka I. Effect of Modification of Poly(ε-Caprolactone) on its Biodegradation in Natural Environments. International Polymer Science and Technology 2002; 29(11): 77-84.
[http://dx.doi.org/10.1177/0307174X0202901116]

[112] Sekiguchi T, Saika A, Nomura K, *et al.* Biodegradation of aliphatic polyesters soaked in deep seawaters and isolation of poly(ε-caprolactone)-degrading bacteria. Polym Degrad Stabil 2011; 96(7): 1397-403.
[http://dx.doi.org/10.1016/j.polymdegradstab.2011.03.004]

[113] Lu B, Wang GX, Huang D, *et al.* Comparison of PCL degradation in different aquatic environments: Effects of bacteria and inorganic salts. Polym Degrad Stabil 2018; 150: 133-9.

[http://dx.doi.org/10.1016/j.polymdegradstab.2018.02.002]

[114]   Alvarez-Zeferino JC, Beltrán-Villavicencio M, Vázquez-Morillas A. Degradation of Plastics in Seawater in Laboratory. Open Journal of Polymer Chemistry 2015; 5(4): 55-62.
[http://dx.doi.org/10.4236/ojpchem.2015.54007]

[115]   Wang G, Huang D, Ji J, Völker C, Wurm F. Seawater-Degradable Polymers: Seawater-Degradable Polymers-Fighting the Marine Plastic Pollution (Adv. Sci. 1/2021). Adv Sci 2021; 8(1):2170004

[116]   Wang GX, Huang D, Ji JH, Völker C, Wurm FR. Seawater-Degradable Polymers-Fighting the Marine Plastic Pollution. Adv Sci (Weinh) 2021; 8(1): 2001121.
[http://dx.doi.org/10.1002/advs.202001121] [PMID: 33437568]

[117]   Bagheri AR, Laforsch C, Greiner A, Agarwal S. Fate of So-Called Biodegradable Polymers in Seawater and Freshwater. Glob Chall 2017; 1(4): 1700048.
[http://dx.doi.org/10.1002/gch2.201700048] [PMID: 31565274]

[118]   Rodriguez EJ, Marcos B, Huneault MA. Hydrolysis of polylactide in aqueous media. J Appl Polym Sci 2016; 133(44): 1-11.
[http://dx.doi.org/10.1002/app.44152]

[119]   Urbanek AK, Rymowicz W, Mirończuk AM. Degradation of plastics and plastic-degrading bacteria in cold marine habitats. Appl Microbiol Biotechnol 2018; 102(18): 7669-78.
[http://dx.doi.org/10.1007/s00253-018-9195-y] [PMID: 29992436]

[120]   Grande PM, Domínguez de María P. Enzymatic hydrolysis of microcrystalline cellulose in concentrated seawater. Bioresour Technol 2012; 104: 799-802.
[http://dx.doi.org/10.1016/j.biortech.2011.10.071] [PMID: 22101072]

[121]   Janik H, Sienkiewicz M, Przybytek A, Guzman A, Kucinska-Lipka J, Kosakowska A. Novel Biodegradable Potato Starch-based Compositions as Candidates in Packaging Industry, Safe for Marine Environment. Fibers Polym 2018; 19(6): 1166-74.
[http://dx.doi.org/10.1007/s12221-018-7872-1]

[122]   Muhammadi S, Shabina , Afzal M, Hameed S. Bacterial polyhydroxyalkanoates-eco-friendly next generation plastic: Production, biocompatibility, biodegradation, physical properties and applications. Green Chem Lett Rev 2015; 8(3-4): 56-77.
[http://dx.doi.org/10.1080/17518253.2015.1109715]

[123]   Jedliński Z, Kurcok P, Lenz RW. First Facile Synthesis of Biomimetic Poly-(*R*)-3-hydroxybutyrate *via* Regioselective Anionic Polymerization of (*S*)-β-Butyrolactone. Macromolecules 1998; 31(19): 6718-20.
[http://dx.doi.org/10.1021/ma980663p]

[124]   Tang X, Chen E. Chemical synthesis of perfectly isotactic and high melting bacterial poly(3-hydroxybutyrate) from bio-sourced racemic cyclic diolide. Nature Comm 9(1)2018;

[125]   Yu J, Plackett D, Chen LXL. Kinetics and mechanism of the monomeric products from abiotic hydrolysis of poly[(R)-3-hydroxybutyrate] under acidic and alkaline conditions. Polym Degrad Stabil 2005; 89(2): 289-99.
[http://dx.doi.org/10.1016/j.polymdegradstab.2004.12.026]

[126]   Miller RA, Brady JM, Cutright DE. Degradation rates of oral resorbable implants (polylactates and polyglycolates): Rate modification with changes in PLA/PGA copolymer ratios. J Biomed Mater Res 1977; 11(5): 711-9.
[http://dx.doi.org/10.1002/jbm.820110507] [PMID: 893490]

[127]   Brannigan RP, Dove AP. Synthesis, properties and biomedical applications of hydrolytically degradable materials based on aliphatic polyesters and polycarbonates. Biomater Sci 2017; 5(1): 9-21.
[http://dx.doi.org/10.1039/C6BM00584E] [PMID: 27840864]

[128]   Blomqvist J, Mannfors B, Pietilä LO. Amorphous cell studies of polyglycolic, poly(l-lactic), poly(l,d-lactic) and poly(glycolic/l-lactic) acids. Polymer (Guildf) 2002; 43(17): 4571-83.

[http://dx.doi.org/10.1016/S0032-3861(02)00312-9]

[129] Amass W, Amass A, Tighe B. A review of biodegradable polymers: uses, current developments in the synthesis and characterization of biodegradable polyesters, blends of biodegradable polymers and recent advances in biodegradation studies. Polym Int 1998; 47(2): 89-144.
[http://dx.doi.org/10.1002/(SICI)1097-0126(1998100)47:2<89::AID-PI86>3.0.CO;2-F]

[130] Danso D, Schmeisser C, Chow J, Zimmermann W, Wei R, Leggewie C, *et al. et al.* New Insights into the Function and Global Distribution of Polyethylene Terephthalate (PET)-Degrading Bacteria and Enzymes in Marine and Terrestrial Metagenomes. Environ Microbiol 84(8)2018;

[131] Sangroniz A, Zhu J, Tang X, Etxeberria A, Chen E, Sardon H. Packaging materials with desired mechanical and barrier properties and full chemical recyclability 10(1)2019;

[132] Chiellini E, Corti A, D'Antone S, Solaro R. Biodegradation of poly (vinyl alcohol) based materials. Prog Polym Sci 2003; 28(6): 963-1014.
[http://dx.doi.org/10.1016/S0079-6700(02)00149-1]

[133] Julinová M, Vaňharová L, Jurča M. Water-soluble polymeric xenobiotics – Polyvinyl alcohol and polyvinylpyrrolidon – And potential solutions to environmental issues: A brief review. J Environ Manage 2018; 228: 213-22.
[http://dx.doi.org/10.1016/j.jenvman.2018.09.010] [PMID: 30223180]

[134] Guo M, Trzcinski AP, Stuckey DC, Murphy RJ. Anaerobic digestion of starch–polyvinyl alcohol biopolymer packaging: Biodegradability and environmental impact assessment. Bioresour Technol 2011; 102(24): 11137-46.
[http://dx.doi.org/10.1016/j.biortech.2011.09.061] [PMID: 22001054]

[135] Baker MI, Walsh SP, Schwartz Z, Boyan BD. A review of polyvinyl alcohol and its uses in cartilage and orthopedic applications. J Biomed Mater Res B Appl Biomater 2012; 100B(5): 1451-7.
[http://dx.doi.org/10.1002/jbm.b.32694] [PMID: 22514196]

[136] Guo M, Murphy RJ. Is There a Generic Environmental Advantage for Starch–PVOH Biopolymers Over Petrochemical Polymers? J Polym Environ 2012; 20(4): 976-90.
[http://dx.doi.org/10.1007/s10924-012-0489-3]

[137] Fredi G, Dorigato A. Recycling of bioplastic waste: A review. Advanced Industrial and Engineering Polymer Research 2021; 4(3): 159-77.
[http://dx.doi.org/10.1016/j.aiepr.2021.06.006]

[138] Marušincová H, Husárová L, Růžička J, *et al.* Polyvinyl alcohol biodegradation under denitrifying conditions. Int Biodeterior Biodegradation 2013; 84: 21-8.
[http://dx.doi.org/10.1016/j.ibiod.2013.05.023]

[139] Corti A, Solaro R, Chiellini E. Biodegradation of poly(vinyl alcohol) in selected mixed microbial culture and relevant culture filtrate. Polym Degrad Stabil 2002; 75(3): 447-58.
[http://dx.doi.org/10.1016/S0141-3910(01)00247-6]

[140] Nogi Y, Yoshizumi M, Miyazaki M. *Thalassospira povalilytica* sp. nov., a polyvinyl-alcoho--degrading marine bacterium. Int J Syst Evol Microbiol 2014; 64(Pt_4): 1149-53.
[http://dx.doi.org/10.1099/ijs.0.058321-0] [PMID: 24408523]

[141] Mayer J, Kaplan D, Stote R, Dixon K, Shupe A, Allen A, *et al.* Biodegradation of Polymer Films in Marine and Soil Environments. ACS Symposium Series. 59-170.
[http://dx.doi.org/10.1021/bk-1996-0627.ch013]

[142] Herzberger J, Niederer K, Pohlit H, *et al.* Polymerization of Ethylene Oxide, Propylene Oxide, and Other Alkylene Oxides: Synthesis, Novel Polymer Architectures, and Bioconjugation. Chem Rev 2016; 116(4): 2170-243.
[http://dx.doi.org/10.1021/acs.chemrev.5b00441] [PMID: 26713458]

[143] Bernhard M, Eubeler JP, Zok S, Knepper TP. Aerobic biodegradation of polyethylene glycols of different molecular weights in wastewater and seawater. Water Res 2008; 42(19): 4791-801.

[http://dx.doi.org/10.1016/j.watres.2008.08.028] [PMID: 18823927]

[144] Bauer KN, Liu L, Wagner M, Andrienko D, Wurm FR. Mechanistic study on the hydrolytic degradation of polyphosphates. Eur Polym J 2018; 108: 286-94.
[http://dx.doi.org/10.1016/j.eurpolymj.2018.08.058]

[145] Lin H, Wolf T, Wurm FR, Kelland MA. Poly(alkyl ethylene phosphonate)s: A New Class of Non-amide Kinetic Hydrate Inhibitor Polymers. Energy Fuels 2017; 31(4): 3843-8.
[http://dx.doi.org/10.1021/acs.energyfuels.7b00019]

[146] Shah AA, Hasan F, Hameed A, Ahmed S. Biological degradation of plastics: A comprehensive review. Biotechnol Adv 2008; 26(3): 246-65.
[http://dx.doi.org/10.1016/j.biotechadv.2007.12.005] [PMID: 18337047]

[147] Mohee R, Unmar GD, Mudhoo A, Khadoo P. Biodegradability of biodegradable/degradable plastic materials under aerobic and anaerobic conditions. Waste Manag 2008; 28(9): 1624-9.
[http://dx.doi.org/10.1016/j.wasman.2007.07.003] [PMID: 17826972]

[148] https://www.din.de/de/mitwirken/normenausschuesse/navp/wdc-beuth:din21:101923058 [Internet] Din.de. 2020 [cited 28 September 2020]. Available from: .

[149] Fritz J, Sandhofer M, Stacher C, Braun R. Strategies for detecting ecotoxicological effects of biodegradable polymers in agricultural applications. Macromol Symp 2003; 197(1): 397-410.
[http://dx.doi.org/10.1002/masy.200350734]

[150] Witt U, Einig T, Yamamoto M, Kleeberg I, Deckwer WD, Müller RJ. Biodegradation of aliphatic–aromatic copolyesters: evaluation of the final biodegradability and ecotoxicological impact of degradation intermediates. Chemosphere 2001; 44(2): 289-99.
[http://dx.doi.org/10.1016/S0045-6535(00)00162-4] [PMID: 11444312]

[151] Rudnik E, Milanov N, Matuschek G, Kettrup A. Ecotoxicity of biocomposites based on renewable feedstock – Preliminary studies. Chemosphere 2007; 70(2): 337-40.
[http://dx.doi.org/10.1016/j.chemosphere.2007.06.026] [PMID: 17669461]

[152] Green DS, Boots B, Sigwart J, Jiang S, Rocha C. Effects of conventional and biodegradable microplastics on a marine ecosystem engineer (Arenicola marina) and sediment nutrient cycling. Environ Pollut 2016; 208(Pt B): 426-34.
[http://dx.doi.org/10.1016/j.envpol.2015.10.010] [PMID: 26552519]

[153] Besseling E, Wang B, Lürling M, Koelmans AA. Nanoplastic affects growth of S. obliquus and reproduction of D. magna. Environ Sci Technol 2014; 48(20): 12336-43.
[http://dx.doi.org/10.1021/es503001d] [PMID: 25268330]

[154] Cole M, Lindeque P, Fileman E, Halsband C, Galloway TS. The impact of polystyrene microplastics on feeding, function and fecundity in the marine copepod *Calanus helgolandicus*. Environ Sci Technol 2015; 49(2): 1130-7.
[http://dx.doi.org/10.1021/es504525u] [PMID: 25563688]

[155] Volova TG, Boyandin AN, Vasiliev AD, *et al.* Biodegradation of polyhydroxyalkanoates (PHAs) in tropical coastal waters and identification of PHA-degrading bacteria. Polym Degrad Stabil 2010; 95(12): 2350-9.
[http://dx.doi.org/10.1016/j.polymdegradstab.2010.08.023]

[156] Rutkowska M, Krasowska K, Heimowska A, *et al.* Environmental Degradation of Blends of Atactic Poly[(R,S)-3-hydroxybutyrate] with Natural PHBV in Baltic Sea Water and Compost with Activated Sludge. J Polym Environ 2008; 16(3): 183-91.
[http://dx.doi.org/10.1007/s10924-008-0100-0]

CHAPTER 12

Role of Alternate Fuels (Bioethanol and Biodiesel) in Preventing Environmental Degradation

Bhupinder Dhir[1,*]

[1] *School of Sciences, Indira Gandhi National Open University, New Delhi, India*

Abstract: The diminishing quantity of fossil fuels and environmental degradation lead to the search for renewable and environmentally friendly fuels that can substitute petroleum. The burning of petroleum products releases gases that pollute the environment, hence need for alternate fuels was realized. Biofuels such as biodiesel and bioethanol derived from food crops, biomass, algae, vegetable oil, animal fats, or lignocellulosic materials are renewable, biodegradable and non-toxic. They possess low quantities of sulfur, polycyclic aromatic hydrocarbons, and metals and are considered eco-friendly. Biotechnological methods have been adapted to increase the production of crop plants that are used in the production of biofuels. Genes encoding for enzymes that degrade lignin, an important component of food crops,have also been inserted in food crops so that processing can be made easier for getting increased production of biofuels.

Keywords: Biodiesel, Bioethanol, Biofuels, Biotechnology, Eco-friendly, Lignin.

INTRODUCTION

Depleting reserves of fossil fuels due to the rise in population has generated interest in the search for alternate renewable sources of fuels. Liquid or gaseous fuel is derived from renewable feedstocks, and biomass is referred to as biofuel. Biofuels mainly include bioethanol, biodiesel, alkanes, and various other hydrocarbon mixtures. These fuels are environmentally friendly and are expected to reduce dependency on fossil fuels such as petroleum. Biofuels reduce the effect of global warming as they do not produce greenhouse gas emissions hence protecting the environment. Hence biofuels have emerged as a safe, clean, eco-friendly, sustainable solution to energy. Biofuels offer sustainable alternatives to petroleum [1].

Biodiesel and bioethanol are the primary biofuels [2, 3]. Biodiesel acts as an alternate fuel to diesel while bioethanol can be used instead of petrol. Corn

* **Corresponding author Bhupinder Dhir:** School of Sciences, Indira Gandhi National Open University, New Delhi, India; E-mail: bhupdhir@gmail.com

ethanol and biodiesel are referred to as first-generation biofuels. They are largely made from food crops such as cereals, sugar crops, and oil seeds [4]. To overcome limitations such as environmental and social concerns, "next-generation," *i.e.*, second- and third-generation biofuels are being developed from non-edible lignocellulosic materials [5]. Woody biomass, wood wastes, crop residues, municipal wastes, and energy crops such as switchgrass and algae are some of the lignocellulosic feedstocks. Agricultural, forestry residues, municipal solid wastes, industrial wastes, and terrestrial and aquatic crops grown solely for energy purposes form the major resources for biomass.

Biodiesel

Biodiesel is a fuel obtained from vegetable oils and animal fats. Some of the oilseed crops used in biodiesel production include rape, sunflower, palm and soybean [6, 7]. Algae and cyanobacteria as also used as a source of biodiesel. Feedstocks from waste animal fats, rapeseed oil, sunflower oil and palm oil are also used in biodiesel production [8]. Biodiesel (B100) is composed of mono-alkyl esters of long-chain fatty acids. Transesterification of oils or fats obtained from plants/animals or alcohols such as methanol, ethanol derived from vegetable oils or animal fats produces biodiesel. Fuel made up of 100% esters of fatty acids is called pure biodiesel (B100). Esters of fatty acids when mixed with diesel in the ratio of 20% form B20 (*i.e.*, 20% B100 and 80% diesel). Similarly, B5 (5% B100) and B2 (2% B 100) can also be formed.

Biodiesel is produced *via* the transesterification of oils along with alcohols (such as methanol or ethanol) or by the esterification of fatty acids. A chemical process called transesterification helps in the production of biodiesel. A catalyst such as an alkali or acid helps in transesterification, and glycerol is formed as a byproduct. The process of transesterification converts fats and oils, *i.e.*, triglycerides, to their corresponding fatty acid methyl esters (FAME) or methyl esters in a very short time. The process of transesterification is used for the production of biodiesel on a commercial scale [9].

Vegetable oils are the main feedstocks used for biodiesel production in the United States of America. Feedstocks such as raw vegetable oils, used cooking oils, yellow grease and animal fats are also used for transesterification. About 2.6 billion gallons of biodiesel were produced by the US in 2018. Oilseed rape is used as the main source of biodiesel in UK. Biofuels make up to 5% of petroleum blends in UK. The United States produced about 1.7 billion gallons of B100 in 2019. Europe is also one of the biggest producers of biodiesel. Fleet vehicles in the European Union and the United States currently use E-diesels (blends of ethanol in diesel) as fuel. In E-diesel, ethanol is immiscible in diesel over a wide range of temperatures, becoming its major drawback.

Biodiesel is a renewable, safe, biodegradable and nontoxic fuel. It leads to less pollution in comparison to conventional petroleum or diesel. The use of biodiesel reduces the emission of gases, such as CO_2, carbon monoxide, nitrogen oxides (NO_x), particulate matter and other chemicals such as hydrocarbons, aromatic hydrocarbons, alkenes, aldehydes and ketones. Hence proves very useful in environmental protection. It also reduces the emission of SO_2 because it is low in sulfur content. A reduction in the release of exhaust emissions by about 45% has been noted after the use of a 20% blend of biodiesel, *i.e.*, B20. A significant reduction in greenhouse gases (~ 86 percent), hydrocarbon emissions (~67 percent), particulate matter (~47 percent) and smog has been noted after the use of biodiesel [10]. Biodiesel can be used in any engine without modification. It acts as an alternative to fuel additives such as methyl tertiary butyl ether (MTBE). Advantages such as good economic potential, low emissions and high efficiency make biodiesel a good option for the replacement of fossil fuels.

Bioethanol

Bioethanol is an environmentally friendly alternative petrol additive/substitute. It is produced from starch, feedstocks based on sugar crops (such as corn, and sugar cane) or cellulosic feedstocks (wood, straw, and even household wastes). It is produced from sucrose or simple sugars *via* the process of alcoholic fermentation [11]. Starch-rich plant material gets converted to sugars and fermented to produce bioethanol. In another method, cellulose from the cell walls of stems or leaves gets converted into sugars, and fermented to produce bioethanol (Fig **1**) [9]. At present, bioethanol is being produced by the fermentation of sugar by microbes. Microbes use sugar as a source of food and convert it to ethanol [12, 13]. Hence bioethanol is derived mainly from biological feedstocks that contain a good amount of sugars that easily get fermented. The plant material is treated with sulphuric acid and heated to convert cellulose into sugars, or cellulose is treated with enzymes. Lignocellulosic materials are subjected to biochemical and thermochemical treatment that breakdown biomass into cellulose and other polymers, which undergo hydrolysis and fermentation [14, 15].

A pretreatment step of chemical or enzymatic hydrolysis is done to the lignocellulosic feedstock to remove the lignin present within it [16]. In the method of chemical hydrolysis, acid is used to break sugar molecules of the feedstock, whereas, in enzymatic hydrolysis, enzymes help in the breakdown of lignocelluloses of cellulose. The process of chemical pretreatment is fast, and conversion can be done on a large scale. Hydrolysis results in the pollution of land, and the traces of acid present in ethanol can make the fuel corrosive [17].

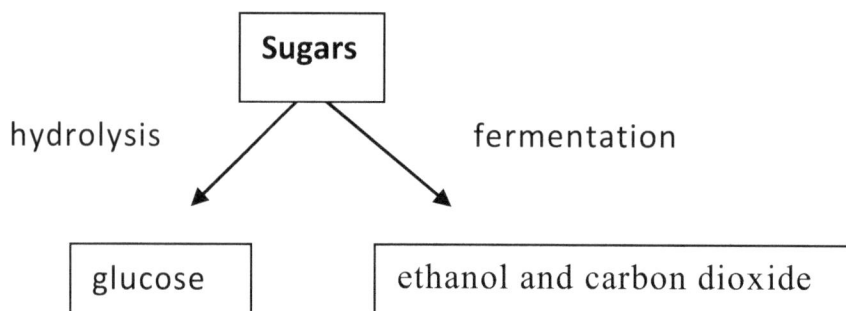

Fig. (1). Breakdown of sugars *via* different processes.

Enzymatic hydrolysis is carried out either by using soluble or immobilized enzymes. The soluble enzymes are conventionally used in hydrolysis. Recently, in many industries, immobilization has been used where an enzyme can be reused, thereby decreasing the cost.

Bioethanol forms about 99% of all biofuels produced in the United States. Brazil, France, the USA, the UK, Argentina and South Africa are some of the countries that use bioethanol as a fuel. Brazil and the United States are leading countries that produce about 90% of world's ethanol using sugarcane (*Saccharum* L.) and corn (*Zea mays* L.) as the primary feedstock.

BIOTECHNOLOGICAL ADVANCES IN BIOFUEL PRODUCTION

Biofuels are mainly produced from crop plants. Biodiesel is produced mainly from seed crops such as soybean and palm oil. Bioethanol is a fuel produced from food-based crops such as corn. Studies have predicted that the production of food crops needs to be increased to almost double to get the biofuels required in the next 50 years to fulfill the demand for fuel. An increase in crop biomass production for more fuel production can be achieved by the technique of genetic engineering [21].

Biofuels are produced mainly from lignocellulosic raw materials that contain cellulose and hemicellulose polymers composed of long chains of sugar monomers bound together by lignin. The plant cell wall is a complex structure made up of polysaccharides (mainly cellulose and hemicellulose) which are cross-linked with lignin, phenolic acids (ferulic and p-coumaric acid), pectins and structural proteins [18, 19]. Plant feedstocks act as a good source of biomass that can be broken down for the conversion to biofuels. Lignin is a complex polysaccharide having carbon-carbon (C-C) cross-linkages or ether linkages. The extensive cross-linking of cell walls by lignin restricts the production of biofuels from biomass. Due to its rigid structure and high crystallinity, cellulose is resistant to chemical or enzymatic hydrolysis. Hence it becomes very difficult to break it into simple sugars [20].The enzymatic hydrolysis of cellulose is a slow process and is influenced by the structural properties of biomass substrate, such as crystallinity, surface area, degree of polymerization and porosity [21 - 23]. The complex lignocellulosic structure is recalcitrant to biodegradation and needs to be disrupted for the production of biofuel from cellulosic materials [24 - 27]. The modification of biomass attributes can improve the efficiency and efficacy of the enzymatic hydrolysis process so that cellulose of biomass can have easy access to cellulase enzymes. Cellulolytic enzymes produced by various fungal and bacterial microorganisms are required for the enzymatic hydrolysis of cellulose [20]. Cellulolytic enzymes act on cellulose chains and their synergistic action randomly cleave bonds of the cellulose chain [27]. Easy degradation of lignocellulose can enhance bioethanol production. Dissolution of hemicellulose leads to the loosening of the structure of raw material, hence acid pretreatment is required for the production of reducing sugars. This biochemical conversion of biomass to ethanol requires pretreatment that reduces the recalcitrance of biomass [20].

Chemical catalysts are also used in the production of biofuels, such as ethanol. Transesterification using lipase has emerged as an alternative for biodiesel fuel production because glycerol produced as a by-product can easily be recovered, and the purification of fatty methyl esters is easy and simple [28]. The cost of lipase is the main limitation of this process [29]. The cost of lipase production can be lowered when recombinant gene technology is used. Microorganisms with high expression levels of lipases and stability towards methanol can be developed. Efforts have been made to increase the activity of enzymes and microbes used in biofuel production in the past decade. Enzymatic processing of biofuels is a sustainable option for easy processing and achieving significant savings.

The cell wall structure and/or lignocellulosic composition of the plant can be improved using biotechnology techniques [30, 31]. The changes in the cell wall composition of plants and the increased fermentation capacity of microbes contribute to increased bioethanol production. This can improve yield of cellulosic ethanol. Hydrolytic enzymes are not able to break the plant cell wall and this limits the conversion of plant biomass to biofuel. Exposure to hydrolytic enzymes and hydrolysis of cell wall polysaccharides is required for production of biofuel from plant biomass. Transgenic approaches have proven useful in developing plants with tolerance and reduction in lignin content without reduction in biomass [31 - 33]. It is suggested that reduction in lignin content could reduce the requirement for pretreatment and need of biomass for biofuel production. The reduction in lignin content could be due to alteration in composition of lignin composition. A significant reduction in lignin content is likely to have negative effect on plant growth and development [34] and any alteration in composition of lignin can improve properties of biomass without harmful pleiotropic effects. Plants that can tolerate reduction in lignin content without severe reduction in biomass can be developed *via* genetic engineering technique [16].

An increase in glucose levels in corn grain can lead to high levels of ethanol production. Corn varieties with better tolerance for insects or herbicides can be used in biofuel production. The genes for pest resistance, herbicide resistance, and glyphosate tolerance can be introduced in soybean and corn plants [30]. Soybean varieties with glyphosate tolerance need to be developed for use in biodiesel production. In plants, glyphosate strongly binds to enzyme 5-enolpyruvylshikimate–3- phosphate (EPSP), hence blocking its activity. In contrast, in a mutated version of the enzyme, a serine residue substitutes the proline and the enzyme becomes insensitive to glyphosate. The introduction of mutated EPSP made the survival of tobacco plants easy on exposure to glyphosate. In addition, genes for abiotic stress tolerance (such as drought and salinity) also need to be introduced in such crops.

Pretreatment of biomass effectively hydrolyzes hemicellulose or removes lignin fraction, which increases the accessibility of cellulose to enzymatic hydrolysis [35]. Pretreatment is done using acids such as dilute sulfuric acid in a concentration range of 0.22–6% w/w at temperatures of 100–200 °C [36]. Sodium hydroxide [37] and ammonia can also be used for this. Ionic liquids have shown the capacity to completely solubilize biomass and hence proved themselves useful in the pretreatment of biomass [38]. The chemicals used in pre-treatment show certain limitations. Acids are corrosive, bases are expensive and require the use of expensive equipment metallurgy and also require neutralization steps. Ionic liquids are much more expensive and can prove toxic at low concentrations. All of these factors make the process environmentally unfriendly and economically unattractive. Therefore, environmentally safe low-cost pretreatment technologies have been studied. It is essential to understand the chemical, biochemical and biological nature of the structural components of plant cell walls. Thus biotechnology approaches can prove useful in modifying lignin content and composition in plants so that steps such as severe pretreatment to improve biomass conversion to ethanol can be avoided.

The approaches, such as a change in lignin composition without reduction in lignin content, can improve biomass properties so that biofuel production can be increased. The hydrolysis and fermentation of starch are involved in the process. Another source of fuel is the cellulosic material of plants that is broken down to produce sugar. Hydrolysis of cellulose is comparatively a difficult process as acid, steam or AFEX (Ammonia Fiber Explosion) pretreatment is required to destabilize lignin and hemicellulose and make cellulose available to cellulase enzymes. Hydroxycinnamic acids, ferulic and sinapic acids, the by-products of lignin biosynthesis, are produced by the oxidation of coniferaldehyde and sinapaldehyde *via* aldehyde dehydrogenase in *Arabidopsis* [39]. Coniferaldehyde dehydrogenase converts coniferaldehyde to ferulic acid. This has been noted in plants, including grasses such as maize [39].

There is also a strong interest in increasing the cellulose to hemicellulose concentration in biomass to improve ethanol yields. It has been suggested that increasing cellulose content in plants, especially grasses, may also improve plant standability [40]. Over-expression of cellulose synthase genes in maize showed improved cellulose synthesis. Alternatively, a reduction in hemicellulose content may improve the ratio of cellulose to hemicellulose content.

Cellulase enzymes from *Thermomonospora fusca* have also been successfully expressed in plants. Genes E2 and E3 encode for an endoglucanase and exoglucanase respectively [41 - 43]. Plants such as tobacco, alfalfa, and potato have been transformed using these. Besides cellulases, other cell wall hydrolyzing enzymes have also been expressed in plants. Endoxylanases hydrolyze xylan, the primary hemicellulose found in woody plants and monocot cell walls. This results in the release of pentose sugar xylose. In an alternate strategy, biomass can be harvested with grain, separated, and processed using the plant-produced enzyme that acts like microbial cellulases that hydrolyze biomass. The gene for amylase isolated from a thermophilic bacterium was cloned. The enzyme remained stable and active at high temperatures (70-80°C) and was commonly used for starch hydrolysis. The incorporation of cell wall hydrolyzing enzymes into plant biomass increased the chances of biomass conversion into ethanol or other alcohol fuels at a much faster rate [44].

Recombinant technology enhances biofuel yield and reduces environmental damage from the feedstock. Efforts can be made to develop biodiesel from nonedible oils, like *Jatropha, Pongamia, Argemone*, castor, sal, *etc.* The development of transgenic plants with enhanced biofuel conversion is expected to be the most rapid and efficient solution for the production of fuels from lignocellulosic biomass [31].

Researchers are also conducted using various processes that aim to improve the quality and economical *via*bility of biofuel. Micro algae [45] and *Jatropha* [46] have been explored as potential sources of oil for biodiesel production. Efforts are also made to use the hydrothermal reaction method for the treatment of organic wastes in bioethanol and biogas production [47]. Diesterol (mixture of fossil diesel fuel, biodiesel and bioethanol) has emerged as a new environment-friendly fuel.

Biotechnology aims to speed up the selection of varieties that are well suited to biofuel production, *i.e.*, they possess properties such as increased biomass per hectare, increased oil content (biodiesel crops), fermentable sugars (ethanol crops) and /or improved processing characteristics that facilitate their conversion to biofuels. Studies have also revealed the genome sequences of several first-generation feedstocks including maize, sorghum and soybean. Apart from this, marker-assisted selection and genetic modifications has also been applied.

The yeast *Saccharomyces cerevisiae* cannot directly ferment starch. The biomass needs to be first be broken down (hydrolysed) into fermentable sugars by enzymes called amylases. These enzymes are produced using genetically modified micro-organisms. Studies related to the development of efficient yeast strains that can produce amylases themselves so that hydrolysis and fermentation can be done. *Agrobacterium tumefaciens*has been used to raise transgenic plants produced *via* the transformation of a range of crop plants, including maize, rice, switchgrass, sorghum and wheat [48 - 51].

CONCLUSION

Biotechnology improves the production of better biomass feedstocks and converts the biomass to biofuels for getting high bioenergy production. Enzymes such as cellulases help in enhancing the process of biofuel production. Studies related to the production of second-generation bioethanol from lignocellulosic materials such as cereal straw and corn stover have been conducted. Biomass feedstocks have been developed *via* genetic engineering by altering plant cell walls *via* manipulating the cell wall polymer biosynthesis pathway. Research studies related to the development of cell walls with reduced recalcitrance to increase the production of fuel crops have been conducted. Genetic modification and development of transgenic prove efficient in producing second-biofuel plants with desirable characteristics. The production of such eco-friendly fuels can prove very useful in curtailing the problem of air pollution, reducing greenhouse effect, global warming and helping to maintain a sustainable environment.

REFERENCES

[1]   Dovì VG, Friedler F, Huisingh D, Klemeš JJ. Cleaner energy for sustainable future. J Clean Prod 2009; 17(10): 889-95.
       [http://dx.doi.org/10.1016/j.jclepro.2009.02.001]

[2]   Fukuda H, Kondo A, Noda H. Biodiesel fuel production by transesterification of oils. J Biosci Bioeng 2001; 92(5): 405-16.

[http://dx.doi.org/10.1016/S1389-1723(01)80288-7] [PMID: 16233120]

[3]   Demirbas A. The Importance of Bioethanol and Biodiesel from Biomass 177-85.2008;

[4]   Kumar A, Sharma S. An evaluation of multipurpose oil seed crop for industrial uses (*Jatropha curcas* L.): A review. Ind Crops Prod 2008; 28(1): 1-10.
[http://dx.doi.org/10.1016/j.indcrop.2008.01.001]

[5]   Sims REH, Mabee W, Saddler JN, Taylor M. An overview of second generation biofuel technologies. Bioresour Technol 2010; 101(6): 1570-80.
[http://dx.doi.org/10.1016/j.biortech.2009.11.046] [PMID: 19963372]

[6]   Kinney AJ, Clemente TE. Modifying soybean oil for enhanced performance in biodiesel blends. Fuel Process Technol 2005; 86(10): 1137-47.
[http://dx.doi.org/10.1016/j.fuproc.2004.11.008]

[7]   Kumar Tiwari A, Kumar A, Raheman H. Biodiesel production from jatropha oil (*Jatropha curcas*) with high free fatty acids: An optimized process. Biomass Bioenergy 2007; 31(8): 569-75.
[http://dx.doi.org/10.1016/j.biombioe.2007.03.003]

[8]   Pinto AC, Guarieiro LLN, Rezende MJC, *et al.* Biodiesel: an overview. J Braz Chem Soc 2005; 16(6b): 1313-30.
[http://dx.doi.org/10.1590/S0103-50532005000800003]

[9]   Balat M, Balat H, Öz C. Progress in bioethanol processing. Pror Energy Combust Sci 2008; 34(5): 551-73.
[http://dx.doi.org/10.1016/j.pecs.2007.11.001]

[10]   Hill J, Nelson E, Tilman D, Polasky S, Tiffany D. Environmental, economic, and energetic costs and benefits of biodiesel and ethanol biofuels. Proc Natl Acad Sci USA 2006; 103(30): 11206-10.
[http://dx.doi.org/10.1073/pnas.0604600103] [PMID: 16837571]

[11]   Chapotin SM, Wolt JD. Genetically modified crops for the bioeconomy: meeting public and regulatory expectations. Bioethanol. Nat Biotechnol 2006; 24: 725.

[12]   Zhao J, Xia L. Ethanol production from corn stover hemicellulosic hydrolysate using immobilized recombinant yeast cells. Biochem Eng J 2010; 49(1): 28-32.
[http://dx.doi.org/10.1016/j.bej.2009.11.007]

[13]   Das S, Bhattacharya A, Haldar S, *et al.* Optimization of enzymatic saccharification of water hyacinth biomass for bio-ethanol: Comparison between artificial neural network and response surface methodology. Sustainable Materials and Technologies 2015; 3: 17-28.
[http://dx.doi.org/10.1016/j.susmat.2015.01.001]

[14]   Valentine J, Clifton-Brown J, Hastings A, Robson P, Allison G, Smith P. Food vs. fuel: the use of land for lignocellulosic 'next generation' energy crops that minimize competition with primary food production. Glob Change Biol Bioenergy 2012; 4(1): 1-19.
[http://dx.doi.org/10.1111/j.1757-1707.2011.01111.x]

[15]   Bayrakci AG, Koçar G. Second-generation bioethanol production from water hyacinth and duckweed in Izmir: A case study. Renew Sustain Energy Rev 2014; 30: 306-16.
[http://dx.doi.org/10.1016/j.rser.2013.10.011]

[16]   Sánchez ÓJ, Cardona CA. Trends in biotechnological production of fuel ethanol from different feedstocks. Bioresour Technol 2008; 99(13): 5270-95.
[http://dx.doi.org/10.1016/j.biortech.2007.11.013] [PMID: 18158236]

[17]   Jegannathan KR, Chan ES, Ravindra P. Harnessing biofuels: A global Renaissance in energy production? Renew Sustain Energy Rev 2009; 13(8): 2163-8.
[http://dx.doi.org/10.1016/j.rser.2009.01.012]

[18]   Grabber JH, Ralph J, Lapierre C, Barrière Y. Genetic and molecular basis of grass cell-wall degradability. I. Lignin–cell wall matrix interactions. C R Biol 2004; 327(5): 455-65.

[http://dx.doi.org/10.1016/j.crvi.2004.02.009] [PMID: 15255476]

[19]   Houston K, Tucker MR, Chowdhury J, Shirley N, Little A. The Plant Cell Wall: A Complex and Dynamic Structure As Revealed by the Responses of Genes under Stress Conditions. Front Plant Sci 2016; 7: 984.
[http://dx.doi.org/10.3389/fpls.2016.00984] [PMID: 27559336]

[20]   Himmel ME, Ding SY, Johnson DK, *et al.* Biomass recalcitrance: engineering plants and enzymes for biofuels production. Science 2007; 315(5813): 804-7.
[http://dx.doi.org/10.1126/science.1137016] [PMID: 17289988]

[21]   Raman Jega K, Chan ES, Ravindra P. Biotechnology in Biofuels-A Cleaner Technology. J Appl Sci (Faisalabad) 2011; 11(13): 2421-5.
[http://dx.doi.org/10.3923/jas.2011.2421.2425]

[22]   Casler MD, Buxton DR, Vogel KP. Genetic modification of lignin concentration affects fitness of perennial herbaceous plants. Theor Appl Genet 2002; 104(1): 127-31.
[http://dx.doi.org/10.1007/s001220200015] [PMID: 12579437]

[23]   Chen H, Qiu W. Key technologies for bioethanol production from lignocellulose. Biotechnol Adv 2010; 28(5): 556-62.
[http://dx.doi.org/10.1016/j.biotechadv.2010.05.005] [PMID: 20546879]

[24]   Chen F, Dixon RA. Lignin modification improves fermentable sugar yields for biofuel production. Nat Biotechnol 2007; 25(7): 759-61.
[http://dx.doi.org/10.1038/nbt1316] [PMID: 17572667]

[25]   Yoshida M, Liu Y, Uchida S, *et al.* Effects of cellulose crystallinity, hemicellulose, and lignin on the enzymatic hydrolysis of *Miscanthus sinensis* to monosaccharides. Biosci Biotechnol Biochem 2008; 72(3): 805-10.
[http://dx.doi.org/10.1271/bbb.70689] [PMID: 18323635]

[26]   Hall M, Bansal P, Lee JH, Realff MJ, Bommarius AS. Cellulose crystallinity - a key predictor of the enzymatic hydrolysis rate. FEBS J 2010; 277(6): 1571-82.
[http://dx.doi.org/10.1111/j.1742-4658.2010.07585.x] [PMID: 20148968]

[27]   Horn SJ, Vaaje-Kolstad G, Westereng B, Eijsink V. Novel enzymes for the degradation of cellulose. Biotechnol Biofuels 2012; 5(1): 45.
[http://dx.doi.org/10.1186/1754-6834-5-45] [PMID: 22747961]

[28]   Shah S, Sharma S, Gupta MN. Biodiesel preparation by lipase-catalyzed transesterification of jatropha oil. Energy Fuels 2004; 18(1): 154-9.
[http://dx.doi.org/10.1021/ef030075z]

[29]   Tamalampudi S, Talukder MR, Hama S, Numata T, Kondo A, Fukuda H. Enzymatic production of biodiesel from *Jatropha* oil: A comparative study of immobilized-whole cell and commercial lipases as a biocatalyst. Biochem Eng J 2008; 39(1): 185-9.
[http://dx.doi.org/10.1016/j.bej.2007.09.002]

[30]   Fernandez CJ, Caswel MF. The First Decade of Genetically Engineered Crops in the United States. USDA-ERS Economic Information Bulletin 2006.SSRN https://ssrn.com/abstract=899582

[31]   Gressel J. Transgenics are imperative for biofuel crops. Plant Sci 2008; 174(3): 246-63.
[http://dx.doi.org/10.1016/j.plantsci.2007.11.009]

[32]   Pichon M, Deswartes C, Gerentes D, *et al.* Variation in lignin and cell wall digestibility in caffeic acid O-methyltransferase down-regulated maize half-sib progenies in field experiments. Mol Breed 2006; 18(3): 253-61.
[http://dx.doi.org/10.1007/s11032-006-9033-2]

[33]   Lee D, Chen A, Nair R. Genetically engineered crops for biofuel production: regulatory perspectives. Biotechnol Genet Eng Rev 2008; 25(1): 331-62.
[http://dx.doi.org/10.5661/bger-25-331] [PMID: 21412361]

[34]   Franke R, McMichael CM, Meyer K, Shirley AM, Cusumano JC, Chapple C. Modified lignin in tobacco and poplar plants over-expressing the *Arabidopsis* gene encoding ferulate 5-hydroxylase. Plant J 2000; 22(3): 223-34.
[http://dx.doi.org/10.1046/j.1365-313x.2000.00727.x] [PMID: 10849340]

[35]   Anterola AM, Lewis NG. Trends in lignin modification: a comprehensive analysis of the effects of genetic manipulations/mutations on lignification and vascular integrity. Phytochemistry 2002; 61(3): 221-94.
[http://dx.doi.org/10.1016/S0031-9422(02)00211-X] [PMID: 12359514]

[36]   Hendriks ATWM, Zeeman G. Pretreatments to enhance the digestibility of lignocellulosic biomass. Bioresour Technol 2009; 100(1): 10-8.
[http://dx.doi.org/10.1016/j.biortech.2008.05.027] [PMID: 18599291]

[37]   N'Diaye S, Rigal L. Pulping of fibre sorghum in a twin screw extruder (extrudeur bi-vis) J. Soc. Ouest-Afr Chim 2009; (27): 55-65.

[38]   Shill K, Padmanabhan S, Xin Q, Prausnitz JM, Clark DS, Blanch HW. Ionic liquid pretreatment of cellulosic biomass: Enzymatic hydrolysis and ionic liquid recycle. Biotechnol Bioeng 2011; 108(3): 511-20.
[http://dx.doi.org/10.1002/bit.23014] [PMID: 21246505]

[39]   Nair DM, Purdue PE, Lazarow PB. Pex7p translocates in and out of peroxisomes in *Saccharomyces cerevisiae*. J Cell Biol 2004; 167(4): 599-604.
[http://dx.doi.org/10.1083/jcb.200407119] [PMID: 15545321]

[40]   Dhugga KS. Maize Biomass Yield and Composition for Biofuels. Crop Sci 2007; 47(6): 2211-27.
[http://dx.doi.org/10.2135/cropsci2007.05.0299]

[41]   Dai Z, Hooker BS, Anderson DB, Thomas SR. Improved plant-based production of E1 endoglucanase using potato: expression optimization and tissue targeting. Mol Breed 2000; 6(3): 277-85.
[http://dx.doi.org/10.1023/A:1009653011948]

[42]   Dai Z, Hooker BS, Quesenberry RD, Thomas SR. Optimization of *Acidothermus cellulolyticus* endoglucanase (E1) production in transgenic tobacco plants by transcriptional, post-transcription and post-translational modification. Transgenic Res 2005; 14(5): 627-43.
[http://dx.doi.org/10.1007/s11248-005-5695-5] [PMID: 16245154]

[43]   Oraby H, Venkatesh B, Dale B, *et al.* Enhanced conversion of plant biomass into glucose using transgenic rice-produced endoglucanase for cellulosic ethanol. Transgenic Res 2007; 16(6): 739-49.
[http://dx.doi.org/10.1007/s11248-006-9064-9] [PMID: 17237981]

[44]   Sticklen M. Plant genetic engineering to improve biomass characteristics for biofuels. Curr Opin Biotechnol 2006; 17(3): 315-9.
[http://dx.doi.org/10.1016/j.copbio.2006.05.003] [PMID: 16701991]

[45]   Chisti Y. Biodiesel from microalgae. Biotechnol Adv 2007; 25(3): 294-306.
[http://dx.doi.org/10.1016/j.biotechadv.2007.02.001] [PMID: 17350212]

[46]   Machmudah S, Shotipruk A, Goto M, Sasaki M, Hirose T. Extraction of astaxanthin from *Haematococcus pluvia*lis using supercritical CO_2 and ethanol as entrainer. Ind Eng Chem Res 2006; 45(10): 3652-7.
[http://dx.doi.org/10.1021/ie051357k]

[47]   He Y, Pang Y, Liu Y, Li X, Wang K. Physicochemical characterization of rice straw pretreated with sodium hydroxide in the solid state for enhancing biogas production. Energy Fuels 2008; 22(4): 2775-81.
[http://dx.doi.org/10.1021/ef8000967]

[48]   Somleva MN, Tomaszewski Z, Conger BV. Agrobacterium mediated genetic transformation of switchgrass. Crop Sci 2002; 42(6): 2080-7.
[http://dx.doi.org/10.2135/cropsci2002.2080]

[49]   Zhao Z, Gu W, Cai T, *et al.* High throughput genetic transformation mediated by *Agrobacterium tumefaciens* in maize. Mol Breed 2002; 8(4): 323-33.
[http://dx.doi.org/10.1023/A:1015243600325]

[50]   Mittal A, Decker SR. Special issue: Application of biotechnology for biofuels: transforming biomass to biofuels. 3 Biotech 2013; 3(5): 341-3.
[http://dx.doi.org/10.1007/s13205-013-0122-8] [PMID: 28324334]

[51]   Hood EE, Love R, Lane J, *et al.* Subcellular targeting is a key condition for high-level accumulation of cellulase protein in transgenic maize seed. Plant Biotechnol J 2007; 5(6): 709-19.
[http://dx.doi.org/10.1111/j.1467-7652.2007.00275.x] [PMID: 17614952]

CHAPTER 13

Remediation of Heavy Metals Using Biochar and its Modified Forms

Akanksha Bhardwaj[1], **Puneeta Pandey**[1] and **Jayaraman Nagendra Babu**[2,*]

[1] *Department of Environmental Science and Technology, Central University of Punjab, VPO-Ghudda, Punjab 151401, India*

[2] *Department of Chemistry, Central University of Punjab, VPO- Ghudda, Punjab 151401, India*

Abstract: Heavy metal contamination has affected various life forms on earth due to their toxic, carcinogenic and bio-assimilative nature. Heavy metals are rapidly transported by various water bodies in our environment. Thus, the remediation of heavy metals in water bodies is essential for sustaining our ecosystems. The treatment technologies available for treating the heavy metals undergoing dynamic biochemical transformations in the environment are a challenge as well as an opportunity for developing alternate cost-effective technologies. Adsorption has emerged as an environment-friendly and cost-effective technology. Biochar, a sustainable and low-cost adsorbent, has shown encouraging results for the remediation of these environmental contaminants. It stands out as a promising adsorbent due to chelating functional moieties apart from high surface area and porosity. These physicochemical attributes of biochar can be modulated using various physicochemical treatments to achieve higher heavy metal removal efficiencies. Biochar is a carbon-neutral material, which can be regenerated and disposed-off easily in an adsorption-based remediation process. This chapter brings out the modifications characteristic of biochar, a comparative statement of properties *vis-a-vis* biochar and their use in the adsorption of heavy metals, and various mechanisms accounting for their removal.

Keywords: Adsorption, Biochar-modifications, Heavy metals, Remediation.

INTRODUCTION

Heavy metal contamination has now spread to almost every part of the ecosphere, affecting various biotic and abiotic components. The main sources of these contaminants include anthropogenic sources, such as industry, mining, pesticide application, metal processing and natural sources. Due to their widespread use, heavy metals can easily enter aquatic systems, soil and atmosphere, thereby impacting human health. These pollutants are toxic, non-biodegradable and have a

---

[*] **Corresponding author Jayaraman Nagendra Babu:** Department of Chemistry, Central University of Punjab, VPO-Ghudda, Punjab 151401, India; E-mail:nagendra.babu@cup.edu.in

Bhupinder Dhir (Ed.)

tendency to bioaccumulate in nature, thus are placed in the category of extremely dangerous contaminants. Various technologies are available for the environmental treatment/remediation of heavy metals. Some of these include adsorption, chemical precipitation, membrane techniques, ion exchange, reverse osmosis and electrochemical methods [1, 2]. However, these methods have many drawbacks and restrictions, such as high cost, production of secondary pollutants, sludge production and disposal problems, skilled labour requirements, and excessive use of chemicals. Of these processes, adsorption offers advantages over high operational costs and environmental risks posed by other technologies. It is highly effective, easy to operate and an efficient process. Flexibility in design makes it one of the most used processes for heavy metal removal. By definition, adsorption is a physicochemical process used for the removal of organic and inorganic pollutants from water and wastewater with the help of a substance called 'adsorbent'. Various adsorbents like zeolites, clay, activated carbon and biomass have been used to remove contaminants. Activated carbon is frequently used for adsorption. It is considered an ideal adsorbent, but its high cost restricts its usage for large-scale applications. So, the focus has shifted to low-cost and renewable materials derived from biomass. Biochar is one such material that has gained worldwide attention among researchers. It is an effective and cheap adsorbent. Properties like well-developed pore structure, abundant functional groups and large surface area make it a favorable material for the adsorption of water-based contaminants [3, 4].

BIOCHAR

The concept of 'biochar' is supposed to have come into existence around 2500 years back with the discovery of Terra Preta in the Amazon basin. These are dark-coloured, carbon-rich, and highly fertile soils of anthropogenic origin found at a depth of two meters. It is believed that civilizations that inhabited the Amazon basin used to smolder the biomass in deep earthen pits instead of burning it, which gave rise to Terra Preta soils. This technique eventually led the way to the production of biochar.

Research on these soils highlighted the benefits of biochar on soil systems [5], and gradually, it emerged as an environment management tool [6]. In modern science, biochar is produced by the thermal degradation of biomass under oxygen-limited conditions with temperatures ranging from 300-700°C and is a highly aromatic, carbon-rich and stable product [7].

Biochar can be produced from a variety of lignocellulosic materials like straw, leaves, wood chips, grass, roots and kitchen and garden waste using various biomass conversion technologies like pyrolysis, hydrothermal carbonization, torrefaction and gasification. Among these methods, pyrolysis is a relatively simpler and most common technique which yields solid product (char), condensable vapours (bio-oil), and non-condensable gaseous fraction. The process involves heating biomass in the absence of oxygen. The pyrolysis process is mainly of two types- fast and slow pyrolysis.

Fast pyrolysis operates at higher temperatures (~500°C) and high heating rates (100-1000°C/s) followed by fast cooling to allow rapid quenching of the vapours formed during the process so as to minimize secondary reactions that can form additional products [8]. The main product of fast pyrolysis is a viscous liquid called bio-oil which consists of organics and water (20-25 wt %). However, biochar production from fast pyrolysis is not preferred as it leads to a larger fraction of biomass converted to bio-oil resulting in poor yield of biochar.

Slow pyrolysis works at low temperatures (400-700°C), longer residence time (>1hr) and heating rates between 5-7 °C/min. Around half of the total carbon present in the feedstock is retained in biochar produced *via* slow pyrolysis [9]. The process also yields bio-oil and gases like H_2, CO_2, CO, and CH_4. Slow pyrolysis can be carried out in reactors like kilns, retorts and converters [10]. Reactors based on heat pipe technology have been used recently for pyrolysis [11, 12]. Constant temperature at any point in the heat pipe makes it better technology than other existing technologies. Depending upon the different pyrolysis conditions like heating rate, residence time, and temperature, the final desired product can be obtained.

The quality and yield of biochar depend upon feedstock type and pyrolysis temperature. Researchers emphasized the influence of both these parameters on biochar, particularly for their application in adsorption studies [13, 14]. Lignocellulosic biomass is mainly composed of lignin, cellulose and hemicellulose. A higher yield of biochar is obtained from biomass with higher lignin content, like wood, as compared to herbaceous biomass. On the other hand, the surface area of wood biochars is higher (62 to 240 m^2/g) and contains lesser functional groups compared to crop residue biochars (0.17 to 68.5 m^2/g) when pyrolyzed at the same pyrolysis temperature [15]. Among crop residues, straw-derived biochars have high ash content due to the high inorganic content in feedstock. It is further observed that the inorganic content of the biochar, particularly alkali and alkaline metals, catalyzes biochar formation [16]. The moisture content of biomass also has an effect on its production, as biomass with higher moisture content yields a higher proportion of biochar under high-pressure conditions [17]. Another property that varies depending on feedstock is pH, which is higher for crop residue-derived biochars than wood biochars because of the higher concentration of alkali and alkaline metals in their precursors [15].

Similarly, pyrolysis temperature also has a profound effect on biochar properties. Within the range of pyrolysis conditions, an increase in temperature leads to lesser biochar yields, and higher fixed carbon content, alkalinity and ash content, which is attributed to the volatilization of organic components and enrichment with ash (Fig. **1**) [15, 18]. Besides, pyrolysis temperature also affects pore size and surface area.

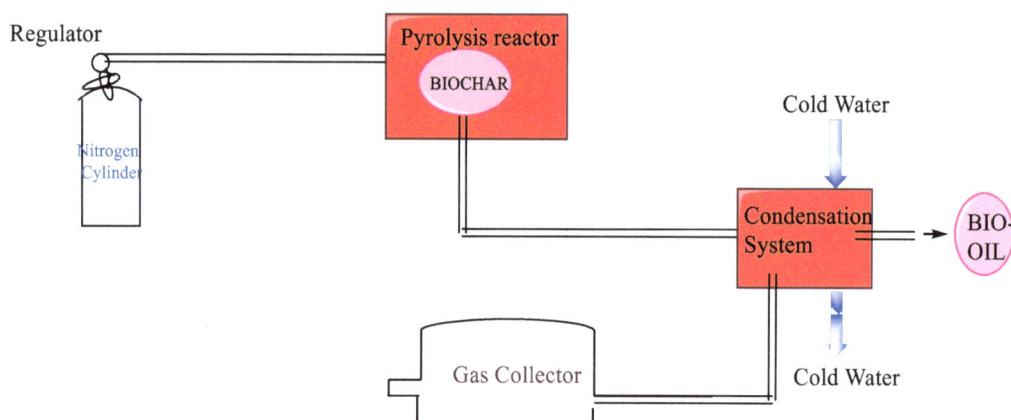

Fig. (1). Pyrolysis Scheme.

Properties of Biochar

Heavy metal sorption capacity of biochar is governed by the various physico-chemical properties of biochar, which are in turn dependent upon pyrolysis conditions and feedstock type, as is already mentioned in the previous section. This section explains the various physicochemical characteristics of biochar, produced under different conditions from various feedstocks, and their significance in the adsorption-based remediation of heavy metals.

Physical and Structural Properties

Porosity of Biochar

When biomass is thermally treated under a certain inert condition, a carbon microstructure having significant porosity is formed. This porous structure resembles a cellular structure of parent material [19]. The appearance of pores in wheat straw biochar is depicted as aligned honeycomb-like groups that resemble the carbon skeleton from the biological capillary structure of biomass [20]. When pores are formed, they get partially blocked by the tars formed during the thermal decomposition of biomass. However, an increase in pyrolysis temperature leads to the volatilization of these compounds and pores formation being accelerated, resulting in higher surface areas [21]. The development of pores with an increase in pyrolysis temperature from 200 to 800°C was also observed in biochar produced from tree waste [22]. Also, as the pyrolysis temperature increases, the widening of pores due to the destruction of walls between adjacent pores was observed upon increasing the pyrolysis temperatures increases [23]. This causes a reduction in micropore volume and an increase in the total pore volume. Enhanced pore developments in biochar were observed when pyrolysis temperature increased from 250°C to 500 °C because of the increased evolution of volatiles, which was most pronounced in a temperature range of 250-400°C due to volatilization of cellulose and lignin. Micropore volume increased from 0.078 cm^3/g to 0.222 cm^3/g with the corresponding increase in total pore volume from 0.193 cm^3/g to 0.466 cm^3/g when the temperature increased from 250°C to 500°C. On the other hand, when the pyrolysis temperature further increased from 500°C to 800°C, there was a reduction in the total pore volume. This is attributed to the deposition of melt in the pores of biochar formed as a result of the decomposition and softening of volatiles [24]. Similar results were obtained for the sewage sludge biochar pyrolyzed between 400-950°C. The pore volume increased drastically from 400 to 600°C while the increase between 600 to 950°C was nearly negligible [25]. Pore size also affects heavy metal adsorption as particles with small pore sizes cannot trap large-sized adsorbates irrespective of their charges or polar nature [26].

Surface Area

The porosity of biochar contributes to its total surface area. The pore formation increases with an increase in pyrolysis temperature on biochar's surface, and hence the total surface area increases [6]. Pores in the range of less than 2 nm (micropores) contribute the greatest to the total surface area. Jindo *et al.* [27] found that with an increase in temperature, the surface areas of biochars increased in biochar derived from different lignocellulosic material. However, the higher ash content in rice residue biochars blocked access to micropores leading to relatively low surface areas at high temperatures. The higher surface areas thus obtained as a function of pyrolysis temperatures in various studies can be attributed to the release of volatiles from biochar which thus enhances the micropore volume. Soybean stover and peanut shell biochar prepared at 700° C were characterized by the surface area of 420 m^2/g and 448 m^2/g compared to biochars prepared at 300°C where the surface area was 6 m^2/g and 3 m^2/g, respectively [18]. This variation was attributed to the loss of O and H-containing functional groups associated with esters, phenols, carbonyl and alkyl chains at a higher temperature which increased the surface area. Feedstock also directly affects pore formation and surface area, *e.g.,* at a given pyrolysis temperature, the surface area of wood and grass-derived biochars was 347 m^2g^{-1} and 139 m^2g^{-1}, respectively [28]. Another variable that affects the surface area is the particle size of the biochar. The surface area was found to decrease by 54% for biochar obtained from 2 mm particle-sized biomass compared to 0.5 mm-sized biomass [29]. The large surface area might facilitate higher adsorption of metal ions, particularly if the surface organic moiety can adsorb positively charged metals. The surface of biochar is negatively charged under acidic conditions due to the pyrolysis-liberated carboxylic acids and is thus suitable for accommodating positively charged metals by electrostatic forces [26].

Zeta Potential

Zeta potential is the charge potential developed in the sliding plane of the colloidal particles, the numerical values and signs of which are related to the surface charge of biochar particles. It develops as a result of the protonation and deprotonation of functional groups, which thus create a net charge on the surfaces of the particles forming an electrical double-layer in the solution phase near the surfaces [30, 31]. Studies have been carried out to measure zeta potential as a function of pH [30, 32], where the negative values were reported throughout the pH range studied. The values of zeta potential become more negative with an increase in pH because of increased neutralization of oxygen-containing acidic functional groups. For *e.g.,* the zeta potential values ranged from -17.9mV (pH=3.5) to -47 mV (at pH=6) [33] for peanut straw biochar prepared at 400°C. The observed surface charge on the biochar is either due to organic functional moieties like -COO$^-$ and -OH or mineral components present in it. The low zeta potentials under acidic pH could be attributed to the interaction between the functional groups (-COO$^-$ and -O$^-$) with H$^+$ which caused the lowering of the negative charge[30]. On the other hand, pyrolysis temperature negatively affects the zeta potential. High temperature chars generally have lower zeta potential. This could be attributed to lesser density of surface functional group in high temperature chars [30].

However, the values of the zeta potential of biochars changed from negative to positive upon *in situ* and *ex-situ* modification with aluminium [34]. The change in values of zeta potential was greater in magnitude in the case of *ex-situ* modification with Al as compared to *in situ* modification. The reduction in negative charge on biochars' surface could be either due to 1) Formation of complexes between Al and functional groups, or 2) Physical coverage of functional moieties by aluminum oxides [34]. On the other hand, the zeta potential values became more negative under the pH range of 3 to 5.5 on the treatment of rice straw biochar with KOH compared to raw biochar [35].

Chemical Properties

The chemical composition of biochar is strongly influenced by the feedstock's elemental concentration [36]. Biochar contains inorganic and organic components in various redox states. It contains elements like C, H, O, K, Na, Ca, Mg, N, P, K, S and Si. Among these elements percentage of C is the highest, followed by H and O. During pyrolysis, these elements are released as gases and vapours, which change the elemental composition of organics, *i.e.*, C, H, N, O and S, hence leading to varying O/C and H/C ratios. These ratios are directly linked to aromaticity, the polarity of biochar and biodegradability [37, 38]. High-temperature biochars have lower O/C and H/C ratios compared to those produced at lower temperatures because, with an increase in temperature, C content increases while H and O decrease [39]. A higher O/C ratio also suggests the presence of extra functional groups on the surface, which in turn signifies higher IEC (Ion Exchange Capacity) values for the biochar [30, 40]. Van Krevelen diagrams have been used to represent the changes in the elemental ratios of biochar as a function of temperature [28]. Fig (**2**) shows the Van Krevelen plot of elemental ratios for wood and grass chars, where the line in the graph signifies the dehydration reactions (Fig. **2**). This elemental variation as a function of temperature leads to a change in atomic ratios. Inorganic components exist mainly as ash in biochar. Doumer *et al.* [41] reported high ash content in water hyacinth biochar compared to other biochars prepared at the same temperature owing to high Ca, Mg and K content. The mineral components like K, Ca and Mg can undergo precipitation or exchange with heavy metals, reducing their availability in solution and aiding in the adsorption process. The amount of minerals concentrated in biochar increases with pyrolysis temperature. For example, increased pyrolysis temperature enhanced inorganic groups in rice straw biochar while the organic moieties declined [42].

pH

pH of the biochar varies with feedstock and the pyrolysis temperature. While the pyrolysis temperature is positively correlated with pH, the dependency upon the feedstock varies with the inorganic content [43]. The increase in pH with temperature is accounted for an increase in inorganic like carbonates, phosphates, and ash, reduction in acidic groups like carboxyl and phenolics with heating and formation of basic surface oxides [4, 30, 43]. This increase in pH with heating is consistent with the results of Boehm titration. For *e.g.,* Al-Webel *et al.* [22] showed an increase in the content of basic groups from 0.15 to 3.55 mmol/g and a decrease in acidic groups from 4.17 to 0.22 mmol/g in biochar with an increase in pyrolysis temperature. Rehrah *et al.* [43] observed a transition in the pH of biochar from weakly acidic (~6) to basic (~10) pH with an increase in temperature from 300 to 750°C because of the release of base cations at high temperatures. Chen *et al.* [44] found that the pH of hardwood obtained at 450°C was weakly acidic while that of straw-based char obtained at 600°C was alkaline. The production of acid and phenolic moiety as a result of the decomposition of cellulose and hemicellulose at 200-300°C lower the pH, and above 300°C, the alkali/alkaline metals begin to expose and increase the pH of biochar, which continues up to the temperature of 600°C after which the pH gets stable [44, 45]. Pyrolysis heating rate and residence time do not have any effect on pH of the resulting biochar [43, 46]. Similarly, feedstock composition and mineral content also have a direct effect on pH. The pH of dairy manure biochar (9.81) was higher than rice husk biochar (8.01), when both were prepared at the same charring temperature [47]. The composition of dairy manure biochar is primarily composed of minerals which raise the pH after separating from biochar at treatment temperature (~300 °C), while the rice husk biochar, a lignocellulosic material while decomposing at the same temperature, produces organic acids and phenolics which lower the pH [45, 47, 48]. Shen *et al.* [29] reported a slightly acidic pH of Salisbury biochar which was attributed to the retention of higher amounts of carboxylic and other acidic groups [49].

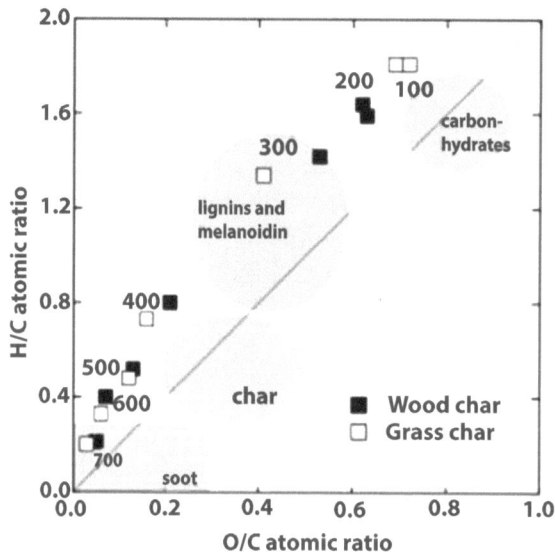

Fig. (2). Van Krevelen plot of elemental ratios for wood and grass chars [28].

Surface Functional Groups

The surface of biochars consists of many hydrophobic and hydrophilic functional groups like carboxyl, phenolic, hydroxyl, and carbonyl, which are acidic and basic in nature. Acidic/basic functional groups on biochar's surface render polarity to the material, which contributes towards the adsorption of polar compounds [50].

FTIR and Solid state ^{13}C Nuclear Magnetic Resonance (NMR) techniques are normally employed to study the functional groups on biochar surfaces. Functional groups like hydroxyl and carboxyl enhance the removal of heavy metals [51]. The density of these functional groups decreases on the surface of biochar with increasing pyrolysis temperature, which is attributed to dehydration reactions and molecular rearrangement [52]. Ding *et al.* [52] reported a reduction in peak intensity at 1722 and 1595 cm^{-1} related to –C=O and disappearance of peak at 1061 cm^{-1} corresponding to –C-O in bagasse biochar obtained at 500°C when compared with biochar obtained at 250°C. Acidic functional groups on biochar surfaces are characterized by Boehm titration [53]. This titration method is used to quantify the oxygen containing surface functional groups on the surface of biochar. Han *et al.* [54] quantified acidic functional groups on biochar obtained at two different temperatures. Carboxylic acid moieties were found to be higher in biochar pyrolyzed at 500°C than at 700°C. It was further established that there was a decrease in the carboxylic and lactone moieties, while phenolic content increased upon increasing the pyrolysis temperature.

Likewise, ssNMR (solid-state nuclear magnetic resonance) has also been used by many researchers for biochar characterization in terms of functional groups and structural changes while undergoing pyrolysis [20, 55]. While studying NMR spectra for biochar pyrolyzed at 300, 500 and 600°C, it was observed that biochar retained cellulosic components as indicated by peaks observed at δ 50-100 ppm and 150 ppm. The peaks corresponded to carbohydrates and phenolic groups, respectively and showed a dominant peak at 130 ppm (Ar-C) for high-temperature chars [55].^{13}C ssNMR analysis of wheat straw pyrolyzed at 200°C, 400°C and 600°C showed a decrease in alkyl C intensity at δ 20-80 ppm and an increase in the intensity at δ 90-130 ppm which is designated to the aryl C moieties with an increase in temperature [20]. High-temperature biochars have the most dominant signal at δ130 ppm in the NMR spectrum, which corresponds to aromatic carbon, whereas the signal corresponding to alkyl C at δ 0-45 ppm was absent. This destruction of the alkyl chain is also supported by the decrease in H/C ratios. Also, signals related to carbohydrate content (45-100ppm) registered a decrease in intensity for the biochars prepared at higher pyrolysis temperatures. Similarly, the amount of oxygenated C was 12.80% w.r.t. the total C intensity recorded, which was way less than 55.83% for the wheat straw biochar. Thus as the pyrolysis temperature is increased, the oxygenated carbon content of biochar is lowered due to the dehydration of carbohydrates.

Methods of Synthesis of Modified Biochars

Modification of biochar improves the metal uptake efficiency as it leads to augmentation in its properties like surface area, functional groups, and pore structure. Tuned biochars have a greater number of surface functional groups and exhibit higher porosity. The treatment/modification can be carried out by-activation (physical and chemical) and loading/impregnation by foreign material [56]. Physical and chemical activation can be bifurcated into i) Activation of biomass followed by pyrolysis and ii) Activation of biochar. Apart from activation, loading/impregnation is another way to change the characteristics of biochar. Loading/Impregnation can be carried out by minerals (hematite, magnetite, bentonite), metal salts/oxides (MnO, CaO), nanoparticles (nZVI, nano-metals oxide/hydroxide), *etc.* These methods of modification of biochars can broadly be classified as *in-situ* (during pyrolysis of biomass or pretreatment of biomass) and *ex-situ* (modification after pyrolysis process, *i.e.*, onto biochar) processes. Activation processes enhance the surface area of biochar by creating pores and enhance surface functional groups, whereas the properties of biochar are completely changed, *e.g.,* new functional groups, as a consequence of reduction, oxidation of functionalization, are attached on its surface when it acts as a platform for embedment of foreign material and forms a composite [56]. However, if raw biomass is treated instead of biochar, then after drying, subsequent pyrolysis is carried out. Physical activation is carried out using steam or CO_2 after performing biomass pyrolysis. The biochar is subjected to gas under high-temperature conditions (800-900°C). Physical activation creates more pores on the biochar surface and enhances 'O' containing functional groups. On the other hand, chemical activation can be carried out majorly using oxidants (acids, bases, oxidizing agents like H_2O_2, $KMnO_4$) and reductants (polyethyleneimine) [57]. Treatment with chemicals is carried out by soaking biochar in their solution for a given duration and temperature in a particular biochar to acid/alkali ratio. If leaching tests like Toxicity Characteristic Leaching Procedure (TCLP) are to be performed with biochar, then the solutions are kept in an overhead shaker. This procedure is followed by the washing and drying of adsorbents. However, if raw biomass is treated instead of biochar, biomass pyrolysis is carried out after drying. Chemical activation leads to the formation of many oxygenated groups like carbonyl, peroxide, carboxyl, phenolic, lactones on the surface of biochar.

In-situ Production

Various modifications are carried out prior to pyrolysis or during the pyrolysis process to get biochar with high adsorptive properties. For example, *in-situ* pore formation *via* catalytic pyrolysis is one of the methods suggested for the alteration of the pore structure of biochar. *In- situ* catalytic pyrolysis involves modification where pyrolysis is carried out in the presence of catalysts like metals and zeolites. In this process, both catalyst and biomass are mixed in a pyrolyser [58]. Various catalysts used in this process are- KOH, NaOH, Al_2O_3, ZnO, CaO, TiO_2, etc [59].

Grounded peels of pomelo fruit were immersed in H_3PO_4 in 1:2.5 (biomass to acid ratio) and were then pyrolyzed at 450°C in order to study their effect on chromium removal. The modified biochar showed well-developed pores as a result of the etching of biochar by H_3PO_4 oxidizing into carbonate and intercalating the phosphorus compounds [60]. FTIR also revealed that the modification was achieved by the interaction between –OH of the biochar and H_3PO_4 molecules.

Magnetized biochar prepared using $FeCl_3$ hydrolysis & ageing on biomass followed by pyrolysis, was used for Cr (VI) adsorption [61]. The hydrolysis and precipitation of Fe^{3+} ions were enhanced by the ageing process at 70°C. γ-Fe_2O_3 particles developed in magnetized biochar significantly increased the removal of Cr(VI) in acidic pH *via* electrostatic attraction between chromate ions and protonated -OH. Another, MnO_x/BC composite was prepared by immersing the crushed pine wood in a manganese chloride ($MnCl_2$) solution, followed by pyrolysis [62].

Ex-situ Modification

Ex-situ modification or post-treatment of biochar can be done after the biomass is pyrolyzed using chemicals or a physical mode of activation. Post-treatment of biochar/hydrochar has been carried out using various activating agents like KOH [63 - 65]. Steam activation of biochar was done by replacing the nitrogen gas with steam after the pyrolysis process. After activation of biochar, N_2 gas was supplied to cool the biochar to a low temperature [66]. The surface of biochar was greatly enhanced upon steam activation as a result of the formation of new pores. Treatment with $KMnO_4$ leads to immobilization of MnOx particles which in composite form is suitable for effective adsorption of heavy metals. MnO_x particles enhance the pore properties and surface functional groups on biochar. A novel biochar was prepared by modification through magnetic treatment before pyrolysis, and treated with $KMnO_4$ after carbonization to study the adsorption of Pb(II) and Cd(II) [67]. A schematic flow chart of the methodology is given in Fig. (**3**).

A similar type of MnO_x/BC composite was prepared where mineral *birnessite* was used for the modification of biochar, which was synthesized by the $KMnO_4$ precipitation method [62, 67]. The birnessite/ BC composite thus formed was used for the adsorption of Pb(II) ions, where birnessite particles served as an effective binding site for the binding of lead. Another study reported ZnO/BC nanomaterial formed by impregnation of zinc nitrate solution to biochar followed by calcination at 380°C where ZnO loading in biochar was varied from 10 to 50% [68]. Ma *et al.* [69] used modified biochar for remediation of Cr(VI) where it was given acid/alkali treatment followed by the addition of polyethylenimine/methanol solution and 1% glutaraldehyde solution. The modification introduced a large number of amine groups, as depicted by the FTIR study, which significantly enhanced the removal of Cr(VI) onto biochar.

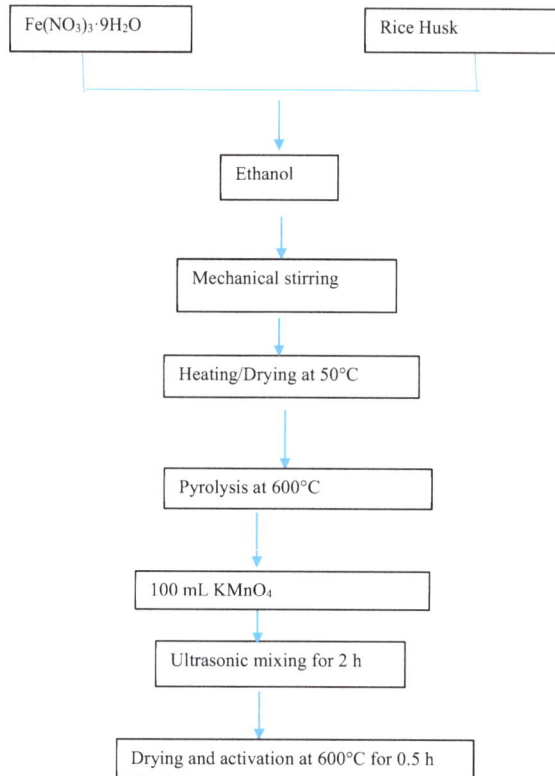

Fig. (3). Flow chart of the methodology of *ex-situ* biochar preparation [67].

nZVI/BC complexes are also known for better removal of heavy metals because of the combined effect of biochar's adsorption and nZVI's reducing properties. Shang *et al.* [70] prepared nZVI/BC adsorbent by dissolving biochar in a ferrous sulphate solution followed by the addition of sodium borohydride. The solution was continuously stirred and then given ethanol washed to avoid oxidation. A similar methodology was adopted while preparing biochar-supported nZVI, except that KBH_4 was used for the reduction of iron [42]. FTIR and XRD results indicated the role of organic moieties like carbonyl and silicate particles within biochar for binding of nZVI particles. The material prepared in earlier reference was impregnated with Pd to form a bi-metallic nanoparticle [71], where it was observed that the bimetallic composite exhibited higher adsorption compared to nZVI/BC and nZVI-Pd composites.

Hydrochar

Hydrochar, is yet another product that resembles biochar, but the difference lies in the production processes and properties. It is produced by a process called hydrothermal carbonization, where the biomass submerged in water is heated at 180-260°C in a confined system under saturated pressure and a residence time of 5 minutes to 12 hours [72]. Solid product (hydrochar) forms the highest fraction of all the products produced. The transformation reactions from biomass to hydrochar include hydrolysis, dehydration, decarboxylation, condensation, polymerization and aromatization [73]. Initially being investigated as a means of energy generation, applications of hydrochar have recently been explored for sorption purposes.

Additionally, hydrochar has been functionalized by adding certain chemicals after production or by means of *in-situ* modifications for tailoring its properties as an adsorbent. Hydrochar obtained from orange peel was given thermal treatment under air/CO_2 environment and chemical treatment with H_3PO_4 (post-treatment). Pretreatment of biomass was also carried out with H_3PO_4 prior to carbonization. The yield of both the acid-treated hydochars was high due to the occurrence of crosslinking reactions that help in the escape of low-weight species onto the carbon matrix [73]. Also, acidic post-treatment and thermal treatment under air environments yield a highly porous structure suitable for its activity as an adsorbent [73].

A new method for the production of hydrochar was proposed using the catalytic effect of H_2SO_4 [74]. In this study, the pretreatment of raw material (rice husk) with the acid was carried out, where the hydrolysis reactions led to the degradation of macromolecules to small molecules. The hydrolysis solution containing raw material was heated at 95°C for about 6 h, which led to hydrochar production following various reactions onto the sugars derived from acid hydrolysis. In another study by Wang *et al.* [75] porous carbons were prepared from the material produced in the previous study [74] by impregnating it with 85% H_3PO_4 in different ratios, and subsequently heating the material to 300-700°C. The resultant material exhibited excellent surface properties like higher surface areas and larger pore volume [75].

In another treatment, the functionalization of glucose was carried out with acrylic acid during hydrothermal carbonization, which led to a higher number of carboxylic acid moieties on the surface of biochar [76]. Therefore, the use of catalytic agents like acids, salts or oxidants lowers the energy required for the activation of biomass' hydrolysis and yields hydochars with higher 'O' containing functional groups [77].

The *in-situ* reduction of iron using hydrochar as a precursor was achieved by thermal activation of the solid residue (Fe-hydrochar) to yield nZVI hydrochar composites. The following were observed during the hydrothermal treatment-

1. Pyrolysis is catalyzed effectively to increase the graphitic structure

2. Fe enhances the charring of biochar

3. Reduction of Fe took place

The particles exhibited increased reactivity due to the formation of graphitic structure within the composites catalyzed by Fe particles [78]. The surface area of the Fe-hydrochar composites increased due to an increase in charring as the temperature was increased upto 600°C. The hydrochar obtained at 700°C showed a marginal decrease in the surface area. The decrease in the surface area of Fe-hydrochar composite obtained at high temperatures is accounted for the partial collapse of pores during charring. The surface area of the Fe-hydrochar composite was about 20 times higher than the untreated hydrochar produced at the same temperature, thus highlighting the role of iron salt as an activation agent during the thermal treatment [78]. Formation of the mesoporous structure was observed in Fe-hydrochar composite at higher treatment temperature because of the carbon removal by the *in-situ* reduction of iron oxides to nanoscale zero-valent iron.

GENERAL PROPERTIES OF MODIFIED BIOCHARS

Properties of biochar can be altered by varying the process parameters of production or through different pre- and post- pyrolysis treatment methods. Pyrolysis temperature is the foremost factor affecting the biochar's physico-chemical properties and its role has been discussed in the previous sections.

Physical activation is usually carried out using steam, while acidic or alkaline treatments are given for the chemical activation of biochar[56]. Steam removes the partially combusted particles trapped in the pores, and by generating gases like H_2, CO it oxidizes the carbon surface [57]. Biochar activated by steam treatment is characterized by lesser organic contents than pure biochar and hence it has lower atomic ratios. Therefore, steam-activated biochars can contribute to adsorption due to an increase in the number of pores and pore volume, rather than functional group-based ion exchange type adsorption [57]. Steam activation of biochars produced at 500°C increased the micropore surface area and micropore volume along with a proportionate increase in the micropore structure, thus making the surface area comparable with the biochar produced at 700°C [54]. It was noted that acidic groups (carboxylic, phenolic and lactonic) present on the surface registered a drastic reduction in steam activation, due to expansion and destruction of pore structure under steam activation conditions.

Acid treatment of biochar has been investigated thoroughly. Modification by acids removes mineral elements from biochar hence enhancing its hydrophilic nature; and also increases its oxygen content. Besides this, acid treatment also enhances the porosity of biochar e.g. the treatment of particles with H_3PO_4 prior to or after pyrolysis led to increase in microporosity, surface area and total pore volume compared to untreated biochar [60, 79]. The post- and pre-treatment with H_3PO_4 increased the surface area of pine sawdust biochar to 795 m^2/g and 1148 m^2/g, respectively, from 9.15 m^2/g [79]. This increase in surface area is mainly attributed to the formation of pores upon acid treatment. To understand the mechanism for generation of pores, the wood biomass components (cellulose and lignin) were pyrolyzed and treated under similar conditions. The cellulosic component contributed more towards the micropore development by-

a. Its swelling property, which aids in the penetration of acid into the biochar, and,
b. The chemical structure which offers conditions for the formation of crosslinks during the acid pyrolysis process [79].

On the other hand, treatment with HNO_3 leads to a reduction in surface area due to the degradation of the micropore wall [80].

Alkaline treatment aids in enhancing the surface area and it produces biochars with higher aromaticity and high polarity [57]. The surface of hickory wood biochar increased from 256 m^2/g to 873 m^2/g after alkaline treatment with NaOH, possibly due to an increase in porosity [81]. The same treatment also led to an increase in IEC up to three times due to an increase in surface functional moieties having oxygen like –COOH and –OH [81].

Biochar composites can be formed either by loading material to the feedstock prior to pyrolysis or post-pyrolysis [56]. Impregnation provides new sites for biochar to bind contaminants, thus enhancing adsorption. The surface area of biochar may increase or decrease after the modification. Magnetic treatment of biochar is known to reduce its surface area [82, 83], which can be attributed to the presence of iron oxides with a smaller surface area than biochar. The surface area of biochar after MgO treatment increased from 27.9 to 136.2 m^2/g [84]. Shang *et al.* [70] prepared nZVI/BC composite and observed a 90.23% decrease in surface area compared to the original biochar due to blockage of pores by nZVI particles, while magnetically modified biochar from the same source observed a 46.77% decrease in the surface area [83].

The chemical properties of biochar are also affected by the modification. MgO-coated biochar derived from sugarcane residue had pH two units higher than the original biochar. The reason for this was the formation of MgO crystallite, which increased the pH. Besides, the formation of nano MgO particles led to higher pore volumes in modified biochar [84]. Acidic treatment with refluxing acids can damage the physical structure of biochar [85], as was observed in the case of Sulphuric acid assisted MgO coated biochar where cracks and irregular structures were seen in FE-SEM imagery compared to the original biochar, which had a smooth surface [84].

BIOCHAR AND ITS MODIFICATIONS FOR HEAVY METALS REMOVAL

Biochar, a carbon-neutral material, tends to have a good adsorption tendency for heavy metal ions Tables **2**, **3** and **4**. Modifications by different methods improved the removal capacity compared to the original biochar as studied from the physical properties, namely charge/zeta potential, surface area and pores of the biochar as surface adsorbent. Although these modifications seem to have a varying effect on the adsorption of metal ions [4], the details of the biochar and its use in the adsorptive removal of heavy metals have been studied and reported for the five primary heavy metals, namely Cr, Pb, Cd, Cu and Zn Tables **2**, **3** and **4**.

Chromium

Chromium is widely used for industrial purposes like in the leather industry, as a wood preservative, electroplating, chemical industry, *etc.* It occurs primarily in two forms- Cr (III) and Cr (VI), the latter being more soluble, toxic and mobile. Cr (III), on the other hand, forms a very stable precipitate and hence is chemically unreactive under environmental conditions [90]. The reduction of Cr(VI) to Cr(III) using biochar has been investigated by many workers [51, 90 - 92]. The pathway involves the transfer of electrons from the inner core to the edges of biochar where hexavalent Cr is adsorbed [90]. The functional groups present on the biochar surface largely govern the adsorption and reduction of chromium. The most favored pH for chromium removal is 2. The biochar under acidic conditions is positively charged due to the protonation of surface functionalities, leading to electrostatic interactions between the negatively charged Cr (VI) and biochar surface [51, 91, 93]. As the pH is increased, deprotonation of biochar occurs and hence the Cr(VI) adsorption decreases. In addition, the presence of inorganic/organic reducing agents like Mg, Si and microbes, tends to enhance the Cr(VI) reduction process and make it more spontaneous [93, 94]. The effect of sugar beet tailing biochar on Cr (VI) removal was investigated [51] with removal of upto 123 mg/g achieved at pH=2. The mechanism of adsorption involves the electrostatic attraction between negatively charged Cr(VI) and protonated functional groups on biochar surface, followed by reduction to lower oxidized form [Cr(III)] and its complexation *via* carboxylic and hydroxyl functional groups. Mohan *et al.* [93] prepared lignin rich oak wood and oak bark biochars (by fast pyrolysis), which contained catechol and its substitutes. These products act as reductants and oxidize to *ortho*-quinone units that help in the binding of metal ions *via* chelation (Reaction 1). The study suggested that modification of biochar by physical or chemical activation could further enhance the surface area of the adsorbents, thereby increasing the removal of Cr(VI) ions.

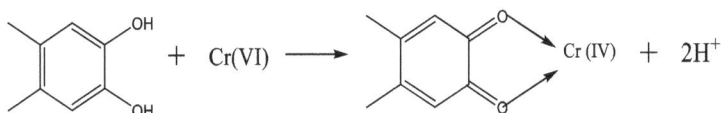

……. Reaction 1(Mohan et al.[25])

An *et al.* [95] studied the adsorption efficiency of biochar modified by $FeCl_3$, citric acid and NaOH, pre and post- pyrolysis. Fe-biochar composite showed the highest removal because of the presence of a large number of functional groups. The functional groups resulted in a more pronounced formation of magnetic particles during pretreatment with $FeCl_3$ (as depicted by XRD), which produced oxygen-containing functional groups on the biochar surface. ZnO-loaded biochar derived from water hyacinth was used for Cr(VI) removal [68], where the maximum adsorption took place at pH=5. XPS analysis of modified biochar after adsorption revealed that 90.6% of the adsorbed Cr was in Cr(III) form. This reduction was achieved because of the photolytic activity of ZnO, due to which it generated electrons to enhance the reduction. Cr (III) got readily adsorbed on inorganic substrates compared to Cr (VI). Also, the precipitation of Cr(III) ions on ZnO nanoparticles is depicted by the formation of zinc chromium oxide hydrate, which led to the removal of chromate ions.

nZVI is known as a strong reductant and adsorbent for various contaminants. However, it has disadvantages like an aggregation of particles in solution which can be overcome by using biochar as immobilizer for nZVI particles because of its properties like high surface area, porous and stable nature [42]. Also, higher removal capacity can be achieved due to the combined effect of two adsorbents. Shang *et al.* [70] used nZVI-supported biochar derived from residues of *Astragalus* sp. for the removal of Cr(VI) ions and were able to achieve 98.71% efficiency under acidic conditions (pH=2). Besides, the effect of common co-existing ions like HCO_3^-, Cl^- and SO_4^{2-} was also investigated in the adsorption process. HCO_3^- ion inhibited the adsorption, while SO_4^{2-} ion enhanced Cr (VI) adsorption significantly. Cr(VI) was absorbed by the mass transfer process of chromate ion to biochar, where the chromate ion was reduced to Cr (III) and afterwards co-precipitated on adsorbent's surface. Shang *et al.* [83] further studied magnetized biochar for the removal of Cr(VI) from the same biochar source and got the removal efficiency of 23.85 mg/g at pH=2 [83]. Qian *et al.* [42] used nZVI supported on rice straw biochar at different temperatures. The maximum removal (40 mg/g) was exhibited by the biochar prepared at 400°C pyrolysis temperature. In another study by Qian *et al.* [71] Cr removal by Pd(0) modified nZVI-supported rice straw biochar prepared at 700°C was investigated. It was observed that adsorption efficiency increased upto 117.1mg/g was observed in 30mg/L Cr(VI) solution, which was attributed to an enhanced reduction of metal ions [71]. Zhao *et al.* [96] used corn straw biochar-based persistent free radicals for the adsorption of Cr (VI). The selective reduction and sorption of Cr(VI) onto corn straw biochar were studied in the presence of competitive ions (salts). The study revealed that in the presence of NaCl, Cr (VI) removal increased from the aqueous solution, whereas ions like HPO_4^{2-}, F^-, HCO_3^-, and CO_3^{2-} suppressed the Cr(VI) removal. However, the selective adsorption of Cr (VI) was enhanced after the treatment of biochar with HNO_3. This was characterized by the presence of Environmentally Persistent Free Radicals (EPFR) on the biochar prepared in the presence of metal oxides. The free radicals act as strong reducing agents and react with metals having higher redox potentials. The free radicals act as strong reducing agents and react with metals having higher redox potentials. The biochar-based EPFR offered a highly selective Cr reduction in a neutral solution contaminated by salts, as revealed by electron spin resonance and free radical quenching studies [96]. In another study, Wang *et al.* [82] used biochar derived from eucalyptus leaf residue and modified by magnetic treatment to remove chromium from electroplating wastewater where removal of up to 97% was achieved for Cr (VI) removal.

Cadmium

Cadmium, one of the most toxic metals, has been regarded as a human carcinogen by various regulatory agencies. It causes anemia, hypertension, itai-itai disease in which a person experiences severe pain and bone disorders. Various researchers have studied the removal of Cd from aqueous solution using biochar derived from various ligno-cellulosic materials [97, 98]. Deng *et al.* [99] did a comparative study for cadmium adsorption using ten different biochars produced from plant (rice straw) and animal-based (swine manure) biomass pyrolyzed at 300, 400, 500, 600 and 700°C. It was observed that rice straw biochar exhibited higher q_{max} compared to swine manure biochar and the removal increased with increasing pyrolysis temperature. The dominant adsorption mechanisms involved were-cation exchange (for low-temperature biochars) and precipitation reactions (for high-temperature biochars), along with cation-interactions and complexation with surface functional groups [99]. Bashir *et al.* [35] studied rice straw biochar prepared by pyrolysis at 500°C for the adsorption of Cd(II) and obtained the maximum removal of 12.17 mg g^{-1}. However, the adsorption capacity nearly tripled on modifying the pure biochar with a 2M KOH solution. This increased Cd(II) adsorption by alkali-activated biochar was attributed to the availability of more functional groups, more negative charge on the surface of biochar and higher surface area [35]. Contrarily, cassava stem biochar activated by KOH showed the highest removal of Cd(II) from aqueous solution by biochar prepared at 300°C compared to 400°C and 500°C owing to its rich surface chemistry [100]. However, Higashikawa *et al.* [86] inferred that feedstock type was more dominant in the adsorption process compared to a pyrolysis temperature. In contrast to these studies, Tran *et al.* [98] concluded that the pyrolysis temperature and pyrolysis time did not affect Cd(II) adsorption. In their study, orange peels were used as a precursor to produce biochar in temperatures range of 400-800°C and residence times of 2h and 6h. Cd(II) adsorption efficiency of 114.69 mg/g was obtained from biochar produced at 700°C with heating time of 6h *via* chemisorption. In Table **1**, the efficiencies of biochars prepared in the study at different temperatures are given for a heating time of 6h. A similar study was carried out using water hyacinth-derived biochars prepared at 250-550°C, amongst which the one prepared at 450°C showed maximum adsorption owing to ion exchange and surface complexation reactions. Liu and Fan [101] investigated the role of mineral ions present in wheat straw biochar prepared at three different temperatures (300°C, 500°C and 700°C) onto Cd(II) removal. Demineralization of biochar was carried out with 1.0 M HCl. Adsorption of Cd(II) increased with increasing pyrolysis temperature and reduced when HCl-treated biochar was used as a sorbent, which clearly reflected the role of mineral ions. For low-temperature biochars, surface complexation and cation-π exchange were the main mechanism operative in the reaction while precipitation with minerals was the dominant mechanism for removal of Cd(II) using high temperature biochar. Pretreatment of biomass followed by pyrolysis is also a common method to obtain modified biochars [4, 102 - 104]. Rapeseed biochar when pretreated with KMnO$_4$ exhibited the highest removal, compared to magnetic and base treated. Also, the kinetic study revealed its ability to reach equilibrium rapidly and efficiently, which highlights its applicability to large scale applications [4]. KMnO$_4$ is a strong oxidant and also produces manganese oxide, therefore, the oxidation reaction can enhance the impregnation of the precursor into the raw material, which adds its effect for metal removal[103]. In a similar study by Wang *et al.* [103] the removal of Cd(II) increased upto seven times from ~4mg/g to 28 mg/g with KMnO$_4$ treatment owing to surface functional moieties like carboxyl, hydroxyl and phenolic, and manganese oxide particles that served as adsorption sites. Treatments do not always enhance the adsorption capacity of biochars. Karunanayake *et al.* [105]

produced magnetized biochar by precipitating it with Fe_3O_4 and a drop in adsorption efficiency was reported in modified biochar compared to pristine biochar. But the adsorption equilibrium was attained within 2 minutes in this study, making it an efficient adsorbent. A removal efficiency of 29 mg/g was observed for pristine grape stalks biochar which was largely due to ion exchange [88]. However, it was quite low compared to magnetized biochar from the same feedstock, where the removal increased upto 64 mg/g [106].

Table 1. Various physico-chemical properties of biochar derived from various feedstocks at varying pyrolysis conditions

Feedstock	Pyrolysis Temperature	Surface area (m^2g^{-1})	Pore volume (cm^3g^{-1})	pH	Elemental composition										References
					C %	H %	O %	N %	K (mg/kg)	Na (mg/kg)	Ca (mg/kg)	Mg (mg/kg)	Al (mg/kg)	P (mg/kg)	
Hardwood	450	0.43	0.00036	5.57	53.41	2.30	5.67	0.07	32.95	5.55	27.22	4.42	19.44	0.59	Chen *et al.* [44]
Corn Straw	600	13.08	0.014	9.54	35.88	1.64	1.86	0.43	208.47	20.08	27.34	26.37	245.17	2.51	
Water hyacinth	250	-	-	7.24	-	-	-	-	34.08	23.12	32.85	2.975	-	2.2	Zhang *et al.* [20]
	350			9.43					-	-	-	-		-	
	450			10.49					81.18	21.12	46	7.625		4.1	
	550			10.46					-	-	-	-		-	
Lotus seedpod	300	2.246	6.189*10-3	-	65.12	2.05	23.52	1.70	-	-	-	-	-	-	Chen *et al.* [50]
-	600	24.15	1.839*10-2	-	73.79	1.96	12.76	1.47							
Rice Straw	500	39.9	-	10.1	49.72	-	-	1.01	-	-	9690	2320	-	-	Higashikawa *et al.* [86]
Sugarcane straw	350	-	-	8.67	60.13	2.44	35.78	1.66	6750	397.19	2910	2280		940	
-	650			9.17	69.37	2.45	26.69	1.50	13650	587.47	6100	3660		2730	
Rice husk	350			8.44	32.79	1.09	66.09	0.04	930	360.36	700	220		BDL	
-	650			8.72	49.48	1.47	49.04	0.02	870	180.43	590	210		BDL	
Sawdust	350			7.59	71.63	3.94	24.35	0.10	240	1510.66	1770	640		BDL	
-	650			7.48	84.60	2.84	12.35	0.22	260	1918.01	2020	840		1050	
Pine wood	300	0.2	0.002	7.10	69.24	4.95	25.17	0.34	200	-	1400	780	280	300	Wang *et al.* [87]
-	450	0.1	0	7.61	80.18	3.31	15.76	0.31	200		2000	1200	360	400	
-	600	209	0.003	7.05	85.68	2.13	11.40	0.33	500		1900	1200	410	400	
Citrus wood	300	0.8	0.005	7.76	64.48	4.73	27.07	1.53	3400		15100	2100	120	1000	
-	450	2.8	0.009	10	67.73	3.61	25.02	1.23	4100		17100	2500	110	1100	
-	600	182	0.013	9.48	78.28	2.08	14.90	1.28	6600		22800	3500	130	1500	
Alfalfa	300	0.6	0.006	8.38	64.72	5.26	24.24	3.10	5100		11900	5600	230	1500	
-	450	0.7	0.005	9.17	69.66	3.01	22.13	2.42	7000		13300	5300	240	1700	
-	600	0.2	0.006	10.35	73.25	1.91	19.43	2.22	7600		15300	6500	250	1900	
Switchgrass	300	1.2	0.01	8.21	59.32	4.84	31.68	2.34	8400		4300	4400	390	2100	
-	450	10	0.02	9.74	64.02	2.87	28.27	2.23	10500		5600	5900	390	3000	
-	600	15	0.024	9.84	68.15	2.21	24.99	1.90	12100		5200	6100	420	3200	

(Table 1) cont.....

Feedstock	Pyrolysis Temperature	Surface area (m^2g^{-1})	Pore volume (cm^3g^{-1})	pH	Elemental composition										References
					C %	H %	O %	N %	K (mg/kg)	Na (mg/kg)	Ca (mg/kg)	Mg (mg/kg)	Al (mg/kg)	P (mg/kg)	
Nut shell	600	465	0.18	8.63	88.2	2.10	7.82	0.31	-	-	-	-	-	-	Trakal *et al.* [88]
Wheat straw	-	364	0.13	9.86	70.4	1.73	6.27	0.97							
Grape stalks	-	72	0.03	10	70.2	1.70	12.5	1.45							
Grape husk	-	77	0.032	9.98	74.9	1.92	10.8	1.93							
Plum stone	-	443	0.172	7.36	81.8	2.01	10.6	0.82							
Burcucumber	300	0.85	0.004	10.86	65.98	5.55	23.09	5.08	-	-	-	-	-	-	Rajapaksha *et al.* [89]
-	700	2.31	0.008	12.32	69.41	1.31	24.45	4.61							
Soybean Stover	300	5.61	-	7.27	68.81	4.29	24.99	1.88	-						Ahmad *et al.* [18]
-	700	420.30	0.19	11.32	81.98	1.27	15.45	1.30							
Sugar beet tailing	300	0.2	-	-	-	-	-	-	18960	2650	28210	986	2950	-	Dong *et al.* [51]
Sugarcane harvest residue	550	27.90	0.04	8.54	-	-	-	-	17.11	6.40	6.70	16.71	-	3.80	Xiao *et al.* [84]
Hickory wood	600	256	-	-	84.7	1.83	11.3	0.30	2800	-	-	2900	400	-	Ding *et al.* [81]

Table 2. Adsorption of heavy metals (Cd, Cr and Cu) using pristine biochars from various sources and pyrolysis temperature.

Heavy Metal	Material	Temperature (°C)	Adsorption Capacity (mg g^{-1})	Isotherm	Kinetic Model	Adsorption pH	Adsorption Mechanism	Reference
Cd	Grape Stalk	600	29	Langmuir	-	5	Ion exchange	Trakal *et al.* [88]
	Rice Straw	500	12.17	Langmuir and Freundlich	-	6.5	Ion exchange	Bashir *et al.* [35]
	Rice Straw	700	66.23	Freundlich	Pseudo-second order	5.5	Cation exchange, Precipitation with minerals, Complexation, Cation-π	Deng *et al.* [99]
	Swine manure	700	46.5	Freundlich	Pseudo-second order	5.5	-	

(Table 2) cont.....

Heavy Metal	Material	Temperature (°C)	Adsorption Capacity (mg g^{-1})	Isotherm	Kinetic Model	Adsorption pH	Adsorption Mechanism	Reference
Cd	Wheat Straw	300	38.4	Langmuir and Freundlich	Pseudo-second order	5	Cation-π interaction, Surface complexation, Mineral precipitation	Liu and Fan [101]
	Wheat Straw	500	52.1	Langmuir and Freundlich	Pseudo-second order	5	Cation-π interaction, Surface complexation, Mineral precipitation	
	Wheat Straw	700	69.8	Langmuir and Freundlich	Pseudo-second order	5	Cation-π interaction, Surface complexation, Mineral precipitation	
	Orange Peel	400	104.99	Langmuir	Pseudo-second order	-	Cπ-cation interactions, surface precipitation	Tran *et al.* [98]
		500	113.66					
		600	111.95					
		700	114.69					
		800	98.60					
	Water hyacinth	250	49.5	Langmuir	Pseudo-second order	-	Ion exchange, Surface Complexation	Zhang *et al.* [102]
		350	69					
		450	70.313					
		550	34					
Cr	Sugar beet tailing	300	123	Langmuir	Pseudo-second order	2	Electrostatic attraction, reduction, complexation	Dong *et al.* [51]
	Corn straw	300	33.33	Langmuir	-	7	Reduction, precipitation	Zhao *et al.* [96]
		500	22.03					
		700	26.18					
	Rice straw	100	2.2	-	-	3	-	Qian *et al.* [42]
		300	3.8	-	-	-	-	
		400	6.4					
		500	6.2					
		700	5.8					

(Table 2) cont.....

Heavy Metal	Material	Temperature (°C)	Adsorption Capacity (mg g⁻¹)	Isotherm	Kinetic Model	Adsorption pH	Adsorption Mechanism	Reference
Cu	Cedar wood	250	2.73	Langmuir	-	-	Electrostatic attraction	Rodríguez-Vila *et al.* [14]
		300	1.50					
		350	12.80					
		400	14.66					
		450	16.72					
		500	21.23					
		550	27.17					
		600	30.58					
		650	23.92					
		700	28.65					
	Pine strip wood	450	0.82	Langmuir			Physisorption	
	Whitewood spruce		1.03					
	Pistachio nut shells		2.98					
	Greenwaste compost		6.55					
	Conifer bark		7.86					
	Bamboo canes		9.36					
	Cedar wood		16.72					
	Farmyard manure		35.84					
	Horse chestnut leaves		56.50					
	Chicken manure		81.30					
	Hardwood	450	6.79	Langmuir	Pseudo-second order	5	Chemisorption	Chen *et al.* [44]
-	Corn straw	600	12.52					

Table 3. Adsorption of heavy metals (Zn and Pb) using pristine biochars from various sources and pyrolysis temperature.

Heavy Metal	Material	Temperature (°C)	Adsorption Capacity (mg g^{-1})	Isotherm	Kinetic Model	Adsorption pH	Adsorption Mechanism	Reference
Zn	Cedar wood	250	1.65	Langmuir	-	-	Electrostatic attraction	Rodríguez-Vila *et al.* [14]
	-	300	1.14					
	-	350	0.75					
	-	400	1.17					
	-	450	10.64					
	-	500	12.01					
	-	550	13.93					
	-	600	13.51					
	-	650	15.20					
	-	700	15.06					
	Pine strip wood	450	0.59	Langmuir	-	-	Physisorption	
	Whitewood spruce	-	0.81					
	Pistachio nut shells	-	2.91					
	Greenwaste compost	-	5.18					
	Conifer bark	-	5.87					
	Bamboo canes	-	6.65					
	Cedar wood	-	10.64					
	Farmyard manure	-	26.04					
	Horse chestnut leaves	-	35.21					
	Chicken manure	-	31.95					
	Hardwood	450	4.54	Langmuir	Pseudo-second order	5	Chemisorption	Chen *et al.* [44]
	Corn straw	600	11			5		

(Table 3) cont.....

Heavy Metal	Material	Temperature (°C)	Adsorption Capacity (mg g^{-1})	Isotherm	Kinetic Model	Adsorption pH	Adsorption Mechanism	Reference
Pb	Bagasse	250	21	Langmuir	Pseudo-second order	-	Complexation, cation exchange	Ding *et al.* [52]
	-	600	6.1			-	Intraparticle diffusion	
	Corn Straw	450	352	Langmuir	Pseudo-second order	7	Electrostatic interaction, Surface complexation, Coordination	Wang *et al.* [67]
	Saw dust	600	27.39	Langmuir	-	4	-	Jellali *et al.* [112]
	Wine lees	300	73.53	Freundlich	Pseudo-second order	6	-	Zhu *et al.* [92]
		400	89.29					
		500	99.01					
		600	104.17					
		700	81.97					
	Pine wood	300	4.25	Langmuir	Pseudo-second order	5	-	Liu and Zhang [78]
	Rice husk	300	2.40					
	Sugarcane bagasse	600	6.48	Langmuir	-	-	Surface adsorption	Inyang *et al.* [108]
	Anaerobically digested Sugarcane bagasse	600	135.4				Precipitation	
	Canola Straw	300	84	Freundlich	Pseudo-second order	-	IEC, Precipitation, Inner sphere complexation	Kwak *et al.* [66]
		500	58					
		700	108					
	Manure pellet	300	49	Freundlich	Pseudo-second order			
		500	96					
		700	68					
	Sawdust	300	43	Freundlich	Pseudo-second order			
		500	62					
		700	43					
	Wheat Straw	300	55	Freundlich	Pseudo-second order			
		500	81					
		700	109					

Table 4. Adsorption of heavy metals using modified biochars

Heavy Metal	Material	Modification	Temperature (°C)	Adsorption Capacity (mg gm^{-1})	Isotherm	Kinetic Model	Adsorption pH	Adsorption Mechanism	Reference
Cd	Cassava stem	KOH	300	24.88	Langmuir	-	5-6	-	Prapagdee *et al.* [100]
	Hickory wood	KMnO$_4$	600	28.1	Langmuir	Ritchie nth order	6	Surface adsorption mechanism involving MnO$_x$particles and 'O; containing functional groups	Wang *et al.* [103]
	Rape straw	NaOH	600	72.369	Langmuir	Pseudo-second order	5	Cation exchange, Cation-π	Li *et al.* [4]
		KMnO$_4$		81.096					
		FeCl$_3$		72.369					
	Rice Straw	KOH	500	41.9	Langmuir and Freundlich	-	6.5	Surface precipitation, Ion exchange	Bashir *et al.* [35]
	Douglas Fir	Fe$_3$O$_4$	-	16 to 11	Langmuir	-	5		Karunanayake *et al.* [105]
Cr	Rice Straw	nZVI	400	40	-	-	4	Reduction, Adsorption	Qian *et al.* [42]
	Herb residue	nZVI	400	99.98	Langmuir	Pseudo-second order	2	Reduction, Co-precipitation	Shang *et al.* [70]
		Magnetic	400	23.85	Langmuir	Pseudo-second order	2	Electrostatic attraction	Shang *et al.* [83]
	Corn Stalk	Pretreatment with FeCl$_3$	500	25.68	Langmuir	Pseudo-second order	1	Fe$_3$O$_4$ coordination and lattice oxygen, Electrostatic attraction, Complexation, Reduction	An *et al.* [95]
		Post treatment with FeCl$_3$	500	23.09	Langmuir	Pseudo-second order	1	Electrostatic attraction, Complexation, Reduction	
	Water hyacinth	ZnO nanoparticles	700	24	Langmuir	Pseudo-second order	5	Reduction, Precipitation	Yu *et al.* [68]
	Rice Straw	Pd-nZVI	700	117.1	Langmuir	-	4	Reduction, Precipitation, Adsorption	Qian *et al.* [71]

(Table 4) cont.....

Heavy Metal	Material	Modification	Temperature (°C)	Adsorption Capacity (mg gm^{-1})	Isotherm	Kinetic Model	Adsorption pH	Adsorption Mechanism	Reference
Pb	Grape pomace	KOH	220	137	Sips model	Pseudo-second order	5	-	Petrovic *et al.* [63]
	Corn Straw	Fe-Mn binary oxide	700	113.7	Langmuir	Pseudo-second order	4	Ion exchange and Hydrogen bonding	Zhang *et al.* [134]
	Tree leaves	CeO$_2$-MoSC and magnetic	600	263.6	Langmuir	Pseudo-second order	-	Electrostatic attraction, Cπ-Pb interaction, Complexation	Li *et al.* [111]
	Maple Wood	Hydrogen Peroxide	500	43.3	Langmuir	-	-	-	Wang *et al.* [135]
	Canola Straw	Steam	300	72	Freundlich	Pseudo-second order	-	IEC, Precipitation, Inner sphere complexation	Kwak *et al.* [66]
			500	93					
			700	195					
	Manure pellet	Steam	300	69	Freundlich	Pseudo-second order			
			500	62					
			700	115					
	Sawdust	Steam	300	58	Freundlich	Pseudo-second order			
			500	41					
			700	69					
	Wheat Straw	Steam	300	57	Freundlich	Pseudo-second order			
			500	87					

Lead

Lead is used as a raw material in printing, battery, pesticides and photographic applications. Lead contamination in the environment is of serious concern because of its biological toxicity. Acidity or alkalinity of biochar affects the equilibrium pH values and hence the removal of Pb (II) ions. In basic solution, a decrease in adsorption capacity occurs due to the precipitation of lead ions to $Pb(OH)_2$ and other Pb complexes [107]. Pb (II) was precipitated as $Pb_3(CO_3)_2(OH)_2$ and $PbCO_3$ on the surface of digested bagasse biochar after adsorption when the pH of the solution was increased. The formation of these complexes was confirmed by XRD and SEM analysis. Further, the FTIR study of biochar after adsorption revealed the disappearance of O=C=O bond (2343 cm^{-1}) thus indicating its role in metal precipitation [108]. Liu and Zhang [78] reported pH 5 to be most favorable for the adsorption of Pb(II) by biochar produced by hydrothermal treatment, after which it was reported to form a hydroxyl complex. The effect of particle size of biochar on Pb (II) adsorption was also studied [29]. Higher removal was obtained using particles of smaller dimension particles (0.15 mm) compared to large particles (2 mm) due to the former comprise of higher surface area as well as higher IEC, which contributed for the exchange of metal ions. Komkiene and Baltrenaite [109] also reported that Silver birch biochar showed higher removal efficiency for heavy metals despite having lower surface area, which was attributed to its much higher IEC than Scots pine biochar. Wang *et al*. [67] obtained high Pb(II) removal from solution (352 mg/g) using corn straw biochar. The FTIR spectra of adsorbent before and after adsorption revealed a significant displacement of functional groups like -OH, -COOH and -CH suggesting complexation and coordination with electrons as primary mechanisms for metal removal. Upon adsorption of Pb(II) onto biochar, the FTIR spectra showed a minor shift in the spectra of –OH band and a significant shift at 1556 cm^{-1} to 1584 cm^{-1} which indicates the role of hydroxyl and –C=C- groups in adsorption of lead ions [110, 111]. Pb(II) adsorption is also affected by the operating conditions of pyrolysis particularly the pyrolysis temperature. Ding *et al*. [52] observed the higher efficiency of low temperature biochars than high temperature biochars prepared from bagasse. The relatively higher content of oxygen-containing-functional groups (hydroxyl, ether) accounted for higher adsorption in low temperature biochars where adsorption occurred *via* complexation and ion exchange mechanisms. Decrease in intensity of C=O group, and a disappearance of C-O band was observed in FTIR spectra of the low temperature biochars upon Pb(II) adsorption. While adsorption in high temperature biochars was attributed to intraparticle diffusion as a result of presence of abundant pores. Besides, some of the adsorptions were also contributed by the precipitation mechanism in biochars that contained significant amounts of phosphate [52]. Contrarily, lead adsorption increased with the increasing temperature of pyrolysis [66, 92]. It was observed that there was an increased formation of alcohol hydroxyl and phenolic hydroxyl groups with pyrolysis temperatures up to 600°C that provided exchangeable protons. However, further increasing the pyrolysis temperature destruction of functional groups affected the adsorption process and the Pb(II) removal efficiency was lowered [92]. Kwak *et al*. [66] reported similar results of an increase in adsorption efficiency for wheat and canola straw biochars produced at 300, 500 and 700°C but the increase was due to higher surface area, IEC, and ash content rather than contribution by the functional groups. Further, steam activation of these biochars enhanced the Pb (II) removal from 108 to 195 mg/g for canola straw biochar because of the increased surface area of biochars achieved due to an increase in micropore volume [66]. Chemical pre-treatment of Cyprus saw dust with MgCl$_2$ followed by subsequent pyrolysis yields Mg-activated biochar, which, when compared with Cyprus biochar enhanced Pb(II) adsorption of ~7.5 folds due to higher zero point charge values, increased surface functional groups, enhanced surface area and well-formed pore structure and increased ion exchange [112]. The effect of temperature during adsorption process is also more pronounced in Pb(II) adsorption as the higher temperature provides energy to metal ions to reach onto the biochar surface and get adsorb [78].

Zinc and Copper

Zinc is released into the environment mainly as a result of industrial activity like mining, iron and steel production, paints and metal coating. Zamani *et al.* [113] used oil palm fruit residues based biochar to remove Zn(II) from solution. Response Surface Methodology (RSM) was employed to study the effect of biochar production conditions on yield and adsorption capacity. Results showed that the optimum biochar produced at 615°C showed the maximum adsorption capacity of 15.18 mg/g for Zn. Bogusz *et al.* [114] studied two types of biochars derived from -*Sidahermaphrodita* and wheat straw for metal ions removal. Maximum removal of Zn(II) took place at pH=7 and *Sidahermaphrodita* biochar showed higher removal (45.62 mg/g) compared to wheat straw biochar (40.18 mg/g) which was due to high aromaticity and polarity of the former and the role of polar groups like –COOH in the binding of metal ions. The removal of metals is also studied in binary and multi-metal solution systems to give a real-time scenario where different metals exist in wastewater and the presence of one metal might affect other's adsorption. In one such study, the adsorption efficiency of Zn(II) in mixed metal solution (Pb^{2+}, Cd^{2+}, Cu^{2+} and Ni^{2+}) increased from 0.71 upto 1.83 mg/g using alkali-modified hickory wood biochar [81]. Chen *et al.* [44] also studied the competitive adsorption using hardwood and corn straw biochars prepared at 450°C and 600°C where Zn(II) adsorption was suppressed in the presence of Cu(II) ions. Over 75% drop was observed in Zn(II) removal efficiency, indicating that Cu(II) can compete with binding sites of Zn(II) ions. But the effect on Cu(II) adsorption was minimal when Zn(II) concentration in solution was increased. Also, a decline in metal removal was observed with an increase in biochar concentration because of the aggregation of particles, however, the total removal increased. Likewise, Park *et al.* [115] while working with sesame straw biochar prepared at 700°C achieved 34mg/g removal for Zn(II) which reduced to 7 mg/g in multi metal solution.

Rodríguez-Vila *et al.* [14] studied the effect of charring temperature and feedstock type on resulting biochars for their application in the removal of Cu(II) and Zn(II). A sigmoidal relationship between temperature and maximum adsorption capacity of these two metals was observed for cedarwood biochar. Cu(II) and Zn(II) adsorption increased with (i) Increasing pyrolysis temperature and (ii) Decreasing C/N of the feedstock. However, the adsorption of Cu(II) lmost doubled compared to Zn(II) adsorption, which was attributed to the higher charge-to-radius ratio and higher electronegativity of Cu(II) due to which it readily gets bonded with functional groups present on the biochar. Zn adsorption was because of cation-π interactions linked to aromaticity observed at higher pyrolysis temperatures [116, 117]. Also, the biochar with the lowest C/N ratio showed the maximum removal efficiency when tested for Zn(II) contaminated mine wastewater.

Copper is released into the environment by the electroplating industry, fertilizers, agro-chemicals, brass manufacturing and mining. Owing to its bioaccumulation and toxicity, copper affects organisms. Tong *et al.* [33] used four straw-based biochars obtained at 300, 400 and 500°C for Cu (II) removal from electroplating industry wastewater. Biochar obtained at 400°C was found to be more efficient for Cu (II) removal. Higher efficiency by high-temperature chars was accounted to their more basic nature, which helps in neutralizing the acidic Cu(II) solution and brings the pH to around pH 5-6, where precipitation reactions occur. The effect of biochar dosage was also clearly visible on solution pH as high dosage led to an increase in pH of the solution and removal of Cu (II) by precipitation. Similarly, adsorption increased with an increase in pH from 4 to 6 in a study involving six types of biochars for Cu(II) adsorption [118]. These biochar samples were divided into two groups – one having Q_{max} higher than 20mg/g characterized by high ash content, P content & high O/C ratios; and the other group having Q_{max} lower than 20 mg/g characterized by low O/C and ash content. The mechanisms involved in adsorption were precipitation, complexation and ion exchange. The distribution of Cu(II) between solid phase and aqueous phase (associated with dissolved organic matter and free Cu(II)) was also studied. Rice husk biochar, olive mill waste biochar, chicken manure biochar and corn cob biochar had ~97% Cu(II) associated with solid fraction, while *Acacia* and *Eucalyptus* biochar had most part associated with organic matter in an aqueous phase, which was, although not toxic but highly mobile [118]. Saleh *et al.* [119] compared the effectiveness of three types of adsorbents- namely sunflower seed husk biomass, its pyrolyzed form (biochar) and activated carbon for the removal of Cu (II) and reported a higher efficiency of biochar compared to the other two adsorbents. BET analysis revealed that there was a 266% increase in micropores in biochar compared to raw feedstock. Though the surface area and pore volume of biochar were comparable with activated carbon, the number of micropores in the former was higher. Cu(II) removal by biochar was independent of the time and temperature of adsorption. In another study, banana peel biochar showed an efficiency of 147 mg/g for the removal of Cu (II), where equilibrium was attained within 30 minutes [120]. Optimum conditions for the maximum removal of 147 mg/g were pH=5.5, the initial metal concentration of 200 mg/L and dose of 1.4 g/L. Five types of adsorption models were applied to the study where the order of fit was Langmuir>D-R>Temkin> Freundlich and Hasley>Harkin-Jura. Unlike most of the studies on Cu(II) removal where equilibrium was attained very fast, Li *et al.* [121] in their study utilizing biochar derived from *Spartina alterniflora* reported equilibrium at 30h owing to chemical interaction which took time to attain the same. Functional groups -COOH and -OH and IEC mechanism was attributed for adsorption evident from IR band shift at 1107 cm^{-1}, and emergence and disappearance of bands at 1459 cm^{-1} and 1386 cm^{-1}, respectively. To further verify the role of the complexation mechanism, three site model involving two acidic and one basic site were proposed as per results given by potentiometric titrations.

Treatment of biochar with oxidants like nitric acid introduced functional moieties like carboxylic groups, which assist in binding heavy metal ions [122, 123]. Hadjittof *et al.* [122] studied the adsorption of Cu (II) using cactus fibers activated by 12 M HNO$_3$. The activation produced –COOH moieties (confirmed by FTIR study) that served as binding sites for Cu (II) ions. Titration curves obtained for activated biochar further confirm the presence of acidic groups against non-activated biochar, especially at pH 3; above which a buffer zone was observed up to 7.2 pH, which denotes the deprotonation of COOH groups. Based on the observations, two different mechanisms were proposed, which led to the formation of outer-sphere complexes and inner-sphere complexes at pH 3 and 6.5, respectively. The following two reactions describe the adsorption mechanism at pH 3, representing a cation exchange reaction (Reaction 2) and another at pH 6 accompanied by hydrolysis where maximum adsorption took place (Reaction 3), leading to the formation of the more stable neutral surface complex.

$$R\text{-}COOH + Cu^{2+} R\text{-}COO\text{-}Cu^+ + H^+ \qquad \dots\dots\text{\textbf{Reaction (2)}}$$

$$R\text{-}COOH + CuOH^+ R\text{-}COO\text{-}CuOH + H^+ \qquad \dots\dots\text{\textbf{Reaction (3)}}$$

A similar study reported the introduction of COOH moieties on to biochar's surface on treatment with 8 M HNO$_3$, where removal of 248 mg/g was achieved *via* the formation of highly stable inner sphere complexes [123]. Peng *et al.* [124] studied the effect of pyrolysis temperature on the adsorption of Cu (II) using sawdust biochar obtained by pyrolysis of sawdust at 200, 350, 500 and 650°C. Maximum adsorption was exhibited by biochar prepared at 650°C due to its high surface area. Treatment with H$_3$PO$_4$ reversed the trend and adsorption decreased with temperature. Further, there was a 12 to 44-fold increase in adsorption upon acid modification. It was attributed to the increased surface 'O' and 'P' containing groups (-COOH, -OH, P=O, P=OOH) *via* surface complexation mechanism, and enhanced surface areas. FTIR study and XPS verified the elevated content of P and O. Similarly, amino modification of saw dust biochar showed enhanced Cu(II) removal in both batch and fixed bed column because of the strong complexation between the amino groups and heavy metal as per XPS and ATR-FTIR analysis [125]. The surface of biochar became protonated when the pH<pH$_{zpc}$ (3.9), which resulted in electrostatic repulsion and low removal of metal ions. However, adsorption was stabilized at pH of 4 to 6. Also, the biochar

showed higher selectivity for the removal of copper in the presence of K^+, Ca^{2+} and Mg^{2+} ions owing to the formation of outer sphere complexes with Cu(II). Physical activation by steam nearly doubled the surface area of biochar but did not improve the adsorption of Cu(II) [126, 127]. Adsorption by biochars is attributed to combination of pseudo first order kinetics and intraparticle diffusion (Two compartment model), which led to the inference that fast sorption was dominant in pristine biochar *via* surface complexation, whereas slow sorption was dominant for steam-activated char because of its higher surface area [126]. Lou *et al.* [127] prepared four types of biochar to study the effect of pyrolysis temperature and steam activation on metal adsorption. While high adsorption was achieved for biochar prepared at 550°C compared to biochar prepared at 300°C, steam activation did not affect the adsorption process at all, despite doubling the surface area of pristine biochar produced at 550°C. Adsorption mechanisms like precipitation and ion exchange with a minor contribution of metal-π interactions were responsible for Cu(II) adsorption on biochar prepared at high pyrolysis temperature [127]. Kim *et al.* [128] explained the role of the ion exchange mechanism for the Cu (II) removal. Biochar derived from green algae was used for adsorption, which had high alkali and alkaline metal content due to its sea water origin. The biochar on physical activation increased the ash content and hence exchangeable ions, which enhanced the adsorption of metal ions, which was further supported by the presence of a significant quantity of cations on the surface of the adsorbent upon adsorption. On the other hand, activation by KOH decreased the removal because the base removed most of the ash content from biochar, despite increasing its specific surface area to a higher amount compared to physically activated biochar [128]. The role of ash content in biochar was further demonstrated by using biochar from two different sources- pine chips and peanut straw, followed by the washing of biochar with deionized water and acid solution (HCl and HF) to remove ash content. Unlike in previously mentioned studies, the role of cations was observed to suppress the Cu (II) removal probably because of decreased competition by cations for adsorption while the presence of anions enhanced the adsorption *via* precipitation in peanut shell biochar. While for pine chips, biochar Cu(II) removal increased with the removal of ash [129].

ECONOMIC FEASIBILITY OF ADSORPTION

The cost of operation of the adsorption process is mainly dependent upon the cost of production of the adsorbent. Biochar is derived from biomass (mainly waste materials or discarded agricultural residue), and studies discussed in the previous section have shown its adsorptive potential for the removal of heavy metals. The cost of biochar production is evaluated in terms of the collection of precursor materials and pyrolysis, apart from taking into account the transportation cost

[130]. For its environmental impact and economic feasibility, Alhashimi and Aktas [131] compared biochar with activated carbon (the most widely used adsorbent). It was reported that biochar has low environmental impact than activated carbon when assessed on the basis of energy demand, transportation to long distances and global warming potential (GWP) [131]. In terms of economic performance, the cost of adsorption operation for the removal of Cr and Zn was found to be lower than activated carbon, while for Cu and Pb, it was comparable. Importantly, the studies suggest that if the biochar is tuned precisely for properties, on the basis of application, it could be more viable and quite cheap in comparison to activated carbon [131 - 133]. In addition, biochar regeneration is another process, making it more economical for its use as an adsorbent [133 - 135].

CONCLUDING REMARKS

Biochar is a 'green' renewable energy source and is mainly used in the remediation of heavy metals, in particular Cr(VI), Pb(II), Cd(II) Zn(II) and Cu(II) ions. Properties like high surface area, porous structure, and the presence of surface functional groups make it an efficient adsorbent for the uptake of metal ions. Thus, biochar can prove to be a very useful material that can be used as a fuel and also an adsorbent for the removal of various contaminants such as heavy metals. Modified biochar can be prepared by activation process (physical or chemical) or impregnation, either *in-situ* or *ex-situ*. The prepared biochars show significant differences in physico-chemical characteristics compared to their raw forms (biomass and biochar), which assists in higher heavy metal uptake. Various adsorption mechanisms involved in the processes are- complexation with surface functional groups, cation exchange, cation-π interactions and precipitation reactions on the surface of biochar as well as the modified biochar.

From the above review, it is quite clear that there are few methods investigated for the functionalization of biochar at its surface. Studies related to the use of biochar for the adsorption of heavy metal ions are scarce. The present review emphasizes the state-of-the-art knowledge existent and paves a way forward for opportunities for surface scientists to explore possibilities of functionalization and use of modified biochars in heavy metal adsorption. This promising investigation would bring the sustainable development goal alive, particularly with biomass which has an opportunity for management.

REFERENCES

[1]   Demirbas A. Heavy metal adsorption onto agro-based waste materials: A review. J Hazard Mater 2008; 157(2-3): 220-9.
[http://dx.doi.org/10.1016/j.jhazmat.2008.01.024] [PMID: 18291580]

[2]   Barakat MA. New trends in removing heavy metals from industrial wastewater. Arab J Chem 2011; 4(4): 361-77.
[http://dx.doi.org/10.1016/j.arabjc.2010.07.019]

[3]   Park J, Hung I, Gan Z, Rojas OJ, Lim KH, Park S. Activated carbon from biochar: Influence of its physicochemical properties on the sorption characteristics of phenanthrene. Bioresour Technol 2013; 149: 383-9.
[http://dx.doi.org/10.1016/j.biortech.2013.09.085] [PMID: 24128401]

[4]   Li B, Yang L, Wang C, *et al.* Adsorption of Cd(II) from aqueous solutions by rape straw biochar derived from different modification processes. Chemosphere 2017; 175: 332-40.
[http://dx.doi.org/10.1016/j.chemosphere.2017.02.061] [PMID: 28235742]

[5]   Ahmad M, Rajapaksha AU, Lim JE, *et al.* Biochar as a sorbent for contaminant management in soil and water: A review. Chemosphere 2014; 99: 19-33.
[http://dx.doi.org/10.1016/j.chemosphere.2013.10.071] [PMID: 24289982]

[6]   Lehmann J, Joseph S. Biochar for environmental management: science, technology and implementation. Routledge 2015.
[http://dx.doi.org/10.4324/9780203762264]

[7]   Wang L, Wang Y, Ma F, *et al.* Mechanisms and reutilization of modified biochar used for removal of heavy metals from wastewater: A review. Sci Total Environ 2019; 668: 1298-309.
[http://dx.doi.org/10.1016/j.scitotenv.2019.03.011] [PMID: 31018469]

[8]   Bridgwater AV, Carson P, Coulson M. A comparison of fast and slow pyrolysis liquids from mallee. Int J Glob Energy Issues 2007; 27(2): 204-16.
[http://dx.doi.org/10.1504/IJGEI.2007.013655]

[9]   Gaunt JL, Lehmann J. Energy balance and emissions associated with biochar sequestration and pyrolysis bioenergy production. Environ Sci Technol 2008; 42(11): 4152-8.
[http://dx.doi.org/10.1021/es071361i] [PMID: 18589980]

[10]   Emrich W. Handbook of charcoal making: the traditional and industrial methods. Springer Science & Business Media 2013; Vol. 7.

[11]   Hoslett J, Ghazal H, Ahmad D, Jouhara H. Removal of copper ions from aqueous solution using low temperature biochar derived from the pyrolysis of municipal solid waste. Sci Total Environ 2019; 673: 777-89.
[http://dx.doi.org/10.1016/j.scitotenv.2019.04.085] [PMID: 31003106]

[12]   Jouhara H, Ahmad D, Czajczyńska D, *et al.* Experimental investigation on the chemical characterisation of pyrolytic products of discarded food at temperatures up to 300 °C. Therm Sci Eng Prog 2018; 5: 579-88.
[http://dx.doi.org/10.1016/j.tsep.2018.02.010]

[13]   Sun Y, Gao B, Yao Y, *et al.* Effects of feedstock type, production method, and pyrolysis temperature on biochar and hydrochar properties. Chem Eng J 2014; 240: 574-8.
[http://dx.doi.org/10.1016/j.cej.2013.10.081]

[14]   Rodríguez-Vila A, Selwyn-Smith H, Enunwa L, Smail I, Covelo EF, Sizmur T. Predicting Cu and Zn sorption capacity of biochar from feedstock C/N ratio and pyrolysis temperature. Environ Sci Pollut Res Int 2018; 25(8): 7730-9.
[http://dx.doi.org/10.1007/s11356-017-1047-2] [PMID: 29288302]

[15]   Wang Y, Hu Y, Zhao X, Wang S, Xing G. Comparisons of biochar properties from wood material and crop residues at different temperatures and residence times. Energy Fuels 2013; 27(10): 5890-9.
[http://dx.doi.org/10.1021/ef400972z]

[16]   Yaman S. Pyrolysis of biomass to produce fuels and chemical feedstocks. Energy Convers Manage 2004; 45(5): 651-71.
[http://dx.doi.org/10.1016/S0196-8904(03)00177-8]

[17] Manyà JJ. Pyrolysis for biochar purposes: a review to establish current knowledge gaps and research needs. Environ Sci Technol 2012; 46(15): 7939-54.
[http://dx.doi.org/10.1021/es301029g] [PMID: 22775244]

[18] Ahmad M, Lee SS, Dou X, *et al.* Effects of pyrolysis temperature on soybean stover- and peanut shell-derived biochar properties and TCE adsorption in water. Bioresour Technol 2012; 118: 536-44.
[http://dx.doi.org/10.1016/j.biortech.2012.05.042] [PMID: 22721877]

[19] Chia CH, Downie A, Munroe P. Characteristics of biochar: physical and structural properties. Biochar for environmental management: Science, technology and implementation 89-109.2015;

[20] Zhang J, Liu J, Liu R. Effects of pyrolysis temperature and heating time on biochar obtained from the pyrolysis of straw and lignosulfonate. Bioresour Technol 2015; 176: 288-91.
[http://dx.doi.org/10.1016/j.biortech.2014.11.011] [PMID: 25435066]

[21] Bansal RC, Donnet J, Stoeckli F. A review of "Active Carbon". New York: Marcel Decker. Inc. 1988; p. 482.

[22] Al-Wabel MI, Al-Omran A, El-Naggar AH, Nadeem M, Usman ARA. Pyrolysis temperature induced changes in characteristics and chemical composition of biochar produced from conocarpus wastes. Bioresour Technol 2013; 131: 374-9.
[http://dx.doi.org/10.1016/j.biortech.2012.12.165] [PMID: 23376202]

[23] Zhang T, Walawender W, Fan L, Fan M, Daugaard D, Brown R. Preparation of activated carbon from forest and agricultural residues through CO activation. Chem Eng J 2004; 105(1-2): 53-9.
[http://dx.doi.org/10.1016/j.cej.2004.06.011]

[24] Lua AC, Yang T, Guo J. Effects of pyrolysis conditions on the properties of activated carbons prepared from pistachio-nut shells. J Anal Appl Pyrolysis 2004; 72(2): 279-87.
[http://dx.doi.org/10.1016/j.jaap.2004.08.001]

[25] Bagreev A, Bandosz TJ, Locke DC. Pore structure and surface chemistry of adsorbents obtained by pyrolysis of sewage sludge-derived fertilizer. Carbon 2001; 39(13): 1971-9.
[http://dx.doi.org/10.1016/S0008-6223(01)00026-4]

[26] Ahmedna M, Marshall WE, Husseiny AA, Rao RM, Goktepe I. The use of nutshell carbons in drinking water filters for removal of trace metals. Water Res 2004; 38(4): 1062-8.
[http://dx.doi.org/10.1016/j.watres.2003.10.047] [PMID: 14769427]

[27] Jindo K, Mizumoto H, Sawada Y, Sanchez-Monedero MA, Sonoki T. Physical and chemical characterization of biochars derived from different agricultural residues. Biogeosciences 2014; 11(23): 6613-21.
[http://dx.doi.org/10.5194/bg-11-6613-2014]

[28] Keiluweit M, Nico PS, Johnson MG, Kleber M. Dynamic molecular structure of plant biomass-derived black carbon (biochar). Environ Sci Technol 2010; 44(4): 1247-53.
[http://dx.doi.org/10.1021/es9031419] [PMID: 20099810]

[29] Shen Z, Jin F, Wang F, McMillan O, Al-Tabbaa A. Sorption of lead by Salisbury biochar produced from British broadleaf hardwood. Bioresour Technol 2015; 193: 553-6.
[http://dx.doi.org/10.1016/j.biortech.2015.06.111] [PMID: 26141669]

[30] Yuan JH, Xu RK, Zhang H. The forms of alkalis in the biochar produced from crop residues at different temperatures. Bioresour Technol 2011; 102(3): 3488-97.
[http://dx.doi.org/10.1016/j.biortech.2010.11.018] [PMID: 21112777]

[31] Hunter RJ. Zeta potential in colloid science: principles and applications. Academic press 2013; Vol. 2.

[32] Xu R, Xiao S, Yuan J, Zhao A. Adsorption of methyl violet from aqueous solutions by the biochars derived from crop residues. Bioresour Technol 2011; 102(22): 10293-8.
[http://dx.doi.org/10.1016/j.biortech.2011.08.089] [PMID: 21924897]

[33] Tong X, Xu R. Removal of Cu(II) from acidic electroplating effluent by biochars generated from crop

straws. J Environ Sci (China) 2013; 25(4): 652-8.
[http://dx.doi.org/10.1016/S1001-0742(12)60118-1] [PMID: 23923773]

[34]   He X, Jiang J, Hong Z, Pan X, Dong Y, Xu R. Effect of aluminum modification of rice straw–based biochar on arsenate adsorption. J Soils Sediments 2020; 20(8): 3073-82.
[http://dx.doi.org/10.1007/s11368-020-02595-2]

[35]   Bashir S, Zhu J, Fu Q, Hu H. Comparing the adsorption mechanism of Cd by rice straw pristine and KOH-modified biochar. Environ Sci Pollut Res Int 2018; 25(12): 11875-83.
[http://dx.doi.org/10.1007/s11356-018-1292-z] [PMID: 29446023]

[36]   Gaskin JW, Steiner C, Harris K, Das K, Bibens B. Effect of low-temperature pyrolysis conditions on biochar for agricultural use. Trans ASABE 2008; 51(6): 2061-9.
[http://dx.doi.org/10.13031/2013.25409]

[37]   Oliveira FR, Patel AK, Jaisi DP, Adhikari S, Lu H, Khanal SK. Environmental application of biochar: Current status and perspectives. Bioresour Technol 2017; 246: 110-22.
[http://dx.doi.org/10.1016/j.biortech.2017.08.122] [PMID: 28863990]

[38]   Crombie K, Mašek O, Sohi SP, Brownsort P, Cross A. The effect of pyrolysis conditions on biochar stability as determined by three methods. Glob Change Biol Bioenergy 2013; 5(2): 122-31.
[http://dx.doi.org/10.1111/gcbb.12030]

[39]   Tan X, Liu Y, Zeng G, *et al.* Application of biochar for the removal of pollutants from aqueous solutions. Chemosphere 2015; 125: 70-85.
[http://dx.doi.org/10.1016/j.chemosphere.2014.12.058] [PMID: 25618190]

[40]   Yan J, Han L, Gao W, Xue S, Chen M. Biochar supported nanoscale zerovalent iron composite used as persulfate activator for removing trichloroethylene. Bioresour Technol 2015; 175: 269-74.
[http://dx.doi.org/10.1016/j.biortech.2014.10.103] [PMID: 25459832]

[41]   Doumer ME, Rigol A, Vidal M, Mangrich AS. Removal of Cd, Cu, Pb, and Zn from aqueous solutions by biochars. Environ Sci Pollut Res Int 2016; 23(3): 2684-92.
[http://dx.doi.org/10.1007/s11356-015-5486-3] [PMID: 26438367]

[42]   Qian L, Zhang W, Yan J, *et al.* Nanoscale zero-valent iron supported by biochars produced at different temperatures: Synthesis mechanism and effect on Cr(VI) removal. Environ Pollut 2017; 223: 153-60.
[http://dx.doi.org/10.1016/j.envpol.2016.12.077] [PMID: 28110906]

[43]   Rehrah D, Reddy MR, Novak JM, *et al.* Production and characterization of biochars from agricultural by-products for use in soil quality enhancement. J Anal Appl Pyrolysis 2014; 108: 301-9.
[http://dx.doi.org/10.1016/j.jaap.2014.03.008]

[44]   Chen X, Chen G, Chen L, *et al.* Adsorption of copper and zinc by biochars produced from pyrolysis of hardwood and corn straw in aqueous solution. Bioresour Technol 2011; 102(19): 8877-84.
[http://dx.doi.org/10.1016/j.biortech.2011.06.078] [PMID: 21764299]

[45]   Abe I, Iwasaki S, Iwata Y, Kominami H, Kera Y. Relationship between Production Method and Adsorption Property of Char. TANSO 1998; 1998(185): 277-84.
[http://dx.doi.org/10.7209/tanso.1998.277]

[46]   Angın D. Effect of pyrolysis temperature and heating rate on biochar obtained from pyrolysis of safflower seed press cake. Bioresour Technol 2013; 128: 593-7.
[http://dx.doi.org/10.1016/j.biortech.2012.10.150] [PMID: 23211485]

[47]   Xu X, Cao X, Zhao L. Comparison of rice husk- and dairy manure-derived biochars for simultaneously removing heavy metals from aqueous solutions: Role of mineral components in biochars. Chemosphere 2013; 92(8): 955-61.
[http://dx.doi.org/10.1016/j.chemosphere.2013.03.009] [PMID: 23591132]

[48]   Cao X, Harris W. Properties of dairy-manure-derived biochar pertinent to its potential use in remediation. Bioresour Technol 2010; 101(14): 5222-8.
[http://dx.doi.org/10.1016/j.biortech.2010.02.052] [PMID: 20206509]

[49]   Ronsse F, van Hecke S, Dickinson D, Prins W. Production and characterization of slow pyrolysis biochar: influence of feedstock type and pyrolysis conditions. Glob Change Biol Bioenergy 2013; 5(2): 104-15.
[http://dx.doi.org/10.1111/gcbb.12018]

[50]   Chen H, Xie A, You S. In *A* review: advances on absorption of heavy metals in the waste water by biochar IOP Conference Series: Materials Science and Engineering. 012160.
[http://dx.doi.org/10.1088/1757-899X/301/1/012160]

[51]   Dong X, Ma LQ, Li Y. Characteristics and mechanisms of hexavalent chromium removal by biochar from sugar beet tailing. J Hazard Mater 2011; 190(1-3): 909-15.
[http://dx.doi.org/10.1016/j.jhazmat.2011.04.008] [PMID: 21550718]

[52]   Ding W, Dong X, Ime IM, Gao B, Ma LQ. Pyrolytic temperatures impact lead sorption mechanisms by bagasse biochars. Chemosphere 2014; 105: 68-74.
[http://dx.doi.org/10.1016/j.chemosphere.2013.12.042] [PMID: 24393563]

[53]   Boehm HP. Some aspects of the surface chemistry of carbon blacks and other carbons. Carbon 1994; 32(5): 759-69.
[http://dx.doi.org/10.1016/0008-6223(94)90031-0]

[54]   Han Y, Boateng AA, Qi PX, Lima IM, Chang J. Heavy metal and phenol adsorptive properties of biochars from pyrolyzed switchgrass and woody biomass in correlation with surface properties. J Environ Manage 2013; 118: 196-204.
[http://dx.doi.org/10.1016/j.jenvman.2013.01.001] [PMID: 23454371]

[55]   Kim KH, Kim JY, Cho TS, Choi JW. Influence of pyrolysis temperature on physicochemical properties of biochar obtained from the fast pyrolysis of pitch pine (Pinus rigida). Bioresour Technol 2012; 118: 158-62.
[http://dx.doi.org/10.1016/j.biortech.2012.04.094] [PMID: 22705519]

[56]   Sizmur T, Fresno T, Akgül G, Frost H, Moreno-Jiménez E. Biochar modification to enhance sorption of inorganics from water. Bioresour Technol 2017; 246: 34-47.
[http://dx.doi.org/10.1016/j.biortech.2017.07.082] [PMID: 28781204]

[57]   Ahmed MB, Zhou JL, Ngo HH, Guo W, Chen M. Progress in the preparation and application of modified biochar for improved contaminant removal from water and wastewater. Bioresour Technol 2016; 214: 836-51.
[http://dx.doi.org/10.1016/j.biortech.2016.05.057] [PMID: 27241534]

[58]   Shafaghat H, Rezaei PS, Ro D, *et al. In-situ* catalytic pyrolysis of lignin in a bench-scale fixed bed pyrolyzer. J Ind Eng Chem 2017; 54: 447-53.
[http://dx.doi.org/10.1016/j.jiec.2017.06.026]

[59]   Chen W, Fang Y, Li K, *et al.* Bamboo wastes catalytic pyrolysis with N-doped biochar catalyst for phenols products. Appl Energy 2020; 260114242
[http://dx.doi.org/10.1016/j.apenergy.2019.114242]

[60]   Wu Y, Cha L, Fan Y, Fang P, Ming Z, Sha H. Activated biochar prepared by pomelo peel using H_3PO_4 for the adsorption of hexavalent chromium: performance and mechanism. Water Air Soil Pollut 2017; 228(10): 405.
[http://dx.doi.org/10.1007/s11270-017-3587-y]

[61]   Han Y, Cao X, Ouyang X, Sohi SP, Chen J. Adsorption kinetics of magnetic biochar derived from peanut hull on removal of Cr (VI) from aqueous solution: Effects of production conditions and particle size. Chemosphere 2016; 145: 336-41.
[http://dx.doi.org/10.1016/j.chemosphere.2015.11.050] [PMID: 26692510]

[62]   Wang S, Gao B, Li Y, *et al.* Manganese oxide-modified biochars: Preparation, characterization, and sorption of arsenate and lead. Bioresour Technol 2015; 181: 13-7.
[http://dx.doi.org/10.1016/j.biortech.2015.01.044] [PMID: 25625462]

[63] Petrović JT, Stojanović MD, Milojković JV, *et al.* Alkali modified hydrochar of grape pomace as a perspective adsorbent of Pb2+ from aqueous solution. J Environ Manage 2016; 182: 292-300.
[http://dx.doi.org/10.1016/j.jenvman.2016.07.081] [PMID: 27494605]

[64] Xue Y, Gao B, Yao Y, *et al.* Hydrogen peroxide modification enhances the ability of biochar (hydrochar) produced from hydrothermal carbonization of peanut hull to remove aqueous heavy metals: Batch and column tests. Chem Eng J 2012; 200-202: 673-80.
[http://dx.doi.org/10.1016/j.cej.2012.06.116]

[65] Zuo X, Liu Z, Chen M. Effect of H_2O_2 concentrations on copper removal using the modified hydrothermal biochar. Bioresour Technol 2016; 207: 262-7.
[http://dx.doi.org/10.1016/j.biortech.2016.02.032] [PMID: 26894566]

[66] Kwak JH, Islam MS, Wang S, *et al.* Biochar properties and lead(II) adsorption capacity depend on feedstock type, pyrolysis temperature, and steam activation. Chemosphere 2019; 231: 393-404.
[http://dx.doi.org/10.1016/j.chemosphere.2019.05.128] [PMID: 31146131]

[67] Wang S, Guo W, Gao F, Yang R. Characterization and Pb(II) removal potential of corn straw- and municipal sludge-derived biochars. R Soc Open Sci 2017; 4(9)170402
[http://dx.doi.org/10.1098/rsos.170402] [PMID: 28989751]

[68] Yu J, Jiang C, Guan Q, *et al.* Enhanced removal of Cr(VI) from aqueous solution by supported ZnO nanoparticles on biochar derived from waste water hyacinth. Chemosphere 2018; 195: 632-40.
[http://dx.doi.org/10.1016/j.chemosphere.2017.12.128] [PMID: 29289904]

[69] Ma Y, Liu WJ, Zhang N, Li YS, Jiang H, Sheng GP. Polyethylenimine modified biochar adsorbent for hexavalent chromium removal from the aqueous solution. Bioresour Technol 2014; 169: 403-8.
[http://dx.doi.org/10.1016/j.biortech.2014.07.014] [PMID: 25069094]

[70] Shang J, Zong M, Yu Y, Kong X, Du Q, Liao Q. Removal of chromium (VI) from water using nanoscale zerovalent iron particles supported on herb-residue biochar. J Environ Manage 2017; 197: 331-7.
[http://dx.doi.org/10.1016/j.jenvman.2017.03.085] [PMID: 28402915]

[71] Qian L, Liu S, Zhang W, *et al.* Enhanced reduction and adsorption of hexavalent chromium by palladium and silicon rich biochar supported nanoscale zero-valent iron. J Colloid Interface Sci 2019; 533: 428-36.
[http://dx.doi.org/10.1016/j.jcis.2018.08.075] [PMID: 30172153]

[72] Kambo HS, Dutta A. A comparative review of biochar and hydrochar in terms of production, physico-chemical properties and applications. Renew Sustain Energy Rev 2015; 45: 359-78.
[http://dx.doi.org/10.1016/j.rser.2015.01.050]

[73] Fernandez ME, Ledesma B, Román S, Bonelli PR, Cukierman AL. Development and characterization of activated hydrochars from orange peels as potential adsorbents for emerging organic contaminants. Bioresour Technol 2015; 183: 221-8.
[http://dx.doi.org/10.1016/j.biortech.2015.02.035] [PMID: 25742754]

[74] Wang L, Guo Y, Zhu Y, *et al.* A new route for preparation of hydrochars from rice husk. Bioresour Technol 2010; 101(24): 9807-10.
[http://dx.doi.org/10.1016/j.biortech.2010.07.031] [PMID: 20709533]

[75] Wang L, Guo Y, Zou B, *et al.* High surface area porous carbons prepared from hydrochars by phosphoric acid activation. Bioresour Technol 2011; 102(2): 1947-50.
[http://dx.doi.org/10.1016/j.biortech.2010.08.100] [PMID: 20851598]

[76] Demir-Cakan R, Baccile N, Antonietti M, Titirici MM. Carboxylate-rich carbonaceous materials *via* one-step hydrothermal carbonization of glucose in the presence of acrylic acid. Chem Mater 2009; 21(3): 484-90.
[http://dx.doi.org/10.1021/cm802141h]

[77] Jain A, Balasubramanian R, Srinivasan MP. Hydrothermal conversion of biomass waste to activated

carbon with high porosity: A review. Chem Eng J 2016; 283: 789-805.
[http://dx.doi.org/10.1016/j.cej.2015.08.014]

[78]   Liu Z, Zhang F, Hoekman SK, Liu T, Gai C, Peng N. Homogeneously dispersed zerovalent iron nanoparticles supported on hydrochar-derived porous carbon: simple, in situ ynthesis and use for dechlorination of PCBs. ACS Sustain Chem& Eng 2016; 4(6): 3261-7.
[http://dx.doi.org/10.1021/acssuschemeng.6b00306]

[79]   Chu G, Zhao J, Huang Y, *et al.* Phosphoric acid pretreatment enhances the specific surface areas of biochars by generation of micropores. Environ Pollut 2018; 240: 1-9.
[http://dx.doi.org/10.1016/j.envpol.2018.04.003] [PMID: 29729563]

[80]   Stavropoulos GG, Samaras P, Sakellaropoulos GP. Effect of activated carbons modification on porosity, surface structure and phenol adsorption. J Hazard Mater 2008; 151(2-3): 414-21.
[http://dx.doi.org/10.1016/j.jhazmat.2007.06.005] [PMID: 17644248]

[81]   Ding Z, Hu X, Wan Y, Wang S, Gao B. Removal of lead, copper, cadmium, zinc, and nickel from aqueous solutions by alkali-modified biochar: Batch and column tests. J Ind Eng Chem 2016; 33: 239-45.
[http://dx.doi.org/10.1016/j.jiec.2015.10.007]

[82]   Wang S, Tang Y, Li K, Mo Y, Li H, Gu Z. Combined performance of biochar sorption and magnetic separation processes for treatment of chromium-contained electroplating wastewater. Bioresour Technol 2014; 174: 67-73.
[http://dx.doi.org/10.1016/j.biortech.2014.10.007] [PMID: 25463783]

[83]   Shang J, Pi J, Zong M, Wang Y, Li W, Liao Q. Chromium removal using magnetic biochar derived from herb-residue. J Taiwan Inst Chem Eng 2016; 68: 289-94.
[http://dx.doi.org/10.1016/j.jtice.2016.09.012]

[84]   Xiao R, Wang JJ, Li R, *et al.* Enhanced sorption of hexavalent chromium [Cr(VI)] from aqueous solutions by diluted sulfuric acid-assisted MgO-coated biochar composite. Chemosphere 2018; 208: 408-16.
[http://dx.doi.org/10.1016/j.chemosphere.2018.05.175] [PMID: 29885507]

[85]   Yang X, Zhang S, Ju M, Liu L. Preparation and modification of biochar materials and their application in soil remediation. Appl Sci (Basel) 2019; 9(7): 1365.
[http://dx.doi.org/10.3390/app9071365]

[86]   Higashikawa FS, Conz RF, Colzato M, Cerri CEP, Alleoni LRF. Effects of feedstock type and slow pyrolysis temperature in the production of biochars on the removal of cadmium and nickel from water. J Clean Prod 2016; 137: 965-72.
[http://dx.doi.org/10.1016/j.jclepro.2016.07.205]

[87]   Wang S, Gao B, Zimmerman AR, *et al.* Physicochemical and sorptive properties of biochars derived from woody and herbaceous biomass. Chemosphere 2015; 134: 257-62.
[http://dx.doi.org/10.1016/j.chemosphere.2015.04.062] [PMID: 25957037]

[88]   Trakal L, Bingöl D, Pohořelý M, Hruška M, Komárek M. Geochemical and spectroscopic investigations of Cd and Pb sorption mechanisms on contrasting biochars: Engineering implications. Bioresour Technol 2014; 171: 442-51.
[http://dx.doi.org/10.1016/j.biortech.2014.08.108] [PMID: 25226061]

[89]   Rajapaksha AU, Vithanage M, Ahmad M, *et al.* Enhanced sulfamethazine removal by steam-activated invasive plant-derived biochar. J Hazard Mater 2015; 290: 43-50.
[http://dx.doi.org/10.1016/j.jhazmat.2015.02.046] [PMID: 25734533]

[90]   Rajapaksha AU, Alam MS, Chen N, *et al.* Removal of hexavalent chromium in aqueous solutions using biochar: Chemical and spectroscopic investigations. Sci Total Environ 2018; 625: 1567-73.
[http://dx.doi.org/10.1016/j.scitotenv.2017.12.195] [PMID: 29996453]

[91]   Xu X, Huang H, Zhang Y, Xu Z, Cao X. Biochar as both electron donor and electron shuttle for the

reduction transformation of Cr(VI) during its sorption. Environ Pollut 2019; 244: 423-30.
[http://dx.doi.org/10.1016/j.envpol.2018.10.068] [PMID: 30352357]

[92] Zhu Q, Wu J, Wang L, Yang G, Zhang X. Adsorption characteristics of Pb^{2+} onto wine lees-derived biochar. Bull Environ Contam Toxicol 2016; 97(2): 294-9.
[http://dx.doi.org/10.1007/s00128-016-1760-4] [PMID: 26920696]

[93] Mohan D, Rajput S, Singh VK, Steele PH, Pittman CU Jr. Modeling and evaluation of chromium remediation from water using low cost bio-char, a green adsorbent. J Hazard Mater 2011; 188(1-3): 319-33.
[http://dx.doi.org/10.1016/j.jhazmat.2011.01.127] [PMID: 21354700]

[94] Park D, Yun YS, Hye Jo J, Park JM. Mechanism of hexavalent chromium removal by dead fungal biomass of Aspergillus niger. Water Res 2005; 39(4): 533-40.
[http://dx.doi.org/10.1016/j.watres.2004.11.002] [PMID: 15707625]

[95] An Q, Li XQ, Nan HY, Yu Y, Jiang JN. The potential adsorption mechanism of the biochars with different modification processes to Cr(VI). Environ Sci Pollut Res Int 2018; 25(31): 31346-57.
[http://dx.doi.org/10.1007/s11356-018-3107-7] [PMID: 30194580]

[96] Zhao N, Yin Z, Liu F, *et al.* Environmentally persistent free radicals mediated removal of Cr(VI) from highly saline water by corn straw biochars. Bioresour Technol 2018; 260: 294-301.
[http://dx.doi.org/10.1016/j.biortech.2018.03.116] [PMID: 29631179]

[97] Kim WK, Shim T, Kim YS, *et al.* Characterization of cadmium removal from aqueous solution by biochar produced from a giant Miscanthus at different pyrolytic temperatures. Bioresour Technol 2013; 138: 266-70.
[http://dx.doi.org/10.1016/j.biortech.2013.03.186] [PMID: 23619139]

[98] Tran HN, You SJ, Chao HP. Effect of pyrolysis temperatures and times on the adsorption of cadmium onto orange peel derived biochar. Waste Manag Res 2016; 34(2): 129-38.
[http://dx.doi.org/10.1177/0734242X15615698] [PMID: 26608900]

[99] Deng Y, Huang S, Laird DA, Wang X, Dong C. Quantitative mechanisms of cadmium adsorption on rice straw- and swine manure-derived biochars. Environ Sci Pollut Res Int 2018; 25(32): 32418-32.
[http://dx.doi.org/10.1007/s11356-018-2991-1] [PMID: 30232770]

[100] Prapagdee S, Piyatiratitivorakul S, Petsom A. Activation of Cassava Stem Biochar by Physico-Chemical Method for Stimulating Cadmium Removal Efficiency from Aqueous Solution. Environ Asia 2014; 7(2): 60-9.

[101] Liu L, Fan S. Removal of cadmium in aqueous solution using wheat straw biochar: effect of minerals and mechanism. Environ Sci Pollut Res Int 2018; 25(9): 8688-700.
[http://dx.doi.org/10.1007/s11356-017-1189-2] [PMID: 29322394]

[102] Zhang M, Gao B, Varnoosfaderani S, Hebard A, Yao Y, Inyang M. Preparation and characterization of a novel magnetic biochar for arsenic removal. Bioresour Technol 2013; 130: 457-62.
[http://dx.doi.org/10.1016/j.biortech.2012.11.132] [PMID: 23313693]

[103] Wang H, Gao B, Wang S, Fang J, Xue Y, Yang K. Removal of Pb(II), Cu(II), and Cd(II) from aqueous solutions by biochar derived from $KMnO_4$ treated hickory wood. Bioresour Technol 2015; 197: 356-62.
[http://dx.doi.org/10.1016/j.biortech.2015.08.132] [PMID: 26344243]

[104] Wang S, Gao B, Zimmerman AR, *et al.* Removal of arsenic by magnetic biochar prepared from pinewood and natural hematite. Bioresour Technol 2015; 175: 391-5.
[http://dx.doi.org/10.1016/j.biortech.2014.10.104] [PMID: 25459847]

[105] Karunanayake AG, Todd OA, Crowley M, *et al.* Lead and cadmium remediation using magnetized and nonmagnetized biochar from Douglas fir. Chem Eng J 2018; 331: 480-91.
[http://dx.doi.org/10.1016/j.cej.2017.08.124]

[106] Trakal L, Veselská V, Šafařík I, Vítková M, Číhalová S, Komárek M. Lead and cadmium sorption

mechanisms on magnetically modified biochars. Bioresour Technol 2016; 203: 318-24.
[http://dx.doi.org/10.1016/j.biortech.2015.12.056] [PMID: 26748045]

[107] Mohan D, Kumar H, Sarswat A, Alexandre-Franco M, Pittman CU Jr. Cadmium and lead remediation using magnetic oak wood and oak bark fast pyrolysis bio-chars. Chem Eng J 2014; 236: 513-28.
[http://dx.doi.org/10.1016/j.cej.2013.09.057]

[108] Inyang M, Gao B, Ding W, Pullammanappallil P, Zimmerman AR, Cao X. Enhanced lead sorption by biochar derived from anaerobically digested sugarcane bagasse. Sep Sci Technol 2011; 46(12): 1950-6.
[http://dx.doi.org/10.1080/01496395.2011.584604]

[109] Komkiene J, Baltrenaite E. Biochar as adsorbent for removal of heavy metal ions [Cadmium(II), Copper(II), Lead(II), Zinc(II)] from aqueous phase. Int J Environ Sci Technol 2016; 13(2): 471-82.
[http://dx.doi.org/10.1007/s13762-015-0873-3]

[110] Li R, Liang W, Wang JJ, *et al.* Facilitative capture of As(V), Pb(II) and methylene blue from aqueous solutions with MgO hybrid sponge-like carbonaceous composite derived from sugarcane leafy trash. J Environ Manage 2018; 212: 77-87.
[http://dx.doi.org/10.1016/j.jenvman.2017.12.034] [PMID: 29428656]

[111] Li R, Deng H, Zhang X, *et al.* High-efficiency removal of Pb(II) and humate by a CeO_2–MoS_2 hybrid magnetic biochar. Bioresour Technol 2019; 273: 335-40.
[http://dx.doi.org/10.1016/j.biortech.2018.10.053] [PMID: 30448686]

[112] Jellali S, Diamantopoulos E, Haddad K, Anane M, Durner W, Mlayah A. Lead removal from aqueous solutions by raw sawdust and magnesium pretreated biochar: Experimental investigations and numerical modelling. J Environ Manage 2016; 180: 439-49.
[http://dx.doi.org/10.1016/j.jenvman.2016.05.055] [PMID: 27266649]

[113] Zamani SA, Yunus R, Samsuri A, Salleh M, Asady B. Removal of Zinc from aqueous solution by optimized oil palm empty Fruit bunches biochar as low cost adsorbent. Bioinorganic Chem Applications 2017.
[http://dx.doi.org/10.1155/2017/7914714]

[114] Bogusz A, Oleszczuk P, Dobrowolski R. Application of laboratory prepared and commercially available biochars to adsorption of cadmium, copper and zinc ions from water. Bioresour Technol 2015; 196: 540-9.
[http://dx.doi.org/10.1016/j.biortech.2015.08.006] [PMID: 26295440]

[115] Park JH, Ok YS, Kim SH, *et al.* Competitive adsorption of heavy metals onto sesame straw biochar in aqueous solutions. Chemosphere 2016; 142: 77-83.
[http://dx.doi.org/10.1016/j.chemosphere.2015.05.093] [PMID: 26082184]

[116] McBride MB. Environmental chemistry of soils Oxford University Press. New York 1994.

[117] Xiao X, Chen B, Zhu L. Transformation, morphology, and dissolution of silicon and carbon in rice straw-derived biochars under different pyrolytic temperatures. Environ Sci Technol 2014; 48(6): 3411-9.
[http://dx.doi.org/10.1021/es405676h] [PMID: 24601595]

[118] Arán D, Antelo J, Fiol S, Macías F. Influence of feedstock on the copper removal capacity of waste-derived biochars. Bioresour Technol 2016; 212: 199-206.
[http://dx.doi.org/10.1016/j.biortech.2016.04.043] [PMID: 27099945]

[119] Saleh ME, El-Refaey AA, Mahmoud AH. Effectiveness of sunflower seed husk biochar for removing copper ions from wastewater: a comparative study. Soil Water Res 2016; 11(1): 53-63.
[http://dx.doi.org/10.17221/274/2014-SWR]

[120] Amin MT, Alazba AA, Shafiq M. Removal of copper and lead using banana biochar in batch adsorption systems: isotherms and kinetic studies. Arab J Sci Eng 2018; 43(11): 5711-22.
[http://dx.doi.org/10.1007/s13369-017-2934-z]

[121] Li M, Liu Q, Guo L, *et al.* Cu(II) removal from aqueous solution by Spartina alterniflora derived biochar. Bioresour Technol 2013; 141: 83-8.
[http://dx.doi.org/10.1016/j.biortech.2012.12.096] [PMID: 23317555]

[122] Hadjittofi L, Prodromou M, Pashalidis I. Activated biochar derived from cactus fibres – Preparation, characterization and application on Cu(II) removal from aqueous solutions. Bioresour Technol 2014; 159: 460-4.
[http://dx.doi.org/10.1016/j.biortech.2014.03.073] [PMID: 24718356]

[123] Liatsou I, Constantinou P, Pashalidis I. Copper binding by activated biochar fibres derived from Luffa Cylindrica. Water Air Soil Pollut 2017; 228(7): 255.
[http://dx.doi.org/10.1007/s11270-017-3411-8]

[124] Peng H, Gao P, Chu G, Pan B, Peng J, Xing B. Enhanced adsorption of Cu(II) and Cd(II) by phosphoric acid-modified biochars. Environ Pollut 2017; 229: 846-53.
[http://dx.doi.org/10.1016/j.envpol.2017.07.004] [PMID: 28779896]

[125] Yang GX, Jiang H. Amino modification of biochar for enhanced adsorption of copper ions from synthetic wastewater. Water Res 2014; 48: 396-405.
[http://dx.doi.org/10.1016/j.watres.2013.09.050] [PMID: 24183556]

[126] Shim T, Yoo J, Ryu C, Park YK, Jung J. Effect of steam activation of biochar produced from a giant Miscanthus on copper sorption and toxicity. Bioresour Technol 2015; 197: 85-90.
[http://dx.doi.org/10.1016/j.biortech.2015.08.055] [PMID: 26318926]

[127] Lou K, Rajapaksha AU, Ok YS, Chang SX. Sorption of copper(II) from synthetic oil sands process-affected water (OSPW) by pine sawdust biochars: effects of pyrolysis temperature and steam activation. J Soils Sediments 2016; 16(8): 2081-9.
[http://dx.doi.org/10.1007/s11368-016-1382-9]

[128] Kim BS, Lee HW, Park SH, *et al.* Removal of Cu2+ by biochars derived from green macroalgae. Environ Sci Pollut Res Int 2016; 23(2): 985-94.
[http://dx.doi.org/10.1007/s11356-015-4368-z] [PMID: 25813639]

[129] Zhou D, Ghosh S, Zhang D, *et al.* Role of ash content in biochar for copper immobilization. Environ Eng Sci 2016; 33(12): 962-9.
[http://dx.doi.org/10.1089/ees.2016.0042]

[130] Roberts KG, Gloy BA, Joseph S, Scott NR, Lehmann J. Life cycle assessment of biochar systems: estimating the energetic, economic, and climate change potential. Environ Sci Technol 2010; 44(2): 827-33.
[http://dx.doi.org/10.1021/es902266r] [PMID: 20030368]

[131] Alhashimi HA, Aktas CB. Life cycle environmental and economic performance of biochar compared with activated carbon: A meta-analysis. Resour Conserv Recycling 2017; 118: 13-26.
[http://dx.doi.org/10.1016/j.resconrec.2016.11.016]

[132] Ambaye TG, Vaccari M, van Hullebusch ED, Amrane A, Rtimi S. Mechanisms and adsorption capacities of biochar for the removal of organic and inorganic pollutants from industrial wastewater. Int J Environ Sci Technol 2021; 18(10): 3273-94.
[http://dx.doi.org/10.1007/s13762-020-03060-w]

[133] Ahmed MB, Zhou JL, Ngo HH, Guo W. Insight into biochar properties and its cost analysis. Biomass Bioenergy 2016; 84: 76-86.
[http://dx.doi.org/10.1016/j.biombioe.2015.11.002]

[134] Zhang L, Liu X, Huang X, Wang W, Sun P, Li Y. Adsorption of Pb $^{2+}$ from aqueous solutions using Fe–Mn binary oxides-loaded biochar: kinetics, isotherm and thermodynamic studies. Environ Technol 2019; 40(14): 1853-61.
[http://dx.doi.org/10.1080/09593330.2018.1432693] [PMID: 29364052]

[135] Wang Q, Wang B, Lee X, Lehmann J, Gao B. Sorption and desorption of Pb(II) to biochar as affected by oxidation and pH. Sci Total Environ 2018; 634: 188-94.
[http://dx.doi.org/10.1016/j.scitotenv.2018.03.189] [PMID: 29627541]

CHAPTER 14

Contribution of Green Technologies in Getting Sustainable Environment

Bhupinder Dhir[1,*]

[1] *School of Life Sciences, Indira Gandhi National Open University, New Delhi-110078, India*

Abstract: Green technologies provide an eco-friendly and sustainable alternative to conventional technologies. Conventional technologies used for combating pollution show certain limitations and drawbacks. Green technologies have been accepted worldwide for their advantages, such as easy availability, less environmental harm and sustainability. In recent years, solar, wind, geothermal energy, and alternate fuels, such as biogas and biodiesel, have emerged as eco-friendly alternatives to conventional energy sources and fuels. Green technologies, such as developing eco-friendly and recyclable products, have restricted the release of greenhouse gases, generation of waste, and exploitation of natural resources to a great extent. Green technologies thus provide a sustainable option to prevent environmental degradation and over-exploitation of natural resources. Carbon-neutral alternatives have the potential to meet the needs of present and future generations. The production of clean energy is one of the major approaches to get a sustainable environment. Developing clean and environmentally friendly carbon-neutral alternatives can prove useful in meeting the needs of the fuels of present and future generations.

Keywords: Biodiesel, Biofuels, Geothermal energy, Green products, Solar energy, Wind energy.

INTRODUCTION

Scientific and technological innovations, though, have improved the standard of living but have resulted in the over-exploitation of natural resources and environmental degradation. Global warming, climate change, and frequent occurrence of natural disasters have affected environmental sustainability. Therefore, an urgent need to develop sustainable and eco-friendly technologies was realized. Green technologies have emerged as eco-friendly alternatives to conventional technologies.

---

[*] **Corresponding author Bhupinder Dhir:** School of Life Sciences, Indira Gandhi National Open University, New Delhi-110078, India; E-mail: bhupdhir@gmail.com

A technology that aims at conserving natural resources and the environment is referred to as Green technology (GT). In other words, it is defined as the technology that aims at fulfilling the needs of society by developing products, equipment, and systems that help to conserve the environment and resources without compromising the needs of future generations [1]. The uses of green technologies help us reduce the use of fossil fuels, waste and pollution, conserve natural resources and play an important role in protecting the environment. Green technologies also help monitor, assess, remedy, and restore the environment. Green technology is also known as sustainable technology.

Some green technologies include bio-fuel, eco-forestry, alternative energy sources such as renewable energy, green products, recyclable products (eco-friendly), cleaner fuels, and solid waste management. Green technology aims to minimize resource use, generate waste, conserve natural resources, and reduce human involvement.

The affordability of this technology is the main reason behind wider acceptance, especially in developing nations. Technological innovations possess great potential to maintain environmental sustainability [2].

GREEN TECHNOLOGIES

Some of the major green technologies have been discussed below.

Green Products or Green Manufacturing

Green products are developed to reduce waste and maximize resource efficiency. They are manufactured using toxic-free ingredients and environmentally-friendly procedures. Common examples of green products include solar panels and thermal heating discs. Solar panels heat the sun to generate electricity and can be installed in homes, apartments, and commercial buildings. Thermal heating discs work on the principle of trapping the rays of the sun. Green products provide an alternative to minimizing fossil fuel use [2].

Green chemicals form an important part of green technology. These products do not contain toxic chemicals and reduce water pollution, hence are eco-friendly. Some of the characteristics of green products are that they

- Do not produce toxic chemicals and within hygienic conditions
- Can be recycled, reused and is biodegradable
- Are produced without the use of fewer resources
- Possess less or zero carbon footprint
- Possess less or zero plastic footprint

Examples of such products include cleaning agents and green laundry detergent made of coconut and glycerin, insecticides made from orange or peppermint oil, reusable water bottle development, and microemulsions (aqueous) as cleaning agents for (an alternative to VOCs). Wooden cutlery made with bamboo fiber and handmade watches made from wood is other green products.

Eco-Friendly Products

Eco-friendly products have been developed to maintain sustainable use of materials and save energy. Some of them have been described below.

An E-reader has been developed for people who buy new books frequently. We can reduce the paper that is used in the printing of books. The use of an E-reader helps in reducing the cutting of trees. Rocket book Everlast Reusable Smart Notebook, developed in the US can be used to write and reused many times after wiping and cleaning. Eco Tool's air hairdryer, made from bamboo, recycled aluminum, and plastic, helps dry hair faster (40%) and causes less heat damage. The 100% tree-free packaging paper is also used worldwide. Chic-made bags made from recycled materials also come in the eco-friendly products category. Biodegradable trash bags are normally easy to compost and hence safe for the environment. They are 100% compostable and are BPI certified. Bamboo-made washcloths and makeup removers are now an eco-friendly alternative in houses and parlors. They do not harm sensitive skin and are hypoallergenic and antimicrobial. Reusable makeup remover pads made from organic bamboo fibers are getting more popular daily. Toothbrushes made from bamboo are stronger and cheaper than plastic. LED light bulbs convert 95% of energy into light, and only 5% is wasted as heat. They also use less energy, saving money. Biodegradable Garden Pots made from recycled material decompose easily. They help in reducing the use of plastic and rubber that cause pollution. Recycled Floor Mats made from natural, reclaimed, or recycled rubber material. Ballpoint pens made from recycled plastic bottles. The B2P (Bottle-2-Pen) pens are retractable and refillable. Reusable coffee cups and lids help reduce a significant amount of single-use plastic waste. Eco-friendly phone cases made up of sustainable bioplastics that are completely (100%) compostable. Eco-friendly dishwasher reduces the usage of energy and water to half the amount. This green product benefits the environment and helps consumers save a lot. A programmable thermostat is a green technology product that can set a schedule and automatically adjust the temperature to save energy.

Ecofriendly vehicles that do not emit gases have also been developed. They help in reducing the gases such as carbon dioxide (CO_2), carbon monoxide (CO), nitrogen oxide (NO_x), sulphur dioxide and hydrocarbon compounds (HC) in the atmosphere, hence reducing the negative influence on the environment. Tesla came out with highly efficient electric cars (EVs), which run on clean energy. Hybrid and electric cars have significantly contributed to reducing carbon emissions and created new avenues for innovation and development. Green buildings are designed in a way that they have large windows, allowing fresh air and abundant natural lighting. A solar-powered lamp or Lantern developed by scientists gets charged using the sun's light for five hours and gives bright illumination for five to seven hours and 24 hours of low illumination. A solar charger developed by research scientists can charge devices while traveling. Example- Nekteck 21W portable solar charger has a compact design and weight of just 18 ounces. It is high-performance and reasonably priced and has got attention worldwide. Solar panels are designed to capture energy from the sun. Both electricity and heat can be generated using solar energy. The front of these panels is designed to generate photovoltaic energy, while the rear produces hot water using a heat exchanger.

Green Architecture

Buildings that use existing natural light and ensure adequate insulation to reduce energy consumption are being constructed. Such construction practices reduce energy utilized in lighting, reduce the heat lost outside and remove the need for heating. The construction materials are obtained from urban waste and landfills. This technology ensures that buildings are passive and do not require additional materials for production.

Reclaimed materials, passive solar design, natural ventilation, and green roofing are some environmentally friendly techniques used in green buildings to reduce carbon footprint. Open spaces and natural airflow reduce the need for air conditioning, thus preventing many environmental problems. The creation of green cities with large recycling operations helps to reduce the environmental footprint along with the conservation of resources. Alternative fuel vehicles, known as green vehicles, have been introduced as an alternative that does not cause harm to the environment.

According to EPA, seven 7 components of the green building concept are

- Energy efficiency and use of renewable energy
- Water efficiency
- Environmentally-friendly building materials
- Reduction of toxins
- Indoor air quality

- Sustainable development
- Waste reduction

Most green products are reusable. These products help reduce the risk of producing toxic waste, thus providing environmental protection. Consumers are encouraged to buy products with high energy star certification. This aims to reduce energy consumption, emission of greenhouse gases, and the over usage of non-renewable resources, hence protecting the environment. Using green technology in chemical, petrochemical, pharmaceutical, and automotive industries can help minimize environmental damage. Adoption of clean technologies can minimize waste generation, usage of resources, and emissions.

Advantages of Using Green Products

Green products last longer than conventional products. Moreover, these products consume less energy and other resources. They are cost-effective. Green products result in low maintenance costs if operated and maintained properly. Most eco-friendly products are made from natural materials that are free from harmful chemicals; hence they improve health. According to an estimate, using renewable energy created almost 5 lac new jobs in 2017. It is predicted that with increasing demands for green products, the number of jobs will rise to 16 million by 2030. Thus, the development of green products also boosts economic conditions. Green products reduce the overuse of resources such as fossil fuels. Green products are made from organic and biodegradable materials, hence leading to less generation of greenhouse gases like CFCs, ozone, methane, and toxic wastes. This prevents pollution and deterioration of climatic conditions. In contrast, some green products are costly as they require innovation and investment for research before development. They are not so popular because of the lack of awareness of the masses.

Renewable Energy Sources

Energy sources that replenish themselves naturally are known as renewable sources of energy. Tester [3] defined renewable energy sources as sustainable energy because it deals with meeting the needs of the present generation to preserve them for future generations. They include solar, wind, hydropower, geothermal, ocean (tide and wave), and bioenergy [4 - 7]. The world's growing demand for energy to support the growing population led to the overexploitation of fossil fuel-based energy sources such as coal. Overuse of fossil fuels results in the depletion of their reserves, increased greenhouse gas emissions, and other environmental issues. Renewable energy was explored as one sustainable alternative that helped reduce greenhouse gas emissions [8, 9].

Renewable energy is obtained naturally from ongoing flows of energy in our surroundings. Renewable energy is obtained from natural and persistent energy sources available in our immediate environment [8]. It proved to be a sustainable resource as it is limitless and produces no harmful environmental effects. These sustainable energy sources are clean, affordable, available and accessible [10].

Table 1. List of countries using renewable energy at a large scale.

Rank	Country	Renewable energy used (%)
1	Germany	12.74
2	UK	11.95
3	Sweden	10.96
4	Spain	10.17
5	Italy	8.8
6	Brazil	7.35
7	Japan	5.3
8	Turkey	5.25
9	Australia	4.75
10	USA	4.32

Many countries all over the world have switched to the use of renewable sources of energy at a large scale (Table **1**). Countries using huge amounts of natural energy resources include Germany, UK, and Sweden. According to a 2019 report, about 40% of the power generation in the UK has been achieved using wind energy.

Solar Energy

Electricity is generated from solar energy. Solar irradiance generates electricity using photovoltaic (PV) [11]. Solar photovoltaic technology converts sunlight into electricity using semiconductor modules. Generating electricity from solar energy reduces fossil fuel use, reducing pollution (mainly greenhouse gas emissions). A solar cell directly converts light energy into electrical energy through the process of photovoltaics. The electricity generated through solar energy is used for pumping water in agricultural fields, refrigeration, communication, charging batteries, and lighting streets and houses in rural areas. Solar energy produces low-cost, high-power photovoltaic cells, which makes it one of the most affordable, efficient energy sources [12]. Solar power is also used to generate thermal energy to produce fuels that can be used for transport and other purposes [13].

Solar panels and plants have become common in many households and office premises. Scientists have been working on improving the design of solar panels so that they collect energy in weather conditions such as rain. These are called all-weather solar panels. During rainfall, the solar panel generates electricity from the force of the rain falling on their surface.

Wind Energy

Wind energy is harnessed to produce electricity or mechanical power. Wind turbines of sizes ranging from 900 W to 50 kW installed onshore (land) or offshore (in the sea or freshwater) have been in about 75% of the world and applied off-grid for pumping drinking water, irrigation, homes, schools, and energy supply to large power stations [14]. Energy generation *via* wind turbines reduces carbon emissions.

Wind energy technologies are already being manufactured and deployed on a large scale [13, 15]. According to a 2008 report by the Global Wind Energy Council, a total of 357 megawatts of offshore wind was added, bringing the cumulative total to less than 1.5 GW. The Global Wind Energy Council projects are expected to increase wind volume to 840.9 GW by the end of 2022.

Hydropower

Hydropower is an energy source harnessed from water moving from higher elevations to lower levels. The energy generated by this way is used to turn turbines and generate electricity. Hydropower projects include dams, run-of-river and in-stream projects [16], irrigation and drinking water and navigation [13]. Hydropower does not cause any pollution and can store energy for many hours. According to World Energy Council Report 2013, about 50% of the world's hydropower capacity comes from China, Brazil, Canada, and the USA. Hydropower generation does not produce greenhouse gases but improves the country's socioeconomic status.

Micro and small hydropower projects have been implemented in various nations worldwide. Hydropower plants with 500 kW to 25 MW capacity have been put in countries like Nepal and India. Besides rural electrification, hydropower plants play an important role in the irrigation of agricultural fields.

Geothermal Energy

Energy obtained naturally from the earth's interior is referred to as geothermal energy. Heat is present inside the internal structure of the planet earth, and the physical processes occurring help in the generation of energy. Huge amounts of heat are present in the earth's crust, which can be exploited. Heat is harnessed from geothermal reservoirs. Hot and permeable reservoirs are called hydrothermal reservoirs [13].

Ocean Energy (Tide and Wave)

The energy from waves, tides, and currents is collectively known as ocean energy. The energy is produced through waves created when wind passes over water. Wind, tides and waves obtain energy from oceans and seas [15]. The faster the wind speeds, the greater distance the wind travels, the greater the wave height, and the greater the wave energy produced [12]. Ocean energy stores energy that can meet the worldwide power demand. The units for harnessing tidal energy were installed in the UK-SeaGen and Portugal-Pelamis in 2008.

Tidal energy has a huge potential market because of its predictable and consistent availability. The high-pressure water spins the turbines, which then generate zero-carbon electricity.

Alternate Fuels

Fuels derived from sources other than petroleum are known as alternate fuels. Alternative fuels include hydrogen, propane, ethanol, methanol, butanol, biodiesel, and vegetable and waste-derived oils [17]. Most of them are produced from renewable sources. Bioethanol, biogas and biodiesel are the most widely used biofuels. Many more alternative fuels are under production or development for use in alternative fuel vehicles and advanced technology vehicles. Using alternative fuels and advanced vehicles helps conserve fuels and lower vehicle emissions. These biofuels help in environmental protection as they play a role in producing clean energy and producing less pollution. Green fuels can significantly reduce greenhouse gas emissions by as much as 30%. Biofuels represent sustainable energy sources as they do not increase net CO_2 emissions [10].

Biofuels produced from food or animal feed crops are first-generation biofuels. These biofuels are produced through fermentation, distillation, and transesterification processes. They are also commonly referred to as 'conventional biofuels'. Second-generation biofuels are derived from non-food feedstocks such as energy crops, including *Miscanthus*, switchgrass, short rotation coppice (SRC) and other lignocellulosic plants), agricultural residues, forest residues, municipal solid waste, and other waste. Second- and third-generation biofuels are called 'advanced biofuels'. This is because their production process is still in the research, development, pilot, or demonstration phase. Biodiesel and bioethanol are the most common, *via*ble alternatives to green fuels. Biodiesel is produced mainly from waste cooking of oil, non-edible vegetable oil, animal fat, and microalgae through conventional transesterification or hydro-treatment of algal oil. At the same time, bioethanol is usually derived from forestry waste, lignocellulosic biomass, starchy and sugary vegetable sources, and agricultural residues.

Bioethanol

Ethanol is a renewable liquid fuel made by fermenting and distilling vegetation sources such as corn. In Brazil, sugarcane and starchy materials; in the USA, corn, and Europe, wheat is used as the main source of ethanol production. Lignocellulosic materials and algal biomass are the other raw materials used to synthesize bioethanol. It is blended with gasoline for use in vehicles. It produces less greenhouse gas (GHG) emissions than other conventional fuels. Research is underway to produce ethanol from cellulose.

Biobutanol

Biobutanol is another fuel from feedstocks, sugar-based substrates such as molasses, and starchy materials such as wheat. Low-cost lignocellulosic wastes are also used in the production of biobutanol. Microorganisms such as *Clostridium acetobutylicum* and *C. beijerinckii* are used to produce biobutanol. The process of biobutanol production is cumbersome as the process products, especially butanol, severely inhibit the anaerobic butanol-producing micro organisms. The process of lignobutanol production is in its preliminary stage of research. Laboratory and pilot-scale study problems need to be addressed before the process is implemented at the commercial level.

Biodiesel

Biodiesel is an eco-friendly substitute for conventional fuels. It is a renewable fuel manufactured from vegetable oils and animal fats. It is also synthesized from feedstock such as grapeseed, rapeseed, sunflower, peanut, palm oil, linseed, palm kernel oil, olive, chestnut, soybeans, canola, corn, jatropha, cottonseed, and many other crops. It is produced mainly through the process of transesterification or the use of an enzyme as a catalyst. Biodiesel is diesel derived from vegetable oils and animal fats. It usually produces fewer air pollutants than petroleum-based diesel.

Biodiesel has several advantages: high octane number, inbuilt lubricity, fewer exhaust emissions, and hence no pollution. It is biodegradable. It is widely used in Scandinavian countries and Germany.

Hydrogen

Hydrogen is an alternative fuel produced mainly from domestic resources. Studies have shown that vehicles powered by pure hydrogen do not produce harmful air pollutants.

Natural Gas

Natural gas is a gaseous fuel found abundantly in U.S. It produces less air pollutants and GHGs compared to conventional fuels.

Propane

Propane is a gaseous fuel that has been widely used in vehicles throughout the world. It produces less harmful air pollutants and GHGs than gasoline.

Alternate Sources of Energy

Biogas

Biogas is produced through the anaerobic digestion of organic waste by methanogenic bacteria such as agricultural, municipal, food, and industrial wastes, municipal wastewater sludge, and animal waste. Biogas production utilizes organic agricultural waste and converts it to fuel and fertilizer. Biogas can be produced in simple reactors using a natural consortium of microorganisms such as manure. Biogas production involves hydrolysis, acidogenesis, acetogenesis, and methanogenesis using different types of microbes in a consortium. Biogas production does not need a complicated separation and purification process. Research has been conducted to produce biogas from recalcitrant substrates such as lignocellulose. Germany is one of the countries leading the production of biogas. India is estimated to have about 12 million biogas plants. China generates about 4 billion cubic meters of biogas. The use of biogas results in the saving of fuel wood, agriculture residue, livestock manure, and kerosene.

Biomass

Biomass, such as agricultural residues and wastes, gets converted to electric and thermal energy through combustion, gasification, and cogeneration. Biomass acts as a major of energy production in developing countries.

Bioenergy

Bioenergy is a renewable energy source derived from biological sources [18]. Bioenergy is an important source of energy, which can be used for transport using biodiesel, electricity generation, cooking, and heating. Household waste, vegetable market waste, and waste from the cotton stalks, leather, and pulp; and paper industries can produce useful energy either by direct incineration, gasification, digestion (biogas production), fermentation, or cogeneration. Energy use reductions can be achieved by minimizing energy demand, rational energy use, recovering heat, and using more green energies.

Electricity from bioenergy attracts many different sources, including forest byproducts such as wood residues, agricultural residues such as sugar cane waste, and animal husbandry residue such as cow dung. One advantage of biomass energy-based electricity is that fuel is often a by-product, residue, or waste product from the above sources.

Bioenergy can worsen soil and vegetation degradation related to the overexploitation of forests, exhaustive crop and forest residue removal, and water overuse [19 - 21]. Diversion of crops or land into bioenergy production can induce food commodity prices and food security [22].

Their sustainable development requires a supply of energy resources that are sustainably available at a reasonable cost and can cause no negative societal impacts. Energy resources such as fossil fuels are finite and lack sustainability, while renewable energy sources are sustainable over a relatively long term.

Recycling and Waste Management

Household and industrial solid waste is another resource that can be used to generate energy. Recycling help solve the problem of plastic waste generation. The chemical recycling process uses chemicals to break down post-consumer plastic waste into valuable chemical components. These components can then be used as fuel or converted into new plastic products. Waste materials, when treated, generate energy in the form of steam or electricity. Recycling aims at reducing dependence on energy from hydrocarbons and fossil fuels. Advancements in green technology help to manage and recycle waste material.

Other Approaches to a Sustainable Environment

Smartphones are generally discarded after 3 or 4 years. Companies have been looking to develop an alternative to minimize e-waste. Fairphone is one of the companies that designs and produces smartphones in a way that they have a lower environmental impact. The company developed a smartphone that does not contain conflict minerals and ensures the longer durability of the product. Another sustainable phone is the Teracube, developed in US. Each component of the phone is replaceable, and the chassis is biodegradable. Both smartphones have been developed to have a low supply chain footprint and sustainable design (durable, replaceable, or fixable components).

A passive system that channels sunlight from an external source and transports it through fibre optic cables to illuminate light-deprived rooms has been developed. Since a passive system provides indoor lighting, there is no daytime energy consumption, hence the energy consumption to zero. The technique is known as Sunlight Transport.

Another way developed to get rid of plastics is the formation of "Plastic Roads". In this technique, the roads are made entirely of plastic or mixed with asphalt or only plastic. The raw material consists of prefabricated, hollow, modular elements made from consumer waste plastics. The technique is still in the development phase, and monitoring is going on. Plastic roads can also significantly reduce the carbon footprint (50 to 72%) and have a longer lifespan.

A solar flower is a solar panel system mounted on the ground and shaped like a flower. The structure consists of 12 petals that open up at the beginning of the day with the sun and close with the sunset. These systems are completely portable and ready-to-plug-in. Besides this, the system is self-cleaning, which increases efficiency and durability. It also has a sun tracker to maximize solar energy production. The SmartFlower produces enough to fulfill the average electricity needs of a household. Hence, it is an environment-friendly way to get clean energy.

Plant or Green Walls are vertical structures that can hold enough soil to have different plants or other greens growing on them. These systems have features such as monitoring and self-irrigation, improving their survival, aesthetic, and air purification potential.

Textiles have been developed from the casein found in milk. The cloth is very soft and smooth. It needs chemical processes to get transformed into yarn. These processes include using toxic chemicals (including sulphuric acid and formaldehyde). The German company Qmilk researched it and developed a process that is chemical-free and uses two liters of milk per Kg of fibre. The production of the fiber requires less energy, and the result is natural and smooth fabric as silk.

Engineers at Oregon State University have invented a hybrid electricity generator, which involves the use of wastewater. They have used wastewater to generate electricity *via* microbial fuel cells and reverse electrodialysis. This technique can be scaled up to form the basis of energy and water sustainability in this era of scarcity of natural resources.

Replacement of uranium by thorium in nuclear power can lead to the generation of less radioactive waste. The burning of the agricultural residues in low oxygen conditions can help in reducing the production of greenhouse gases and also produce charcoal, which can be used in many operations.

A plant-based plastic that is recyclable and degradable compared to other packaging materials has been developed by the Dutch company Avantium. Polyethylene furanoate (PEF) is made entirely from bio-based feedstocks (sugars). Sugars used in making it are obtained from sugar beet, sugar cane, wheat, and corn, as well as agricultural residues. The bioplastics developed by Avantium showed less production of greenhouse house gases compared to conventional plastics. Using agricultural residues such as straw as raw material can also provide increased carbon savings in the future.

Photovoltaic (PV), an alternativeto fossil fuel-based electricity, has been developed. They can be directly installed on the roof of a building. Building Integrated Photovoltaics (BIPV) systems are photovoltaic units and solar panels that mimic conventional roofing materials but perform the task of generating electricity. The units can last for two to three decades giving maximum energy output. They produce zero emissions. Electricity production *via* PV panels to meet any electrical energy needs of the systems, carbohydrates (food), liquid fuels, chemical feedstocks, and polymers for fiber manufacture can be produced with minimum water requirements.

Conventional pavements contribute to heat in large urban areas. This is because conventional paving materials such as asphalt and concrete absorb 95 to 60% of the energy reaching them instead of reflecting it into the atmosphere. A road surface has developed using additives or materials that reflect solar radiation. These pavements reflect light and stay cooler in the sun and hence are called "cool pavement". Cool pavements contribute in reducing urban air temperature by 0.6°C (1°F), improve air quality and lower surface temperatures, thus contributing a lot in change in local climate change adaptation.

Carbon Capture and Storage (CCS)

The process of carbon Capture and Storage (CCS) aims to reduce carbon emissions and tackle the problem of global warming. In the process of CCS, carbon dioxide is captured, transported, and stored deep underground. The three main components of the technology are the capture of CO_2, transport of captured CO_2 to storage sites, and injection into the subsurface for storage. This technology has shown the potential to capture more than 90 percent of carbon dioxide (CO_2) emissions. Carbon capture is safe, efficient, and cost-effective. The practice of carbon capture can bring about 14 percent of global greenhouse gas emissions by

2050. In this way, decarbonization in the industrial sector can be achieved. Uncertainties and risks associated with the process, such as storage and leaks, need to be assessed. Twenty-six commercial-scale carbon capture projects are operating around the world. Commercial operation of carbon capture has been demonstrated in coal gasification, production of ethanol, fertilizers and processing of natural gas.

Artificial Photosynthesis

Engineers and scientists are trying to develop a technology that will use sunlight and carbon dioxide to produce energy. Photosynthesis is the process by which trees absorb carbon dioxide from the atmosphere and convert sunlight into energy. The artificial systems for production of fuels from sunlight, based on the principles of natural photosynthesis referred as "artificial leaf".

Artificial photosynthesis systems (APS's) use photocatalysts and biocatalysts to convert and store solar energy for use in the fields of resource, environment, food, and energy. An enzyme bed reactor is used to fix CO_2 in the air, similar to natural photosynthesis. The reactor is run by hydrogen energy and bioelectric transducers. Electrochemical water dissociation into H_2 and O_2 is accomplished by mimicking photosynthesis.

The construction of an artificial photosynthetic system thus requires that one-electron-transfer events can be coupled to catalytic units that are able of mediating the four-electron-four-proton splitting of H_2O. These catalysts may be interfaced with the surface of semiconductor materials to enhance the photoelectrochemical H_2O splitting. The catalysts for the reactions, such as H_2O oxidation and proton reduction, can be attached to a light-harvesting material, or the catalysts can be separately attached to two different semiconductors. The reaction platforms introduced in APS's contribute to the great stability, continuous processing, improve system efficiency and reduce the operation cost. The platforms also improve the photosynthetic performance due to the proximity effect.

In an artificial photosynthetic system, a chromophore (photosensitizer) is first attached to a semiconductor. Upon illumination, a charge-separated state is formed by electron transfer from the photoexcited chromophore to the conduction band of the semiconductor. The charge-separated state allows the chromophore to abstract an electron from the catalyst, thus regenerating the chromophore and a one-electron oxidized catalyst. The generated H_2 could be either used directly as a source of energy or be used to reduce CO_2 to other types of fuels of higher complexity, such as methane or methanol. This chromophore should efficiently absorb and convert the incoming solar energy into an excited state that can transfer an electron to an acceptor for the creation of a charge-separated state, thus generating the required thermodynamic driving force for the desired chemical reactions. Both photoactive molecular dyes and semiconductors can induce electron transfer and have the potential of being used as light-harvesting chromophores in a future artificial device for H_2O splitting.

This technology will reduce carbon dioxide levels and produce renewable fuel. Capturing carbon and efficient conversion of solar energy into electricity will provide success for this technology. Artificial photosynthesis is also supposed to reduce water use significantly.

USE OF GREEN TECHNOLOGY ACROSS THE GLOBE

Many countries have adapted green technology for environmental protection and sustainability. Ireland has been using wind and tidal energy. About 80,000 jobs were created in 2020 and are expected to reach a much higher number in the coming years. Germany emerged as a leader in the renewable energy market. About one-third of the world's solar energy panels have been installed here, and this aims to reduce carbon emissions by 80-95%. Denmark ranks first when the matter of investment in wind energy is concerned. Sweden leads the list of countries that use clean technology (renewable energy) and has a number of clean technology companies. Finland ranks second among the countries that use clean technology. The industry provides jobs to about 50,000 people. Switzerland has various innovative projects, energy and transport infrastructure, and various government supportive policies which support the clean technology industry.

The UK is a country that does innovation in green technology and is currently outperforming in the sector of using wind energy. The USA has invested $ 7.9 billion in research related to implementation of green initiatives aiming at a 30% reduction in carbon emissions by 2030. Iraq has good sources of alternative and clean energy that can be used to generate electricity power generation. Green infrastructure technologies have been adopted in the planning of cities in Iraq to address the environmental problems resulting from the acceleration of desertification and the high temperature. Green building techniques and green architecture have been adopted in the housing sector to achieve sustainability in the housing sector by reducing water and energy consumption. Wind energy is an important source of renewable energy in India. India ranks fifth in the sector of use of wind energy for power generation after China, the United States, Germany, and Spain. Production of 23444MW through wind energy has been planned during the first phase in 2020. India plans to produce 600,000 electric vehicles and about 40 -50 million two-wheeled vehicles and bicycles by 2020 to reduce the burden on conventional fuels and energy sources.

ALTERNATE TECHNOLOGIES AND SUSTAINABLE DEVELOPMENT

Sustainable development is the process in which natural resources are exploited, and the process of technological development and advancement goes in harmony for both current and future generations. Sustainable development aims for economic, social development, and environmental conservation. Sustainability can be attained by harnessing technological advances without negative repercussions on the environment. Modern innovation techniques play a prominent role in achieving sustainable development. Alternative new forms of energy need to be developed to reduce emissions or pollution. These forms include the use of renewable and clean energy and bio-fuels aimed at reducing harmful pollutants and the consumption of fossil resources.

Social and Economic Sustainability

The use of renewable and alternate sources of energy has generated employment from about 2.3 million jobs created worldwide which also has improved health, education, gender equality, and environmental safety [13]. Globally, per capita incomes are positively correlated with per capita energy use, and economic growth can be identified as the essential factor behind increasing energy consumption in the last decades. Technology has been utilized in improving the overall standard of living of people; it is employed in analysis for better allocation of resources.

Environmental Sustainability

Use of renewable energy helps in saving natural resources for future generations. The health, biodiversity and vitality of ecosystems get maintained after the adoption of green technologies. Renewable energy contributes to sustainable development *via* impacts on human development and economic productivity [23, 24]. Renewable energy reduces energy imports and contributes to the economy's stability by enhancing energy security across the globe. The utilization of renewable energy contributes to increasing the reliability of energy services. Diverse forms of energy sources, together with good management and system design, can help to enhance security [13].

CONCLUSION

Renewable energy systems, cleaner energy production, and the development of efficient energy storage devices contribute greatly to a sustainable environment. Green innovations aim to generate high-quality products that can reduce environmental deterioration. Green products, manufacturing, alternative fuels, and renewable energy contribute largely to environmental sustainability. Sustainable green technologies can play a major role in environmental protection and economic development. Green fuels can reduce gaseous emissions, thereby protecting the environment. Green technologies also aim at conserving natural resources. Analysis of the risk involved in the practical implementation of a new technology needs to be done to ensure its acceptability on a global scale.

REFERENCES

[1] Fayomi GU, Mini SE, Fayomi OSI, Odunlami A, Oyeleke OO. Sustainability and Clean Technology: A Technological Perspective. IOP Conference Series of Material Sci Eng.
[http://dx.doi.org/10.1088/1757-899X/1107/1/012076]

[2] Shafiei MWM, Hooman A. The Importance of Green Technologies and Energy Efficiency for Environmental Protection. Int J Appld Environ Sci 2017; 12: 937-51.

[3] Tester JW, Drake EM, Golay MW, Driscoll MJ, Peters WA. Sustainable Energy – Choosing Among Options. Cambridge, Massachusetts, USA: MIT Press 2005; p. 850.

[4] Panwar NL, Kaushik SC, Kothari S. Role of renewable energy sources in environmental protection: A review. Renew Sustain Energy Rev 2011; 15(3): 1513-24.
[http://dx.doi.org/10.1016/j.rser.2010.11.037]

[5] Tabatabaei M, Karimi K, Kumar R, Horváth IS. Renewable energy and alternative fuel technologies. BioMed Res Int 2015; 2015: 1-2.
[http://dx.doi.org/10.1155/2015/245935] [PMID: 25883946]

[6] Owusu PA, Asumadu-Sarkodie S. A review of renewable energy sources, sustainability issues and climate change mitigation. Cogent Eng 2016; 3(1)1167990
[http://dx.doi.org/10.1080/23311916.2016.1167990]

[7] Kumar J, Majid CR. Renewable energy for sustainable development in India: current status, future prospects, challenges, employment, and investment opportunities. Energy Sustain Soc 2020; 10: 2.
[http://dx.doi.org/10.1186/s13705-019-0232-1]

[8] Abbasi T, Premalatha M, Abbasi SA. The return to renewables: Will it help in global warming control? Renew Sustain Energy Rev 2011; 15(1): 891-4.
[http://dx.doi.org/10.1016/j.rser.2010.09.048]

[9] Santiago A, Roxas F. Identifying, developing, and moving sustainable communities through renewable energy. World Journal of Science, Technology and Sustainable Development 2012; 9(4): 273-81.
[http://dx.doi.org/10.1108/20425941211271487]

[10] Twidell J, Weir T. Renewable Energy Resources. 3rd ed., London: Routledge 2015.
[http://dx.doi.org/10.4324/9781315766416]

[11] Asumadu-Sarkodie S, Owusu PA. The potential and economic *via*bility of solar photovoltaic power in Ghana. Energy Sources A Recovery Util Environ Effects 2016; 38(5): 709-16. a
[http://dx.doi.org/10.1080/15567036.2015.1122682]

[12] Jacobson MZ, Delucchi MA. Providing all global energy with wind, water, and solar power, Part I: Technologies, energy resources, quantities and areas of infrastructure, and materials. Energy Policy 2011; 39(3): 1154-69.
[http://dx.doi.org/10.1016/j.enpol.2010.11.040]

[13] Edenhofer O, Pichs-Madruga R, Sokona Y, *et al.* Renewable Energy Sources and Climate Change Mitigation. Cambridge: Cambridge University Press 2011.
[http://dx.doi.org/10.1017/CBO9781139151153]

[14] Asumadu-Sarkodie S, Owusu PA. The potential and economic *via*bility of wind farms in Ghana. Energy Sources A Recovery Util Environ Effects 2016; 38(5): 695-701. b

[http://dx.doi.org/10.1080/15567036.2015.1122680]

[15] Esteban M, Leary D. Current developments and future prospects of offshore wind and ocean energy. Appl Energy 2012; 90(1): 128-36.
[http://dx.doi.org/10.1016/j.apenergy.2011.06.011]

[16] Asumadu-Sarkodie S, Owusu PA, Jayaweera HM. Flood risk management in Ghana: A case study in Accra. Adv Appl Sci Res 2015; 6: 196-201.

[17] Jeswani HK, Chilvers A, Azapagic A. Environmental sustainability of biofuels: a review. Proc- Royal Soc, Math Phys Eng Sci 2020; 476(2243)20200351
[http://dx.doi.org/10.1098/rspa.2020.0351] [PMID: 33363439]

[18] Omer AM. Environment and Development: Bioenergy for Future Arch Chem Eng 11 2019;

[19] Koh LP, Ghazoul J. Biofuels, biodiversity, and people: Understanding the conflicts and finding opportunities. Biol Conserv 2008; 141(10): 2450-60.
[http://dx.doi.org/10.1016/j.biocon.2008.08.005]

[20] Omer AM. Green energies and the environment. Renew Sustain Energy Rev 2008; 12(7): 1789-821.
[http://dx.doi.org/10.1016/j.rser.2006.05.009]

[21] Robertson GP, Dale VH, Doering OC, *et al.* Agriculture. Sustainable biofuels redux. Science 2008; 322(5898): 49-50.
[http://dx.doi.org/10.1126/science.1161525] [PMID: 18832631]

[22] Headey D, Fan S. Anatomy of a crisis: the causes and consequences of surging food prices. Agric Econ 2008; 39: 375-91.
[http://dx.doi.org/10.1111/j.1574-0862.2008.00345.x]

[23] Asumadu-Sarkodie S, Owusu PA. A review of Ghana's energy sector national energy statistics and policy framework. Cogent Eng 2016; 3(1)1155274 c
[http://dx.doi.org/10.1080/23311916.2016.1155274]

[24] Asumadu-Sarkodie S, Owusu PA. Carbon dioxide emissions, GDP, energy use, and population growth: a multivariate and causality analysis for Ghana, 1971–2013. Environ Sci Pollut Res Int 2016; 23(13): 13508-20. d
[http://dx.doi.org/10.1007/s11356-016-6511-x] [PMID: 27030236]

CHAPTER 15

Techniques in Prevention, Detection and Monitoring of Environmental Contaminants

Bhupinder Dhir[1,*]

[1] *School of Sciences, Indira Gandhi National Open University, New Delhi, India*

Abstract: Pollution in various sectors of the environment has produced a threat to human health and aquatic ecosystems. Biosensors play an important role in the detection of toxicants such as heavy metals. Efforts have been made to develop sensitive and efficient sensors for monitoring the presence of contaminants in the environment using nanotechnology and bioengineering techniques. Biosensors, in particular, help in monitoring the presence of pollutants in the environment, protecting our environment. Enzyme, DNA, imuno and whole cell-based biosensors have been developed and work depending on the reaction type, transduction signal, or analytical performance. Advantages such as specificity, low cost, ease of use, and portability establish biosensors as an efficient technique that can be used to detect the presence of various inorganic and organic contaminants.

Keywords: Biosensors, Heavy metals, Electrochemical sensors, Enzymes.

INTRODUCTION

Toxic compounds and contaminants present in soil and water cause threats to ecology and human health [1-3]. Therefore, one needs to develop precise methods for their detection and quantitation. Inorganic, organic and radioactive compounds present in the environment need to be monitored/ traced so that they can be treated effectively and living beings and the environment can be protected from damage and their harmful effects [4 - 7]. Monitoring of air, water and soil (major environmental components) is important to find out the current status of environmental damage/deterioration [8]. Environmental monitoring helps us in assessing the impact of human activities on the environment, and thus enables us to protect the environment.

High-performance liquid chromatography (HPLC), gas chromatography (GC), mass spectroscopy (MS), atomic absorption spectroscopy (AAS), emission spectr-

---

[*] **Corresponding author Bhupinder Dhir:** School of Sciences, Indira Gandhi National Open University, New Delhi, India; E-mail: bhupdhir@gmail.com

Bhupinder Dhir (Ed.)

oscopy, inductively coupled plasma mass spectroscopy, and various chromatographic techniques are some of the analytical methods that have been used in a routine manner to check the presence of toxicants in the environment. These methods are accurate, sensitive and reliable. Sophisticated instrumentation, sample preparation, pretreatment (pre-concentration), and long measuring time of analysis are required for their work. These methods need to be carried out by trained people. Most of these methods cannot assess the cumulative effect of various toxicants present in a sample, therefore need for the development of sensors with high specificity was realized.

Many research studies have been conducted to investigate and develop technologies that aim to reduce or detect the impact of hazardous compounds on the environment. Sensors for the rapid detection of environmental pollutants have been developed [9 - 11]. Ion-selective electrodes, biosensors and voltammetric techniques are some electrochemical methods that are used as an alternative to classical methods. These methods require less instrumentation and take less time for measurement. Biosensors have emerged as a good alternative to conventional techniques [12 - 14]. They do not require any pre-treatment of the sample, hence the analysis carried out by them is very rapid and highly sensitive. They provide real-time, high-frequency monitoring, which is very specific and accurate. Easy portability, low cost, simplicity and selectivity are some of the other advantages reported in biosensors. According to the Environmental protection Agency (EPA), bio-monitoring of pollution using biosensors is important for the implementation of preventive and remedial measures.

A device that provides quantitative or semi-quantitative information about a component using a biological recognition element is termed a biosensor. It performs specific biochemical reactions or interactions with its surrounding environment. Biosensors consist of a recognition element (enzyme, antibody, DNA), a signal-transducing structure (electrical, optical, or thermal), and an amplification/processing element. This analytical device combines information obtained by a biological sensing element, such as an enzyme or an antibody with a physical (optical, mass, or electrochemical) transducer. The target and the bio-recognition molecules interact with each other, and a measurable electrical signal is generated [15]. The biological component is used as a recognition element [16]. The interaction of the target analyte with the biological sensing element, generates a signal in the transducer which gives information related to the concentration of the analyte. The amplification of the signal is displayed by a signal processor.

The detection of an analyte takes place with the help of a recognition component of a sensor. The signal gets converted using a transducer. Various contaminants are monitored according to analytes which include trace metals, radioisotopes, volatile organic compounds, and biological pathogens. Biosensors help in the rapid and accurate detection of hazardous components.

The detection of heavy metals such as lead, cadmium, mercury, and arsenic is done using electrochemical biosensors. The heavy metal ions can also be analysed with biosensors using protein-based (enzyme, metal binding proteins and antibody) and whole cell-based approaches (genetically engineered microorganisms).

BIOSENSORS

Sensors are categorized into different types depending on the type of sensor that is used and environmental factors which are analysed. The classification of biosensors into various groups depends upon signal transduction or recognition principles. Biosensors are categorized as electrochemical (amperometric, and impedance biosensors), optical (optical fibre and surface plasmon resonance biosensors), piezoelectric (quartz crystal microbalance biosensors), or thermal sensors on the basis of the transducing component. Biosensors are classified as immunosensors, apt sensors, genosensors, and enzymatic biosensors (when antibodies, aptamers, nucleic acids, and enzymes are used). This classification is based on the type of their recognition element.

The biosensors mainly used in environmental monitoring are immunosensors and enzymatic biosensors. Apt sensors which have been developed recently, show characteristics such as thermal stability, *in vitro* synthesis, structure designing, easy modification, and capacity to distinguish targets with different functional groups. These features establish them as ideal candidates for environmental monitoring.

Optical biosensors are based on the principle of absorption of light, fluorescence, luminescence, reflectance, Raman scattering and refractive index. These sensors have found many applications in the area of environmental monitoring, food safety, drug development, biomedical research, diagnosis and control of environmental pollution [17, 18]. Enzyme-based biosensors having optical transducers represent autonomous devices that help in environmental monitoring.

Principle of Biosensors

Molecular recognition occurs when the substance that needs to be measured diffuses through the bioactive material. The process of recognition is followed by a biological reaction. The transducer converts the information obtained into a quantitative electrical signal. The signal later gets amplified, and the concentration of the substance is measured.

Characteristics of Biosensors

- Immobilized bioactive substance is used as a catalyst. Enzyme can be used many times.
- Strong specificity, *i.e.*, react with a specific substrate
- The speed of analysis is fast (results are given in one minute)
- High level of accuracy (relative error less than 1%)
- Simple operating system
- Cost-effective

Types of Biosensors

Enzyme-Based Biosensor

Indirect assessment/monitoring of inorganic (such as heavy metals and organic (pesticides) substances can be done using enzyme-based biosensors. These biosensors help in searching for biological molecules that catalyze specific chemical reactions [19]. A wide range of analytes, such as pesticides, heavy metals, and glycoalkaloids (emerging pollutants), can be measured using biosensors that are based on the principle of enzyme inhibition. Enzymatic biosensor measures analyte concentration or product formation during enzymatic reactions *via* a direct mode, while enzyme inhibition due to contact with the analyte of interest can be measured *via* indirect mode biosensors. The enzyme gets immobilized on the electrode surface in an enzyme-based biosensor. The enzyme and the transducer surface come in intimate contact. Immobilization occurs *via* adsorption, entrapment, microencapsulation, covalent binding, and cross-linking with glutaraldehyde. The process of immobilization help in retaining high enzyme activity and stability. Each substrate shows specificity for the substrate. Adsorption, covalent binding to solid surfaces, surface films, entrapment in polymer gels and encapsulation help in the immobilization of enzymes. Immobilization of enzymes on the solid surface increases the sensitivity of the biosensors. Low quantities of immobilized enzymes are required. They help in rapid (less than 5 min) analysis of the sample [20]. Pollutants such as pesticides, phenols, heavy metals, and pharmaceutical compounds can easily be detected using enzyme-based biosensors. They analyze pollutants present in the environment in a quick, efficient, automatized, and economical way [21].

First-generation electrochemical enzymatic biosensors work on the basis of product or co-substrate registration at the electrode surface, while specific redox mediators between the enzyme and the transducer are involved in second-generation biosensors. Direct exchange of electrons between the active site of the enzyme and the material of the electrode forms the basis of the working of third-generation biosensors. High catalytic activity and selectivity are the properties shown by enzymes present in biosensors, even in complex matrices. Inhibitor related to the analyte is responsible for the inhibition of the enzyme. Peroxidases and polyphenol oxidases (laccases and tyrosinases) are functional oxidative enzymes that help in the recognition of analyte.The enzyme activity is inhibited by pollutants.

Enzymes are biomolecules that act as recognition elements in biosensors. Oxidative enzymes can be used as recognition elements in biosensors. Various pollutants such as pesticides, heavy metals, organic compounds such as phenol, and pharmaceutical compounds can be detected with the help of these biosensors. Enzyme sensors help in multi-analyte detection. Enzymes such as acetylcholinesterase, alkaline phosphatase, urease, invertase, peroxidase, lactate dehydrogenase, tyrosinase, and nitrate reductase are used in the detection of heavy metals. Electrochemical or optical measurements are used in the detetction of inhibited immobilized enzyme. Enzyme Cholinesterase (ChE) is inhibited by toxic chemicals, *viz.* organophosphates, and heavy metals, hence is used to monitor toxic compounds [18]. Biosensors based on cholinesterase enzyme (ChE) are used in the detection of toxins. Biosensors having the enzyme acetylcholine esterase can be used for the detection of organophosphorus compounds present in water. Inhibition of the enzyme, butyrylcholine esterase by pesticides and urease by heavy metals help in the simultaneous detection of pesticides and heavy metal ions [22]. In enzymatic biosensor, enzyme toluene orthomonooxygenase and an oxygen-receptive ruthenium-based phosphorescent dye act as a transducer and help in the measurement of toluene in aqueous solution [23, 24]. Organophosphorus pesticide biosensors can be made by the immobilization of acetylcholinesterase on the surface of nanoparticles. Phenol can be detected with the help of the enzyme phenol oxidase-containing biosensor. Polychlorinated biphenyls (PCBs), chlorinated hydrocarbons and other organic compounds can be detected with the help of biosensors.

Immunosensors

The presence of pollutants such as pesticides (aldrin, triazines DDT, glyphosate), herbicides can be detected with the help of immunological techniques (immunotechniques). Immunoelectrodes act as biosensors and detect low concentrations of pollutants. Very low concentrations of pesticides such as triazines, malathion and carbamates have been detected in immunoassays based on the use of specific antibodies. Antibody-antigen interactions also prove useful in the detection of heavy metals. Antibodies that specifically bind to HMs have been developed. In immunosensors, recognition elements are attached to a transduction element.

Electrochemical biosensors

Electrochemical biosensors help in on-site detection and accurate monitoring of environmental conditions. Monitoring and detection of various environmental toxicants, including pesticides and heavy metals, can be done using electrochemical biosensors [25]. Biocatalytic and affinity electrochemical biosensors have been developed in recent years [26]. In affinity biosensors, selective binding between the target analyte and the biological sensing element takes place. Example-DNA or antibody, receptor-based biosensors. In biocatalytic biosensors, an electroactive species is produced *via* interaction with a target analyte [27]. Example-enzymes, whole cells, or tissue-based biosensors.

The biological elements present in the electrochemical biocatalytic sensors recognize a target and induce a response in an electroactive molecule (*e.g.,* enzymes) [28], whereas in electrochemical affinity biosensors, a binding recognition element releases a signal when coupled to the target (*e.g.,* antibodies). Metals, pharmaceutical and pesticide-based compounds can be detected with the use of these biosensors [29, 30] (Table **1**).

Electrochemical biosensors can be potentiometric or amperometric, depending on the electrical signal they measure (potential or current). Electrochemical biosensors have Various advantages, such as real-time monitoring, miniat urization, high selectivity and sensitivity help these biosensors in the accurate monitoring of environmental conditions. These biosensors also allow on-site detection of pollutants [31, 32].

An electrochemical micro-sensor is used for detecting and quantifying trace metal ions present in water [33]. A simple, fast, and portable paper-based dual electrochemical/colorimetric system for the simultaneous detection of gold and iron has been developed. A three-dimensional microfluidic paper-based analytical device for the detection of toxic heavy metals has also been developed [34]. Paper-based electrochemical biosensor device for the detection of Pb(II) ion has been developed [35]. Colorimetric detection of Ni, Fe, Cu, Cr and electrochemical detection of Pb, Cd using a multilayer paper-based device is reported.

Inhibition in the respiratory activity of *Escherichia coli* after exposure to heavy metals, and pesticides, has been monitored using an electrochemical microfluidic paper-based analytical device (μPAD) [36].

Electrochemical sensors possess advantages such as high sensitivity, low cost, less power requirement, easy miniaturization, portability, compatibility with microfabrication and micromachining technologies and suitability for *in situ* analysis [37].

Table 1. Electrochemical Biosensors utilized in the detection of various contaminants

Type of contaminant	Enzyme
Pesticide	
Dichlorvos	ChOx
Paraoxon	Phosphotriesterase
Methyl parathion	Organophosphorus hydrolase
Paraoxon, 2,4-dichlorophenoxyacetic acid, and atrazine	Butyrylcholinesterase, ALP, and tyrosinase
Fenobucarb, temephos, and dimethoate	Glutathione *S*-transferase
Heavy metal(s)	
Cr(VI), Cr(III)	Glucose oxidase/horseradish peroxidase
Hg^{2+}, Cd^{2+}, Pb^{2+}, Cr(VI)	Glucose oxidase
Pb^{2+}, Ni^{2+}, Cd^{2+}	Horseradish peroxidase
Hg^{2+}	Catalase
Cr(VI)	Horseradish peroxidase

Nanobiosensors

Sensors based on nanomaterials help in the detection of environmental contaminants at the nanomolar to sub-picomolar level. Nanobiosensors are composed of a nanomaterial(s), a recognition element for specificity, and a signal transduction device as a means of relaying the presence of the analyte. Sensors can detect a single analyte or multiple analytes, which means they can help in multiple detections. The properties of nanomaterials, such as high surface area-t--volume ratios and easy surface functionalization, make them sensitive so that changes in surface chemistry can be detected even at low detection limits.

Nanomaterial-based sensors show high accuracy in the specific and sensitive detection of environmental contaminants. Nanosensors for the detection of contaminants such as pesticides, heavy metals, and pathogens have been developed [38 - 41]. Heavy metals can be detected using nanomaterials such as nanoclusters, aptamers, antibodies, and quantum dots. Small, uniform size of nanomaterials and high surface-to-volume ratio help them in detecting heavy metals in a better way.

Nanowires or nanotubes have the capacity to act as chemical and biological sensors [42]. Nanowires possess a high surface-to-volume ratio and show high stability because of their crystalline structure, high sensitivity and selectivity. Nanowires are implemented in electrodes to increase the surface-to-volume ratio and augment the electrochemical sensitivity of sensors so that analytes can be detected. SWNTs have shown a high accuracy, faster response and higher sensitivity for the detection of gas molecules such as NO_2 and NH_3. The gas molecules get directly bonded to the surface of SWNTs and thus influence the electrical resistance of the sensor. Though SWNTs can act as good alternatives to nanosensors but show certain limitations.

Electrochemical sensors composed of the nanocomposite materials, such as copper oxide (CuO), along with single-walled carbon nanotubes (SWCNTs), when attached to nanowires, help in the detection of organophosphorus pesticides. Electrochemical sensors composed of nanocomposite materials (CuO-SWCNTs) were highly stable and showed good specificity towards pesticides, inorganic ions, and sugars.

Nanostructured materials having tin dioxide (SnO_2) nanowires and zinc oxide (ZnO) nanorods have been used and developed. They help in the detection of gases. The nanostructures revealed the presence of gases such as ammonia (NH_3), carbon monoxide (CO), hydrogen (H_2), carbon dioxide (CO_2) present in the environment and liquefied petroleum gas (LPG).

A low-cost detection assay was developed to determine reactivity and help in the characterization of selected nanoparticles (Ag, Au, CeO_2, SiO_2 and VO_2) with particle sizes ranging between 5 and 400 nm [43]. Whole-cell biosensors were developed by integrating golTSB genes from *Salmonella enteric serovar typhimurium* to induce Au (I/III) complexes for the detection and exploration of nanoparticles. The fabricated biosensor could identify metal ions, such as Ag (I), Cu (II), Fe (III), Ni (II), Co (II), Zn and Pb (II). Bioassays that involve the use of sulphur-oxidizing bacteria (SOB) have been used for the identification of the presence of Cr (III) and Cr (IV) in water samples. The microbial biosensors employed for the toxicity assessment included *Chlorella* sp., *Chlorella vulgaris*, *Monoraphidium sp., Scenedesmus subspicatus* and *Brachionus calyciflorus sp..* *Nitrosomonas europaea* (an ammonia-oxidizing bacterium) based biosensor has been designed to determine concentrations of allylthiourea and thioacetamide in water. It measures ammonium oxidation rates [44].

Multiple electrodes, coatings, and electrochemical techniques target analytes such as toxic metals, and industrial chemicals (trichloroethylene, methyl-t-butyl ether). Nanoelectrode arrays identify and quantify dissolved metals [45]. Changes in current and voltage are depicted in terms of signals obtained from the electrodes.

In optical fiber nano immunosensor, optics and photonics technology are applied in immunoassay. An antigen or antibody converts to an optical signal when an antigen and antibody combine. An optical fiber nano immunosensor for the detection of BPT (benzopyrene tetrol) has been successfully developed. They used a photomultiplier to record fluorescence produced by the binding of BPT and antibodies. The content of BPT in cells was determined by measuring the change in fluorescence intensity.

Paper-based Electrochemical Biosensors

Paper has been used as a substrate for developing microfluidic devices and biosensors. Paper-based materials show unique characteristics such as porosity, liquid wicking rate, and affinity for numerous analytes. Paper-based models are portable and ideal for on-site and real-time monitoring. They found its applications in clinical, nutraceutical, and environmental areas. Use of paper in a broad spectrum, such as chromatography, filter, and blotting, establishes paper as an ideal candidate for use in sensors. Medina-Sánchez *et al.* [46] developed a flow paper-based sensing device for lead and cadmium ions detection in environmental matrices.

Paper-based electrochemical biosensor for the detection of Pb(II) ion has been developed [35]. A simple, fast, paper-based dual electrochemical sensor for the simultaneous detection of gold and iron has been reported. Studies indicated that a three-dimensional microfluidic paper-based analytical device based on the potential-control technique for the detection of toxic heavy metals (Pb^{2+} and Hg^{2+}) has also been developed [44].

Screen-Printed Electrodes (SPEs)

Screen-Printed Electrodes (SPEs) comprise ink printed circuits on a substrate. A mold, stencil, or mesh is used to cast ink into a substrate. The physical barrier helps in this. The printed ink is cured and fixed with an insulator layer. The substrate comprises materials such as plastics, ceramics, and paper. The basic working principle of these sensors is redox reactions and electrochemical effects. Chemical modifications increase catalytic function, new morphologies increase electrode surface and selectivity is improved by staining electrodes with Biomolecules. All these things improve the sensing mechanism. Martínez-García *et al.* [47] reported an electrochemical enzyme-based biosensor helped in the detection of 3-hydroxybutyrate. It is a SPEs that has been modified using reduced graphene oxide and thionine.

SPEs find their use in testing water quality tests and the detection of organic compounds found in environmental samples (Table **2**). Toxic heavy metals, including lead (Pb^{2+}) and cadmium (Cd^{2+}), have also been determined by SPEs. A mixture of graphene (G)/polyaniline (PANI)/polyestirene (PS) electrospun nanofibers have also been tried. Screen-printed G/PANI/PS electrode has successfully been used for simultaneous detection of Pb^{2+} and Cd^{2+} in real river water samples. An increase in the surface area improved the electrochemical response of electrodes. This happens because of the electrospinning technique for acquiring nanofibers, selection, and mixture of materials. The addition of graphene and other conductive elements to materials used in screen-printed electrodes proved useful in the improvement of the sensor so that detection of contaminants in the environment is easy. Monitoring changes in the concentrations of such pollutants in various environmental conditions is possible and provides an accurate treatment.

Table 2. Screen-printed sensors developed for the detection of organic compounds in environmental samples.

Analyte	Detection Method
Organophosphate	Amperometric
Organophosphate pesticides	Chronoamperometry
Organophosphorus	Amperometric
Organophosphorus and Carbamate Pesticides	Amperometry, flow system
Organophosphorus Pesticide	Amperometry
Organophosphorus Pesticide Dichlofenthion	Photoelectrochemical
Herbicide isoproturon	Amperometric
Herbicide	Amperometric
Picric acid and atrazine	Photo-electrochemical
Chlorsulfuron	Stripping voltammetry
Phenol and catechol	Amperometric measurements
Phenol and pesticide	Electrochemical measurement
Phenol	Amperometric

Whole Cells Biosensors for Heavy Metal Detection

An analytical device that includes the integration of whole cells that are responsible for its selectivity with a physical transducer that generates a signal proportional to the concentration of analytes is referred to as a whole cell-based biosensor. These biosensors are used specifically for the detetction of heavy metals [48, 49].

Natural receptors (noncatalytic or nonimmunogenic proteins and microorganisms such as: bacteria, mosses, algae, yeasts, fungi, and lichens) specifically bind heavy metals [50]. Coupling of cells to transducer help in the conversion of cellular response into detectable signals. These help in the detection of heavy metal toxicity found in environmental media such as soil, sediment, and water.

These biosensors show poor selectivity because of nonspecific cellular response to substrates.

Bioluminescent-Based Sensor

The biosensor is based on the measurement of bioluminescence. The toxicity of heavy metals can be measured as a decrease in light output from the bacterium when exposed to environmental samples. Low concentrations of heavy metals (0.01–0.05 μg/mL) motivate light output. The modifications in the composition of fatty acid present in the cell membrane, synthesis of fatty acids into intracellular media, and disruption in the cell energy production result in stimulation. The production of light from the whole cell depends on the energy derived from the electron transport chain. The measure of luminescence gives information about the metabolic activity.

DNA-Based Biosensor

Nucleic acids find their use in the recognition and monitoring of toxic compounds. DNA is an important intracellular object that undergoes modification or damage when it interacts with endogenous and exogenous factors. This stable, low-cost, and easily adaptable molecule is helpful in targeting heavy metal ions (such as Cu^{2+} and Hg^{2+}). Toxic molecules such as heavy metals interact with nucleic acids and show a high affinity for DNA. The interaction of DNA and heavy metals could damage the replication and transcription of DNA *in vivo*, or cause mutation of the gene. Interaction with DNA helps in monitoring and quantifying levels of HMs (such as Pb, Cd, and Ni).DNA and DNAzymes [single-stranded DNA molecules with catalytic activity] are biodegradable, highly selective, and easily obtained *via in vitro* methods [51].

An electrochemical DNA biosensor is an integrated receptor-transducer device in which DNA acts as a biomolecular recognition element and measures specific binding with DNA through electrochemical transduction. Immobilization of DNA or its components (such as nucleotides, nucleosides, purine, and pyrimidine bases) on the electrode surface is the most important factor for the development of an efficient DNA-based electrochemical biosensor.

Negatively charged phosphate oxygen atoms, ribose hydroxyls, base ring nitrogen molecules, and exocyclic base keto groups are potential DNA sites that bind metal ions. The interaction of small pollutants with the immobilized DNA (drugs, mutagenic pollutants, *etc.*) is monitored in these biosensors. Heavy metal ions interact at more than two different sites on DNA, hence these interactions are complex.

In DNA nanosensors, DNA is the recognition site, and nanomaterials used act as the transducing element. DNA-based nanosensors for various human applications, including the detection of environmental pollutants, have been designed. The nanomaterials used include metal oxides, quantum dots (QDs), platinum (Pt), copper (Cu), magnetic, tungsten disulfide (WS2), mesoporous silica (MSN), graphene, and graphene oxide. Gold (Au) and silica NPs have been considered safe nanomaterials for sensor application.

DNA-based nanosensors have proved easy and cost-effective detection of pathogens, antibiotics, pesticides, and other environmental pollutants. The role of DNA nanosensors in detecting pathogenic virus and fungus is not well studied but has been used to detect genetically modified organisms (GMOs), antibiotics, and pesticides. The features such as specificity, easy design and cost-effectiveness make DNA-based nanosensors useful for the detection of diverse analytes, even in complex mediums.

In an electrochemical DNA biosensor, ssDNA (probe ssDNA) is used for the modifying surface of the working electrode. Electrochemical techniques such as amperometry, voltammetry, or EIS detect hybridization of the probe with a complementary target ssDNA or disruption in its structural integrity. Heavy metal ions bind to specific DNA bases such as thymine (T) or cytosine (C), forming stable DNA duplexes. The development of selective DNA biosensors for heavy metal ion detection has been made possible. A gold electrode surface modified with ssDNA (DNA-1) showed the interaction of sulfhydryl groups and the gold electrode surface in a device.

An electrochemical aptamer (oligonucleotide or peptide molecules bind to specific target molecules) biosensor for the detection of Hg^{2+} ions in water was developed [52]. Apta-sensor was based on $T-Hg^{2+}-T$ coordination chemistry. This sensor could be regenerated by addition of cysteine and Mg^{2+} ions.

Microbial Biosensors

A microbial biosensor is an analytical device in which microorganisms are coupled with a transducer. This enables rapid, accurate and sensitive detection of target analytes. The detection of gases such as sulfur dioxide (SO_2), methane and carbon dioxide has been made possible through microbial biosensors [53, 54]. *Thiobacillus*-based biosensor detect pollutant such as SO_2, while *Methalomonas* based/immobilized biosensor detect methane (CH_4)., A particular strain of *Pseudomonas* is used for monitoring carbon dioxide. A graphite electrode with *Cynobacterium* and *Synechococcus* measures the inhibition of electron transport during photosynthesis due to certain pollutants (such as herbicides).

Microfluidic Sensors

Non-biological contaminants such as pesticides, phosphate, Hg, ammonium ion and As have been detected using micro-scale technologies. Rapid analysis of hazardous metals and other inorganic contaminants in water, soil, and mixed waste sites has been made possible through laser-induced breakdown spectroscopy (LIBS). This technique uses a laser to heat a small area, rapidly generating plasma from the atomic constituents present at the focal point.

Biosensors for Measuring Radioisotopes

RadFET (Radiation field-effect transistor) measures the dose of gamma radiation. It is based on the principle that ionizing radiation promotes high-mobility electrons. High-purity germanium and scintillation crystals such as thallium-doped sodium iodide are commonly used gamma radiation detectors (low energy resolution). Thermoluminescent dosimeter (TLD) uses a photodetector that converts the signal, which is a reading of radiation dose. A thermoluminescent dosimeter is a crystal that absorbs energy when exposed to radiation and promotes electrons into semiconductor holes. The energy trapped in holes is released in the form of light upon heating. An isotope identification gamma detector was developed for the identification of nuclear weapons and radiological dispersal devices. A small neutron generator was developed to be used as a probe to detect the presence of nuclear materials.

Biosensors for Measuring Volatile Organic Compound

Evanescent fiber-optic chemical sensor, grating light reflection spectroscopy (GLRS), miniature chemical flow probe sensor, chemical sensor arrays, gold nanoparticle chemiresistors, the electrical impedance of tethered lipid bilayers on planar electrodes, MicroHound, hyperspectral imaging, and chemiresistor array are some of the sensors/equipments used for detection of volatile organic compounds. The presence of very low levels of organic compounds in aqueous solutions can be detected by polymer optical waves. Quantitative measurements are done using near-infrared (NIR) spectroscopy.

A liquid sample is analyzed using transmission diffraction grating in which an incident light beam is directed onto the grating. Some of the diffracted orders are transformed from traveling waves at certain angles of incidence. The process occurs at a specific wavelength that is a function of the grating period. This technique was used in combination with the electrochemical modulation of a gold-coated metallic spectroscopic grating that is used for the detection of very low amounts of aromatic hydrocarbons. The grating was organized as the working electrode in an electrochemical cell containing water and trace amounts of contaminants such as TNT and a dye.

The analytes react with a reagent and form spectrally distinct products in the chemical flow probe sensor. Target analytes can be volatile organic compounds such as chlorinated halocarbons, and dissolved metals present in air or water. The presence of classes of saturated alkane, aromatic hydrocarbon, chlorinated hydrocarbon, alcohol, ketone, and organophosphonate can be detected using this. The most sensitive sensor is an acoustic sensor that measures a decrease in its active resonant frequency when trace mass is loaded on the active surface.

A chemically sensitive resistor formed of a conductive polymer film deposited on a micro-fabricated circuit is referred to as a chemiresistor sensor. The chemically-sensitive polymer is dissolved in a solvent and mixed with conductive carbon particles. The chemiresistor array detects volatile organic compounds (VOCs), aromatic hydrocarbons (*e.g.,* benzene), chlorinated solvents (*e.g.,* trichloro ethylene (TCE), carbon tetrachloride), aliphatic hydrocarbons (*e.g.,* hexane, iso-octane), alcohols, and ketones (*e.g.,* acetone).

Biosensors for Measuring Pesticides

Enzymatic sensors are based on the inhibition of a selected enzyme. They are mostly used for the determination of organophosphate pesticides such as chlorpyrifos, methyl parathion, and diazinon.

Electrochemical biosensors that possess whole cells are based on the principle of antibody–antigen interaction (immunosensors) and are used to detect and monitor pesticides. Acetylcholine esterase (AChE) enzyme catalyzes the hydrolysis of the neurotransmitter acetylcholine to choline and acetate. An AChE-based electrochemical biosensor for the detection of methyl parathion, an organophosphate pesticide, has been developed. The biosensor was made by cross-linking AChE with glutaraldehyde, immobilization of the cross-linked enzyme on the surface of the modified carbon electrode (using semiconducting single-walled carbon nanotubes) and treatment with bovine serum albumin (BSA) (avoid nonspecific binding). Organophosphates cause phosphorylation of the serine residue present at the enzyme's active center, inhibiting AChE activity. Butyrylcholinesterase (BChe), choline oxidase (ChOx), phosphotriesterase (PTE), and organophosphorus hydrolase (OPH) are other enzymes used for the development of electrochemical biosensors involved in the detection of organophosphate pesticides.

An amperometric biosensor based on the inhibition of the enzyme glutathione *S*-transferase (GST) was developed by Borah *et al.* [55]. The biosensor showed an ability to detect various types of pesticides, including organophosphate (chlorpyrifos, ethion) organochlorine (DDT), and benzimidazole (carbendazim).

Whole cells present in electrochemical biosensors respond to a range of targets. They provide information related to pharmacology, cell physiology, and toxicology and are easily get genetically modified, allowing the detection of specific classes of toxicants. The physiological changes noted after exposure to toxicants, such as changes in respiratory, are used as a means to assess acute biotoxicity in microbial cells. Microbial biosensors have been used for the detection of organochlorine, triazine, and organophosphate pesticides. A microbial biosensor based on the inhibition of the photocurrent generated by the cyanobacterial species *Anabaena variabilis* detection of herbicide, was developed. Photobioelectrocatalytic oxidation of water that occurs in the Photosystem II protein complex located in the thylakoid membrane of photosynthetic microorganisms was responsible for the generation of photocurrent. The bacteria with redox mediator *p*-benzoquinone was immobilized on a carbon electrode with the help of calcium alginate (a trapping polymer). The polymer also incorporated activated carbon to provide a conductive network.

An electrochemical biosensor composed of *Chlorella* sp. algal cells immobilized on an indium tin oxide (ITO) electrode modified with silica-coated ZnO quantum dots, was developed by Pabbi *et al.* [56]. The biosensor measured the inhibition of cell wall alkaline phosphatase (ALP) caused after exposure to acephate, an organophosphate pesticide dephosphorylates. The substrate *p*-nitrophenyl phosphate gets dephosphorylated to electroactive *p*-nitrophenol with the help of ALP, which later gets oxidized at the surface of the electrode.

Antibodies or immunoglobulins are produced by the body in response to the entry of foreign substances known as antigens. They make a family of Y-shaped glycoproteins. Enzyme-linked immunosorbent assays (ELISAs) are the most popular immunosensor. An antigen or antibody is immobilized on solid support (electrode surface) in direct ELISA. In another type, an enzyme-labeled antibody or antigen is introduced, followed by a substrate that is converted to an electroactive product with the signal generated related to the analyte concentration. An unconjugated primary antibody is introduced, followed by an enzyme-labeled secondary antibody that enhances the signal, improving sensitivity in an indirect ELISA. An antigen containing two epitopes interacts with two antibodies, one of which is enzymatically labeled in sandwich ELISA.

An electrochemical immunosensor for the detection of parathion has been developed by Mehta *et al.* [57]. The immunosensor was made by modifying a screen-printed electrode surface with graphene quantum dots, functionalization with NH_2 groups, and (3) bio-interfacing with anti-parathion antibodies. These sensors detect neonicotinoid-type pesticides. Their main target is insect nicotinic acetylcholine receptors (nAChRs), which are widespread in the insect brain and central nervous system.

Detection of Heavy Metals Using Biosensors

Heavy metal ions can be detected by electrochemical and enzymatic biosensors [58]. They have demonstrated the ability to detect heavy metal ions such as Pb^{2+}, Cu^{2+}, Cd^{2+}, Cr^{6+} and Hg^{2+}. Heavy metals inhibit enzyme (urease, horseradish peroxidase, glucose oxidase, and catalase) activity. These biosensors possess advantages such as low detection levels (up to nanomolar concentrations), high selectivity, repeatability, and storage at 4 °C from 14 to 30 days.

Prokaryotic and eukaryotic whole-cell electrochemical biosensors for heavy metal ion detection have been developed. Electrochemical biosensors based on eukaryotic cells have been successfully used to study toxic heavy metals. Living cells, when used as biological sensing elements, make sensing ability more flexible and provide a cheaper alternative to purified enzymes or antibodies. Microbial biosensors also show an ability to detect heavy metal ions [59]. Microbial biosensors show rapid response to toxicants, and are less expensive, more compact and portable. An integrated microbial biosensor in which *E. coli* and the electrochemical redox mediator benzoquinone were co-immobilized within a gelatin/silica hybrid hydrogel on the surface of the carbon electrode was developed by Li *et al.* [60]. This biosensor detected the toxicity of Hg, Cu, and Cd.

Nanomaterials have also been used in the fabrication of enzyme-based biosensors because their enhanced conductivity help to improve the signal-to-noise ratio and signal transduction.

Electroactive biofilms (EABs), a type of microbial biosensor, have found their applications in environmental and bioprocess monitoring. These biofilms can be included in microbial fuel cells (MFCs) and measured biological oxygen demand (BOD). They were used in the monitoring of environmental toxicants. The metabolic processes of microorganisms convert chemical energy contained in the organic matter of MFCs into electrical energy. Microorganisms oxidize organic matter under anaerobic conditions. The presence of electrogenic, organic and inorganic compounds interferes with electron transfer by acting as electron acceptors (metals) or donors (organic matter). This decides the rate of current flow within the MFC.

A graphene oxide quantum dot/carboxylated multiwall carbon nanotube-modified pencil graphite electrode was developed by Zhu *et al.* [61]. It was used to monitor changes in nucleotide catabolism in human hepatoma (HepG2) cells exposed to toxicants. The biosensor detects changes in electrochemical signals due to guanine/xanthine, adenine, and hypoxanthine following exposure to micromolar concentrations of toxicants such as Cd, Hg, and Pb.

DETECTION OF ORGANIC COMPOUNDS USING BIOSENSORS

Benzene, toluene, ethyl benzene and xylene (concentration range 0.5–120 mg/L) are some of the compounds whose presence in water samples has been detected using *Pseudomonas putida* strain TVA8 [62].

MOLECULAR BIOLOGY IN ENVIRONMENTAL MONITORING

Use of molecular probes and immunoassays in detection and monitoring of environmental pollution has increased in the recent years. Semipermeable membrane devices (SPMDs) demonstrated their applicability in sequestering and analyzing a wide array of environmental contaminants particularly organic compounds such as organochlorine pesticides, polychlorinated biphenyls, polycyclic aromatic hydrocarbons, polychlorinated dioxins and dibenzofurans, organophosphate pesticides and pyrethroid insecticides. SPMDs are designed specifically to sample nonpol and hydrophobic chemicals. Use of monoclonal antibodies (MAbs) in the detection and bio-monitoring of environmental pollution is also gaining importance.

Various nanomaterials such as metallic nanoparticles, carbon-based nanomaterials (carbon nanotubes and graphene), carbon coatings, membranes, and some conductive polymers have been used for increasing the working efficiency of electrode surface.

SPEs have been extensively used for testing water quality and the detection of organic compounds in environmental samples. They are also used in the determination of toxic heavy metals, including Pb^{2+} and Cd^{2+} [63]. A mixture of graphene (G)/polyaniline (PANI)/polystyrene (PS) electrospun nanofibers has also been used in biosensors. Screen-printed G/PANI/PS electrode simultaneously detected the presence of Pb^{2+} and Cd^{2+} in real river water samples through anodic stripping voltammetry (ASV).

CONCLUSION

Conventional techniques used for monitoring and detecting contaminants in the environment are accurate and sensitive but expensive, difficult, and require sample pretreatment before analysis. These techniques are not able to evaluate the impact of multiple toxicants in aquatic ecosystems. There is a need to develop precise and quick methods for the detection and quantitation of contaminants present in the environment so that they can be treated at an early rate. Biosensors act as cheap, highly sensitive and easy ways to monitor the presence of various contaminants present in the environment. Immuno, enzymatic, whole-cell, and DNA biosensors for monitoring and detection of pollutants have been developed.

Enzyme, DNA and antibody-based biosensors are easy to use, cost-effective, and show high selectivity toward specific toxicants. Electrochemical biosensors are found suitable for detecting the presence of pesticides and heavy metals. Whole-cell biosensors have been found more suitable for screening toxicants, such as pesticides, heavy metals, organosulfur compounds, and pharmaceuticals. They are also useful in the detection of combined effects of multiple toxicants present in the environment. The performance of biosensors needs to be improved. Efforts have been made to design simple, easy-to-operate, affordable, and user-friendly biosensors with high sensitivity, less volume, and no prior pre-treatment. Research studies are underway to develop biosensors that show high sensitivity, stability, long shelf life and multifunctional approach.

REFERENCES

[1]   Jan A, Azam M, Siddiqui K, Ali A, Choi I, Haq Q. Heavy metals and human health: mechanistic insight into toxicity and counter defense system of antioxidants. Int J Mol Sci 2015; 16(12): 29592-630.
[http://dx.doi.org/10.3390/ijms161226183] [PMID: 26690422]

[2]   Masindi V, Muedi KL. Environmental contamination by heavy metals. Heavy Metals 2018; 10: 115-32.

[3]   Sall ML, Diaw AKD, Gningue-Sall D, Efremova Aaron S, Aaron JJ. Toxic heavy metals: impact on the environment and human health, and treatment with conducting organic polymers, a review. Environ Sci Pollut Res Int 2020; 27(24): 29927-42.
[http://dx.doi.org/10.1007/s11356-020-09354-3] [PMID: 32506411]

[4]   Rice KM, Walker EM Jr, Wu M, Gillette C, Blough ER. Environmental mercury and its toxic effects. J Prev Med Public Health 2014; 47(2): 74-83.
[http://dx.doi.org/10.3961/jpmph.2014.47.2.74] [PMID: 24744824]

[5]   Naidu R, Arias Espana VA, Liu Y, Jit J. Emerging contaminants in the environment: Risk-based analysis for better management. Chemosphere 2016; 154: 350-7.
[http://dx.doi.org/10.1016/j.chemosphere.2016.03.068] [PMID: 27062002]

[6]   Wu X, Cobbina SJ, Mao G, Xu H, Zhang Z, Yang L. A review of toxicity and mechanisms of individual and mixtures of heavy metals in the environment. Environ Sci Pollut Res Int 2016; 23(9): 8244-59.
[http://dx.doi.org/10.1007/s11356-016-6333-x] [PMID: 26965280]

[7]   Zhang H, Reynolds M. Cadmium exposure in living organisms: A short review. Sci Total Environ 2019; 678: 761-7.
[http://dx.doi.org/10.1016/j.scitotenv.2019.04.395] [PMID: 31085492]

[8]   Rodriguez-Narvaez OM, Peralta-Hernandez JM, Goonetilleke A, Bandala ER. Treatment technologies

for emerging contaminants in water: A review. Chem Eng J 2017; 323: 361-80.
[http://dx.doi.org/10.1016/j.cej.2017.04.106]

[9]   Vigneshvar S, Sudhakumari CC, Senthilkumaran B, Prakash H. Recent advances in biosensor technology for potential applications–an overview. Front Bioeng Biotechnol 2016; 4: 11.
[http://dx.doi.org/10.3389/fbioe.2016.00011] [PMID: 26909346]

[10]   Hashemi Goradel N, Mirzaei H, Sahebkar A, *et al.* Biosensors for the Detection of Environmental and Urban Pollutions. J Cell Biochem 2018; 119(1): 207-12.
[http://dx.doi.org/10.1002/jcb.26030] [PMID: 28383805]

[11]   Bano H, Islam S, Noor F, Bhat MA. Biosensors and Bioremediation as Biotechnological Tools for Environmental Monitoring and Protection. Int J Curr Microbiol Appl Sci 2020; 9(10): 3406-25.
[http://dx.doi.org/10.20546/ijcmas.2020.910.394]

[12]   Ho C, Robinson A, Miller D, Davis M. Overview of Sensors and Needs for Environmental Monitoring. Sensors (Basel) 2005; 5(1): 4-37.
[http://dx.doi.org/10.3390/s5010004]

[13]   Rodriguez-Mozaz S, Marco MP, de Alda MJL, Barceló D. Biosensors for environmental applications: Future development trends. Pure Appl Chem 2004; 76(4): 723-52.
[http://dx.doi.org/10.1351/pac200476040723]

[14]   Justino C, Duarte A, Rocha-Santos T. Recent Progress in Biosensors for Environmental Monitoring: A Review. Sensors (Basel) 2017; 17(12): 2918.
[http://dx.doi.org/10.3390/s17122918] [PMID: 29244756]

[15]   Thévenot DR, Toth K, Durst RA, Wilson GS. Electrochemical biosensors: recommended definitions and classification. Biosens Bioelectron 2001; 16(1-2): 121-31.
[PMID: 11261847]

[16]   Pandey CM, Malhotra BD. Biosensors: Fundamentals and Applications. Walter de Gruyter GmbH & Co KG 2019.
[http://dx.doi.org/10.1515/9783110641080]

[17]   Shankaran D, Gobi K, Miura N. Recent advancements in surface plasmon resonance immunosensors for detection of small molecules of biomedical, food and environmental interest. Sens Actuators B Chem 2007; 121(1): 158-77.
[http://dx.doi.org/10.1016/j.snb.2006.09.014]

[18]   Borisov SM, Wolfbeis OS. Optical Biosensors. Chem Rev 2008; 108(2): 423-61.
[http://dx.doi.org/10.1021/cr068105t] [PMID: 18229952]

[19]   Zhu YC, Mei LP, Ruan YF, *et al.* Enzyme-based biosensors and their applications. 2019.
[http://dx.doi.org/10.1016/B978-0-444-64114-4.00008-X]

[20]   Nguyen HH, Lee SH, Lee UJ, Fermin CD, Kim M. Immobilized enzymes in biosensor applications. Materials (Basel) 2019; 12(1): 121.
[http://dx.doi.org/10.3390/ma12010121] [PMID: 30609693]

[21]   Ahmed I, Iqbal HMN, Dhama K. Enzyme-based biodegradation of hazardous pollutants—An overview. J Exp Biol Agric Sci 2017; 5: 402-11.
[http://dx.doi.org/10.18006/2017.5(4).402.411]

[22]   Ligler FS. Perspective on optical biosensors and integrated sensor systems. Anal Chem 2009; 81(2): 519-26.
[http://dx.doi.org/10.1021/ac8016289] [PMID: 19140774]

[23]   Zhong Z, Fritzsche M, Pieper SB, *et al.* Fiber optic monooxygenase biosensor for toluene concentration measurement in aqueous samples. Biosens Bioelectron 2011; 26(5): 2407-12.
[http://dx.doi.org/10.1016/j.bios.2010.10.021] [PMID: 21081273]

[24]   Chen C, Wang J. Optical biosensors: an exhaustive and comprehensive review. Analyst (Lond) 2020;

145(5): 1605-28.
[http://dx.doi.org/10.1039/C9AN01998G] [PMID: 31970360]

[25] Waheed A, Mansha M, Ullah N. Nanomaterials-based electrochemical detection of heavy metals in water: Current status, challenges and future direction. Trends Analyt Chem 2018; 105: 37-51.
[http://dx.doi.org/10.1016/j.trac.2018.04.012]

[26] Ronkainen NJ, Halsall HB, Heineman WR. Electrochemical biosensors. Chem Soc Rev 2010; 39(5): 1747-63.
[http://dx.doi.org/10.1039/b714449k] [PMID: 20419217]

[27] Kimmel DW, LeBlanc G, Meschievitz ME, Cliffel DE. Electrochemical sensors and biosensors. Anal Chem 2012; 84(2): 685-707.
[http://dx.doi.org/10.1021/ac202878q] [PMID: 22044045]

[28] Shyuan LK, Heng LY, Ahmad M, Aziz SA, Ishak Z. Evaluation of pesticide and heavy metal toxicity using immobilized enzyme alkaline phosphatase with an electrochemical biosensor. Asian Journal of Biochemistry 2008; 3(6): 359-65.
[http://dx.doi.org/10.3923/ajb.2008.359.365]

[29] March G, Nguyen TD, Piro B. Modified Electrodes Used for Electrochemical Detection of Metal Ions in Environmental Analysis Academic Editor: Andrew M. Shaw Biosensors 2015; 5(2): 241-275.

[30] Uniyal S, Sharma RK. Technological advancement in electrochemical biosensor based detection of Organophosphate pesticide chlorpyrifos in the environment: A review of status and prospects. Biosens Bioelectron 2018; 116: 37-50.
[http://dx.doi.org/10.1016/j.bios.2018.05.039] [PMID: 29857260]

[31] Hernandez-Vargas G, Sosa-Hernández J, Saldarriaga-Hernandez S, Villalba-Rodríguez A, Parra-Saldivar R, Iqbal H. Electrochemical Biosensors: A Solution to Pollution Detection with Reference to Environmental Contaminants. Biosensors (Basel) 2018; 8(2): 29.
[http://dx.doi.org/10.3390/bios8020029] [PMID: 29587374]

[32] Khanmohammadi A, Jalili Ghazizadeh A, Hashemi P, Afkhami A, Arduini F, Bagheri H. An overview to electrochemical biosensors and sensors for the detection of environmental contaminants. J Indian Chem Soc 2020; 17(10): 2429-47.
[http://dx.doi.org/10.1007/s13738-020-01940-z]

[33] Zhu X, Qin H, Liu J, *et al.* A novel electrochemical method to evaluate the cytotoxicity of heavy metals. J Hazard Mater 2014; 271: 210-9.
[http://dx.doi.org/10.1016/j.jhazmat.2014.02.030] [PMID: 24637447]

[34] Lace A, Cleary J. A Review of Microfluidic Detection Strategies for Heavy Metals in Water. Chemosensors (Basel) 2021; 9(4): 60.
[http://dx.doi.org/10.3390/chemosensors9040060]

[35] Nie Z, Nijhuis CA, Gong J, *et al.* Electrochemical sensing in paper-based microfluidic devices. Lab Chip 2010; 10(4): 477-83.
[http://dx.doi.org/10.1039/B917150A] [PMID: 20126688]

[36] Pérez-Fernández B, Costa-García A, Muñiz AE. Electrochemical (Bio)Sensors for Pesticides Detection Using Screen-Printed Electrodes. Biosensors (Basel) 2020; 10(4): 32.
[http://dx.doi.org/10.3390/bios10040032] [PMID: 32252430]

[37] Yang Y, Fang D, Liu Y, *et al.* Problems analysis and new fabrication strategies of mediated electrochemical biosensors for wastewater toxicity assessment. Biosens Bioelectron 2018; 108: 82-8.
[http://dx.doi.org/10.1016/j.bios.2018.02.049] [PMID: 29501051]

[38] Aragay G, Merkoçi A. Nanomaterials application in electrochemical detection of heavy metals. Electrochim Acta 2012; 84: 49-61.
[http://dx.doi.org/10.1016/j.electacta.2012.04.044]

[39] Govindhan M, Adhikari BR, Chen A. Nanomaterials-based electrochemical detection of chemical

contaminants. RSC Advances 2014; 4(109): 63741-60.
[http://dx.doi.org/10.1039/C4RA10399H]

[40]   Nanda Kumar D, Alex SA, Chandrasekaran N, Mukherjee A. Acetylcholinesterase (AChE)-mediated immobilization of silver nanoparticles for the detection of organophosphorus pesticides. RSC Advances 2016; 6(69): 64769-77.
[http://dx.doi.org/10.1039/C6RA13185A]

[41]   Buledi JA, Amin S, Haider SI, Bhanger MI, Solangi AR. A review on detection of heavy metals from aqueous media using nanomaterial-based sensors. Environ Sci Pollut Res Int 2020; 2020: 1-9.
[PMID: 32036535]

[42]   Ramgir NS, Yang Y, Zacharias M. Nanowire-Based Sensors. Small 2010; 6(16): 1705-22.
[http://dx.doi.org/10.1002/smll.201000972] [PMID: 20712030]

[43]   Kumar P, Kim KH, Bansal V, Lazarides T, Kumar N. Progress in the sensing techniques for heavy metal ions using nanomaterials. J Ind Eng Chem 2017; 54: 30-43.
[http://dx.doi.org/10.1016/j.jiec.2017.06.010]

[44]   Zhang M, Ge L, Ge S, *et al.* Three-dimensional paper-based electrochemiluminescence device for simultaneous detection of Pb^{2+} and Hg^{2+} based on potential-control technique. Biosens Bioelectron 2013; 41: 544-50.
[http://dx.doi.org/10.1016/j.bios.2012.09.022] [PMID: 23058662]

[45]   Ashby CIH, Kelly MJ, Yelton WG, *et al.* Functionalized Nanoelectrode Arrays for In Situ Identification and Quantification of Regulated Chemicals in Water. 2002.

[46]   Medina-Sánchez M, Cadevall M, Ros J, Merkoçi A. Eco-friendly electrochemical lab-on-paper for heavy metal detection. Anal Bioanal Chem 2015; 407(28): 8445-9.
[http://dx.doi.org/10.1007/s00216-015-9022-6] [PMID: 26403238]

[47]   Martínez-García G, Pérez-Julián E, Agüí L, *et al.* An Electrochemical Enzyme Biosensor for 3-Hydroxybutyrate Detection Using Screen-Printed Electrodes Modified by Reduced Graphene Oxide and Thionine. Biosensors (Basel) 2017; 7(4): 50.
[http://dx.doi.org/10.3390/bios7040050] [PMID: 29137135]

[48]   Wan NA, Wan J, Wong LS. Exploring the potential of whole cell biosensor: a review in environmental applications. Int J Chem Environ Biol Sci 2014; 2: 1.

[49]   Gui Q, Lawson T, Shan S, Yan L, Liu Y. The application of whole cell-based biosensors for use in environmental analysis and in medical diagnostics. Sensors (Basel) 2017; 17(7): 1623.
[http://dx.doi.org/10.3390/s17071623] [PMID: 28703749]

[50]   Gupta N, Renugopalakrishnan V, Liepmann D, Paulmurugan R, Malhotra BD. Cell-based biosensors: Recent trends, challenges and future perspectives. Biosens Bioelectron 2019; 141111435
[http://dx.doi.org/10.1016/j.bios.2019.111435] [PMID: 31238280]

[51]   Qu J, Wu L, Liu H, *et al.* A novel electrochemical biosensor based on DNA for rapid and selective detection of cadmium. Int J Electrochem Sci 2015; 10: 4020-8.

[52]   Zeng G, Zhang C, Huang D, *et al.* Practical and regenerable electrochemical aptasensor based on nanoporous gold and thymine-Hg^{2+}-thymine base pairs for Hg^{2+} detection. Biosens Bioelectron 2017; 90: 542-8.
[http://dx.doi.org/10.1016/j.bios.2016.10.018] [PMID: 27825522]

[53]   Leth S, Maltoni S, Simkus R, *et al.* Engineered bacteria based biosensors for monitoring bioavailable heavy metals. Electroanalysis 2002; 14(1): 35-42.
[http://dx.doi.org/10.1002/1521-4109(200201)14:1<35::AID-ELAN35>3.0.CO;2-W]

[54]   Dai C, Choi S. Technology and applications of microbial biosensor. *Open.* Open Journal of Applied Biosensor 2013; 2(3): 83-93.
[http://dx.doi.org/10.4236/ojab.2013.23011]

[55] Borah H, Gogoi S, Kalita S, Puzari P. A broad spectrum amperometric pesticide biosensor based on glutathione S-transferase immobilized on graphene oxide-gelatin matrix. J Electroanal Chem (Lausanne) 2018; 828: 116-23.
[http://dx.doi.org/10.1016/j.jelechem.2018.09.047]

[56] Pabbi M, Mittal SK. An electrochemical algal biosensor based on silica coated ZnO quantum dots for selective determination of acephate. Anal Methods 2017; 9(10): 1672-80.
[http://dx.doi.org/10.1039/C7AY00111H]

[57] Mehta J, Bhardwaj N, Bhardwaj SK, *et al.* Graphene quantum dot modified screen printed immunosensor for the determination of parathion. Anal Biochem 2017; 523: 1-9.
[http://dx.doi.org/10.1016/j.ab.2017.01.026] [PMID: 28161099]

[58] Turdean GL. Design and Development of Biosensors for the Detection of Heavy Metal Toxicity. 2011.
[http://dx.doi.org/10.4061/2011/343125]

[59] Liu Y, Deng Y, Dong H, Liu K, He N. Progress on sensors based on nanomaterials for rapid detection of heavy metal ions. Sci China Chem 2017; 60(3): 329-37.
[http://dx.doi.org/10.1007/s11426-016-0253-2]

[60] Li J, Yu Y, Qian J, Wang Y, Zhang J, Zhi J. A novel integrated biosensor based on co-immobilizing the mediator and microorganism for water biotoxicity assay. Analyst (Lond) 2014; 139(11): 2806-12.
[http://dx.doi.org/10.1039/C4AN00243A] [PMID: 24728093]

[61] Zhu X, Wu G, Lu N, Yuan X. A miniaturized electrochemical toxicity biosensor based on graphene oxide quantum dots/carboxylated carbon nanotubes for assessment of priority pollutants. J Hazard Mater 324: 272-80.2016;

[62] Kuncová G, Pazlarova J, Hlavata A, Ripp S, Sayler GS. Bioluminescent bioreporter Pseudomonas putida TVA8 as a detector of water pollution. Operational conditions and selectivity of free cells sensor. Ecol Indic 2011; 11(3): 882-7.
[http://dx.doi.org/10.1016/j.ecolind.2010.12.001]

[63] Salek Maghsoudi A, Hassani S, Mirnia K, Abdollahi M. Recent Advances in Nanotechnology-Based Biosensors Development for Detection of Arsenic, Lead, Mercury, and Cadmium. Int J Nanomedicine 2021; 16: 803-32.
[http://dx.doi.org/10.2147/IJN.S294417] [PMID: 33568907]

CHAPTER 16

Utility of Biofertilizers for Soil Sustainability

Sekar Hamsa[1]**, Ruby Tiwari**[1,*] **and Chanderkant Chaudhary**[2]

[1] *Department of Genetics, University of Delhi, South Campus, New Delhi, 110021, India*

[2] *Department of Plant Molecular Biology, University of Delhi, South Campus, New Delhi, 110021, India*

Abstract: Modern agriculture is almost entirely reliant on the supply and utilization of agrochemicals, such as fertilizers, pesticides, and insecticides, to maintain and boost agriculture productivity. Heavy use of chemical fertilizers has resulted in numerous adverse effects on the environment and human health. Biofertilizers have emerged as an eco-friendly, inexpensive, and renewable alternative to restore, enhance, and maintain soil fertility, soil health, and crop yield. Biofertilizers are beneficial microbes, including plant growth-promoting rhizobacteria, mycorrhizal fungi, cyanobacteria, and their symbionts. Hence, the importance of biofertilizers in soil management practices for soil and crop sustainability needs to be highlighted in light of their multiple benefits, including augmenting nutrient availability in the rhizosphere, increasing nutrient uptake and recycling, supplementing soil water holding capacity, production of plant growth regulators, and soil reclamation. The challenges regarding the large-scale utilization of biofertilizers need to be emphasized to achieve sustainability in agricultural soils.

Keywords: Biofertilizers, Biotechnology, Nutrient, Phosphorus, Soil he, Sustainable agriculture.

INTRODUCTION

Currently, the global population is still increasing and it is estimated that around 2050 it will reach approximately 9.7 billion people in the world [1].With the increasing human population, food demand grows exponentially, and to cope with this problem, we have major challenges like maintaining sustainable agriculture to create a balance in the ecosystem. The unconscious application of agrochemicals and lack of knowledge regarding biodegradation ability lead to soil chelation with toxic molecules, which disturbs the soil structure, fertility, and water-holding capacity [2]. Over usage of synthetic fertilizers has been directly linked to the eutrophication of water resources [3 - 5] and the toxic build-up of heavy metals

* **Corresponding author Ruby Tiwari:** Department of Genetics, University of Delhi, South Campus, New Delhi, 110021, India; E-mail: ruby12tiwari@gmail.com

such as arsenic, cadmium, and plumbum [6]. The rapid growth in the world population has been the key factor in the explosion of intensive industrialization, urbanization, and agricultural production. The nutritional requirements of the present world population cannot be achieved by traditional agriculture practices, and these methods are incapable of making the countries self-sufficient [7 - 9]. Conventional agricultural practices based on the extensive application of synthetic fertilizers and pesticides are applied to enhance crop productivity and make the crop disease resistant [10]. Unfortunately, both these methods directly or indirectly consist of a high risk to the environment as they expose the water table, air, and soil stratum to these toxic chemicals [11]. The unconscious application of agrochemicals and lack of knowledge regarding biodegradation ability lead to soil chelation with toxic molecules, which disturbs the soil structure, fertility, and water-holding capacity [2]. Over usage of synthetic fertilizers has been directly linked to the eutrophication of water resources [3 - 5] and the toxic build-up of heavy metals such as arsenic, cadmium, and plumbum [6]. The rapid growth in the world population has been the key factor in the explosion of intensive industrialization, urbanization, and agricultural production. The nutritional requirements of the present world population cannot be achieved by traditional agriculture practices, and these methods are incapable of making the countries self-sufficient [7 - 9]. Conventional agricultural practices based on the extensive application of synthetic fertilizers and pesticides are applied to enhance crop productivity and make the crop disease resistant [10]. Unfortunately, both these methods directly or indirectly consist of a high risk to the environment as they expose the water table, air, and soil stratum to these toxic chemicals [11].

Eco-friendly practices like organic farming are a promising method over conventional agriculture, which can reduce the dependence on toxic chemicals and can enable chemical-free food production in addition to maintaining the high quality and biodiversity of soils [12]. Present-day agriculture is shifting more towards organic agriculture, and a considerable amount of growth from 11 million hectares (1999) to 57.8 million hectares (2016) has been recorded in the area of organic agriculture [13, 14]. Biofertilizers seem to be a promising approach in order to attain sustainable agriculture, and it utilizes live microorganisms such as microbial strains to enhance plant growth and development. Microbial strains employed a plethora of mechanisms such as nitrogen fixation, potassium, and phosphorus solubilization, flushing of excess phytohormones, beneficial plant-pathogen interactions, alleviating abiotic and biotic stresses, and sequestration of soil soluble pollutants to enhance soil fertility and crop yield. Present-day agriculture practices involve heavy usage of chemical fertilizers and pesticides, which has created a disorder in the ecological cycle. Microbial inoculants are a promising and non-toxic way to overcome these potent biological hazards.

Biofertilizer application has been showing promising outcomes as it utilizes the strains of microorganisms having beneficial properties to reduce the problems arising from the use of mineral fertilization [15]. These microbial strains have the potential to make agricultural practices sustainable as they enhance plant growth and development by enriching native nutrient (N,P, K, S, Zn) concentrations and bioavailability (Fig **1**) [16]. Moreover, it also minimizes the mycotoxins contamination as these strains pose antibacterial and/or antifungal activity [16]. Previous reports have been suggesting the crucial role of biofertilizers in sustainable agriculture as they enhance the activity of indigenous microorganisms present in the soil [17], activate the plant growth stimulants, induce abiotic and biotic stress tolerance mechanisms, maintain the soil pH, metabolize complex chemical compounds into simple assimilate forms [18] and sequester soil pollutants such as heavy metals [19], atrazine [20] or pesticide mixtures [21]. Various other properties and roles of biofertilizers are represented in Fig. (**1**).

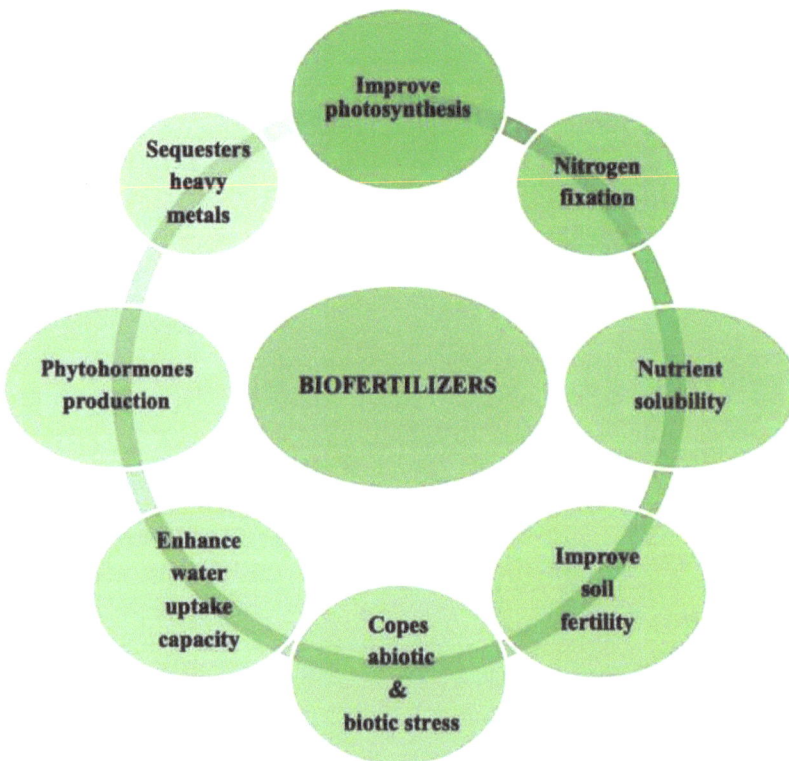

Fig. (1). Diagrammatic overview of various roles and properties of biofertilizers for soil sustainability and sustainable crop production.

Biofertilizers unravel an eco-friendly, affordable, and renewable way to enhance crop productivity in contrast to chemically synthesized fertilizers [22]. This review summarizes the impact and scope of microorganisms in terms of physical, chemical, and biological properties of the soil, sustainable agriculture by using PGPRs, nutrient cycling in soil, and the role of biotechnology and genomics in enhancing the properties of biofertilizers. Several challenges faced by the farmers in applying biofertilizers and the difficulty faced for their commercialization have also been discussed with a brief concluding remark on future solutions.

BIOFERTILIZERS AND AGRICULTURE

The definition of biofertilizers has changed from time to time. It encompasses seaweed extracts, urban wastes, and microbial mixtures with an array of diverse mineral fertilizer products having organic compounds, green manures, animal manures, and plant extracts [23 - 25]. Okon and Labandera-Gonzalez [26] stated that it is a substance that improves the exploitation of nutrients present in the soil, but does not replace them (like mineral fertilizers), so it should not be determined as a biofertilizer but as an inoculant. According to [23], biofertilizer utilizes living microorganisms, which is applied to the plant surfaces, seeds, and soil, or colonizes the rhizosphere and accelerates plant growth by enhancing nutrient availability. This description of biofertilizer does not match with the above-mentioned definition as described by Okon and Labandera-Gonzalez [26], whereas Fuentes-Ramirez and Caballero-Mellado [27] stated biofertilizers as "a product that contains living microorganisms benefiting crop field either directly or indirectly *via* various mechanisms". Somers et al. [28] categorize the microorganisms involved in the synthesis of phytohormones as phyto-stimulators or bio-enhancers and microorganisms carrying out the biodegradation of organic pollutants as rhizoremediators. Therefore, it is impossible to include all microbial inoculants into the category of biofertilizers [29]; instead, the ideal way is to sub-categorize them according to their biological property. Biofertilizer application promotes crop/plant growth without compromising the ecosystem and it acts as an adjuvant in increasing harvest yields [30]. Schütz et al. [31] showed that usage of biofertilizers had increased the crop yield average by 16.2% in comparison with the crops treated without biofertilizers. Microbial biofertilizers are the key players in maintaining soil fertility [32], improving plant-water relations [33], alleviating soil-borne diseases [34], and lowering the insect pests rate [35]. They showed promising results in achieving sustainable agriculture, but at the same time, there are some drawbacks, such as limited shelf life, sensitivity towards high temperature, storage, and transportation challenges [36]. However, new technologies are being developed in order to dilute these challenges [37].

Role of Biofertilizers in Improving Physio-chemical Properties of Soil

The physio-chemical properties of soil are a crucial factor that has a direct impact on crop productivity, nutrient, heavy metal mobilization, and biological composition. The physical properties of the soil include particle shape and size, texture, porosity, and water holding capacity, aeration, color, and density, while the chemical properties include pH, electrical conductivity through cation exchange, nutrient availability, and soil organic matter.

Climate change, land-use change, deforestation and erosion have ruined the quality of soil by reducing soil organic matter, nutrient availability, and electrical conductance, and also by influencing soil hydraulic properties and aggregate stability [38]. Soil aggregation is a major factor having a significant impact on other physical properties of the soil, including porosity and bulk density. Most of the soil's physical and chemical properties are highly correlated with each other in terms of their function [39]. The use of chemical fertilizers was shown to enhance the compaction of soil, further leading to crushed aggregate units, and a reduction in soil volume and porosity which increases the bulk density and penetration resistance [40].

To shift from chemical fertilizers, earlier formulations reported positive use of organic manure and biochar, which is an amalgamation of several organic matters as a potential method to restore soil quality and fertility. However, in the last decade, organic matter in combination with biofertilizers gained importance due to its better results in improving soil quality [41, 42]. Though vermicompost exerts a positive effect on the physio-chemical properties of the soil, vermicompost (organic matter) with a microbe *Paenibacillus azotofoxans* showed the best effect on the stability of both micro and macroaggregates as compared to vermicompost alone. Along with aggregate size, it also shows an increase in the organic carbon content and soil water content, further improving the stability of the soil [43, 44]. Application of biofertilizers (*Bacillus*, *Trichoderma,* and *Pseudomonas*) decreased bulk density and increased water-holding capacity as compared to organic fertilizers, which were more effective in increasing the yield of cotton [45]. The classification of biofertilizers is described in Fig. (**2**).

Biofertilizers are the major drivers to enhance soil properties by decreasing pH, secreting extra polymeric substances, and mobilizing nutrients. Mycorrhiza and PGPRs act as soil conditioners by exerting a positive effect on the physicochemical properties of the soil [46]. Due to the high requirement of water retention in sandy soils, EPS (Extracellular Polymeric substances) secreted by the microbes *Bacillus*, *Pseudomonas, Microbacterium arborescens* and several Phosphate solubilizing microbes (PSMs) are involved in stabilizing the structure of soil by influencing soil aggregation and water holding capacity in these soils [47, 48]. Mycorrhiza, along with PSM inoculated together, enhanced the organic content of the soil, and water holding capacity and caused a significant effect on the ion balance or electrical conductivity of the soil [49]. Cyanobacteria like *Nostoc* reduce erosion and enhance soil structure by secreting exopolysaccharides which form a coating on the macroaggregates and protect them from breaking down into smaller pieces called slaking [50]. They also influence the chemical properties by increasing soil hydraulic conductivity *via* decreasing pH and sodium content in the soil [51, 52]. Soil aggregation is an important factor for heavy metal mobilization in metal-stressed soils. Biofertilizers like PGPR and PSMs have been shown to stimulate the formation of larger aggregates (0.2-2mm) in soils containing lead and cadmium [53, 54].

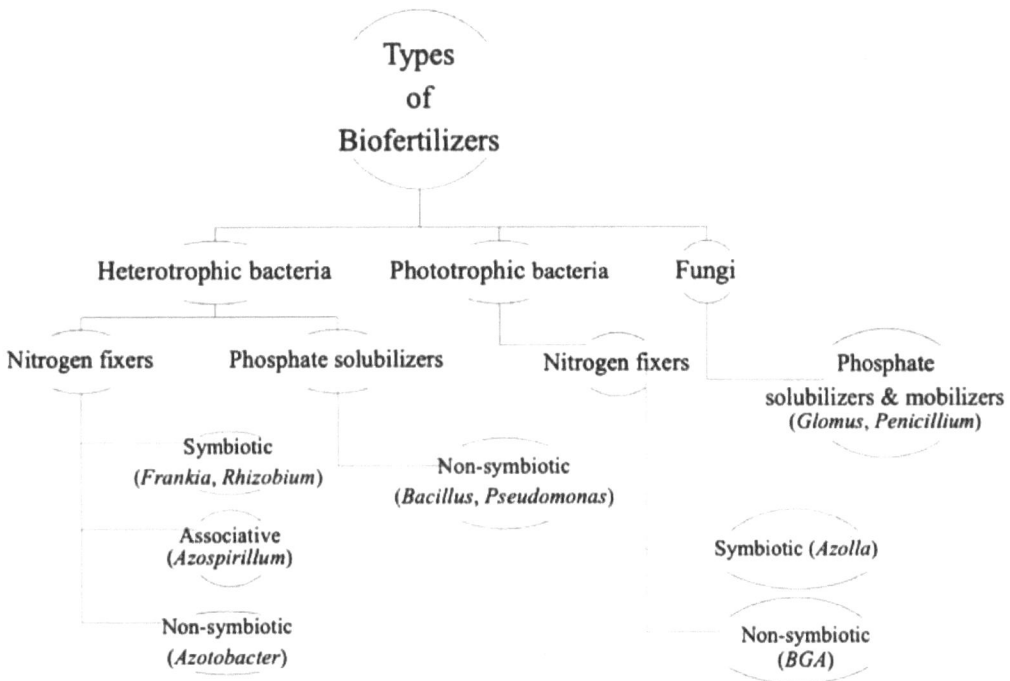

**Types
of
Biofertilizers**

Heterotrophic bacteria   Phototrophic bacteria   Fungi

Nitrogen fixers   Phosphate solubilizers   Nitrogen fixers   Phosphate solubilizers & mobilizers
(*Glomus, Penicillium*)

Symbiotic
(*Frankia, Rhizobium*)   Non-symbiotic
(*Bacillus, Pseudomonas*)

Associative
(*Azospirillum*)   Symbiotic (*Azolla*)

Non-symbiotic
(*Azotobacter*)   Non-symbiotic
(*BGA*)

Fig. (2).   Classification of biofertilizers based on their lifestyle, nature, and functions. (adapted from 206).

A direct correlation was observed between arbuscular mycorrhiza enhancing the water-stable aggregates, which resulted in high microbial numbers. Several other studies showed a positive correlation between physical, chemical, and biological properties [55]. Thus, the physiochemical properties are the main indicators of the quality of the soil in a particular ecosystem and within the constraint of climate.

Integrating them with the biological properties, mainly the interaction with beneficial microbial flora, will enhance our understanding of the sustainability of the soil [56].

ROLE OF NUTRIENT CYCLING IN SOIL IMPROVEMENT

Nutrient cycling is one of the major processes accomplished by soil microbes, thus playing a crucial role in maintaining the fertility of the soil. Nutrients cycled this way are up taken from the roots of the plant from the soil and used as raw materials for various important biological mechanisms for growth and development. Microbes recycle nitrogen, phosphorus and potassium, which are of prime importance, and other nutrients like zinc, manganese, silicon, calcium, *etc*. Over the years, soil microbes have developed and adapted mechanisms to solubilize, fix or chelate nutrients such that they are converted from their unavailable to available forms [57].

Phosphorus

Phosphorus is the second most crucial macronutrient in terms of mineral nutrition. The fixation of phosphate is through absorption or precipitation by free aluminium and iron ions in the acidic soil and by calcium ions in alkaline soil. Though phosphorus is present in surplus, its unavailability in soluble form due to fixation leads to the widespread use of toxic phosphorus fertilizers by farmers, thus polluting the agricultural soils. In the context of excessive use of harmful chemical fertilizers, phosphorus solubilizing microbes (PSMs) have gained widespread attention as a biofertilizer to enhance the uptake of phosphorus by mobilizing and rendering it in the soluble form [58]. Mechanisms of PSMs include solubilization, absorption, and mineralization of unavailable phosphorus complexes either in inorganic forms or organic forms (30-50%).

Solubilization

Production of various low molecular organic acids (OA), *viz*. citric acid, oxalic acid, gluconic acid, aspartic acid, and others lowers the pH of the soil by making it more acidic. This acidification (more H^+ ions), in turn, leads to the liberation of phosphate ions from their respective complexes with Al, Fe, or Ca through the substitution of H^+. Thus, the complex formed is H_2PO_4 which is the major soluble form that can be easily uptaken by the roots. Another mechanism is by chelation and reduction [59], *Trichoderma* T-22 was observed to produce compounds that could effectively chelate Fe^{3+} ions and also causes diffusion of metabolites that causes reduction of Fe^{3+} ions to Fe^{2+} ions. Apart from iron reduction, activity was also seen for manganese and copper ions. Some microbes also produce siderophores which are iron-chelating compounds [60].

Mineralization

Several enzymes are involved to release phosphorus present in the organic form. These include non-specific acid phosphatases, which are involved in the dephosphorylation of phosphate bonds; phytases which cause the release of phosphorus from phytate, which is the major part of organic matter in the soil and phosphatases cleave the C-P bond of organo-phosphonates [59].

Since phosphorus is the most limiting element after nitrogen, PSM has been advocated in agronomical practices efficiently in conjugation with other microorganisms to increase soil fertility and crop production. PSMs have been known to play a major role in legumes, which require more phosphorus than other crops [61]. A combination of organic fertilizers containing rock phosphate and *Bacillus sp* (PSM2) caused a significant increase in the number of nodules, the number of pods, shoot length, and grain yield of chickpeas [62,]. Field inoculations of multiple PSMs *Bacillus, Serratia, Arthrobacter,* and *Pantoea* led to an increase in the yield of rapeseed from 21-40% [63]. PSM-enriched vermicompost in calcareous soils where phosphorus uptake is a major problem resulted in an increase in shoot length of both tomato and wheat [64]. Similarly, *Nocardiopsis alba* and *Streptomyces* were shown to be promising candidates to increase the solubilization of phosphorus and further increase the germination rates and vigor index of wheat crops [65]. Several other PSMs have been characterized and shown to effectively increase the yield of other cereal crops like maize [66] and rice [67].

Nitrogen

Nitrogen is the major essential element and also the most limited in the soil due to runoff, leaching, and soil erosion. To compensate for this loss of nitrogen, heavy chemical fertilizers in the form of nitrates are applied in the form of nitrates which are a direct threat to soil causing nitrogen burn and to water bodies causing eutrophication. As an alternative, biological nitrogen fixation (BNF) through microbes has shown to be a cost-effective and efficient way of fixing nitrogen available to plants and also reducing the use of chemical fertilizers upto 0.160 billion tons per year [68]. BNF has genetically controlled regulatory mechanisms which ensure assimilation of fixed nitrogen into available biomass and the dissipation of the excess nitrogen to avoid its release into the environment [69]. The choice of biofertilizer to be applied depends on the type of soil, and the crops cultivated in that location. Nitrogen-fixing bacteria are classified based on their nature, either free-living or symbiotic and leguminous or non-leguminous based on the crops they colonize. Some of the common BNFs are *Azotobacter, Rhizobium, Azospirillium, Bacillus,* and most important, cyanobacteria, *Nostoc* and *Anabaena* [70].

Mechanism

BNF microbes fix atmospheric oxygen and convert it into ammonia through a specialized nitrogenase complex. Since the oxygen easily degrades nitrogenase, it needs an anaerobic condition for its proper functioning, which is inherently adapted in these microbes. BNF is characterized into three types based on the nature of the microbes [71].

Symbiotic

This is a mutualistic relationship between the microbe, which fixes nitrogen for the plant and gains carbon and energy source from the niche of the rhizosphere. They form specialized structures called nodules in the roots of the plant. The microbes can form associations with both leguminous (*Rhizobium*) and non-leguminous plants (*Frankia*) [72].

Asymbiotic/ Non-symbiotic

These are through free-living bacteria and the fixing of nitrogen is through the reduction process and is independent of any respiration process. Due to their widespread availability, they play an important role in nutrient cycling and maintaining the soil ecosystem. Cyanobacteria fall in this category and are one of the major nitrogen fixers apart from improving soil properties [73].

Associative

They are mainly associated with cereal crops and are mostly present in the rhizosphere and seldom enter the roots. Genus *Azospirillium* is the most common microbe associated with rice, wheat, and maize [74].

Nitrification and Denitrification

The process of conversion of ammonia to nitrate or nitrites is termed nitrification and further conversion of nitrate to molecular nitrogen denitrification. This dinitrogen is then released into the atmosphere is a crucial process for removing excess nitrates from the soil, thus completing the nitrogen cycle. *Alacigens sp, Pseudomonas, and Bradyrhizobium* are common microbes utilized for this process [75].

Several PSBs and Arbuscular mycorrhiza also promote nitrogen fixation in case of low phosphate supply [76, 77]. Apart from nitrogen fixation, these microbes also play a crucial role in phosphate solubilization, enhancing biological properties, increasing water holding capacity, water and mineral uptake by plants, defense against pathogens, and reclamation of eroded and contaminated soils [78]. In the two decades of research, over 200 genes have been identified to play a role in legume nitrogen fixation. A cascade of signalling events and transcriptional control is involved in nitrogen-fixing microorganisms through *nif* genes [79, 80]. Several factors have also been shown to affect the functioning of these microbes, like the addition of fertilizers over a certain load, temperature, precipitation, and the biomes where they are present [81].

Since most of the soils cultivating rice are deficient in nitrogen, BNF is extensively utilized and reported to have a significant increase in grain yield of rice, with Cyanobacteria and *Azolla-Anabena* systems being the most efficient of all nitrogen fixers [82]. The addition of *B. subtilis* in the agricultural soil not only led to the enhanced yield of maize but also reduced the leaching of nitrogen by over 50% by affecting the physicochemical properties as compared to only compost [83]. The application of BNF microbes has been reported to have a profound effect on the yield of various legumes [84, 85] and other crops like sweet potato [86], sugarcane [87] and Solanaceae members [88].

Potassium

Potassium is the third most important macronutrient after phosphorus and nitrogen. It is fixed in the soil in the form of soil minerals (90-98%) like mica, zeolite, and glauconite and is released in the soil solution by weathering or soil water [89]. Though the exchangeable form of potassium can be replenished, Indian agricultural soils are generally limited in potassium due to leaching and excessive crop uptake, which results in the use of heavy fertilizers in the form of potash. Still, most of the unaware farmers do not use potash along with urea and diammonium phosphate leading to the depletion of K in soil and further affecting crop productivity [90]. Since India does not produce sufficient potassium fertilizers, biofertilizers can be a preferable and cost-effective method to convert unavailable forms to available forms. Several microorganisms like *Pseudomonas*, *Bacillus*, *Acidithiobacillus ferroxidans*, *Aspergillus* have been isolated and characterized, of which *Bacillus mucilaginosus* and *B. edaphicus* have been reported to play a major role in unbinding of potassium from minerals [90 - 92]. The mechanisms of K-solubilizing microbes include direct and indirect methods [93, 94].

Direct methods

● Production of various organic acids, mainly tartaric acid, thereby lowering the pH causing solubilization through acidification.

● Proton-promoted mechanisms due to particulate and organic carbon degradation.

● Microbial uptake of PO_4^{3-} and Ca^{2+} ions leads to a gradient in cation-anion concentration.

Indirect methods

● Production of metal complexing ligands like glucuronic acid and alginates.

● Through ion diffusion by secreting various complex polysaccharides.

● Redox reaction by transferring electrons to the metal ions eventually results in destroying metal groups.

● Biofilms/extracellular polymers that accelerate the uptake of potassium and increase the rate of weathering helpful in potassium solubilization.

● Production of HCN, ammonia, and siderophores.

Several KSMs (Potassium solubilizing microbes) have been isolated, which have high efficiency to release potassium from their fixed forms [95, 96]. Many of the KSMs have been characterized in stressed soils like arid soils, saline soils, water deficit soil, and heavy metal-contaminated soils [97 - 99]. It has also been related to improving soil properties as the proportion of soil macro aggregates (>2mm) in soil supplemented with NPK was much higher than in soil supplemented with only NP [100]. KSMs have been shown to increase not only the grain weight but also plant height and dry matter in cereal crops like rice, wheat, and maize by increasing the photosynthetic efficiency [65, 101 - 103]. Since tea plantations grow in acidic soils, they require almost 60Kg K_2O per hectare to increase their productivity and yield; thus, several KSMs have been isolated, of which *Bacillus pseudomycoides* incubation led to the release of high amounts of exchangeable K as compared to control [104]. Biological potassium fertilizer (BSF) is a promising approach for the stimulation of exchangeable K content, however, its efficiency is highly determined by soil type, presence of K-bearing minerals, and differing environmental conditions [105].

Zinc and Iron (Micronutrients)

Though zinc is a micronutrient, it is a key component of various metabolic activities in soil and crops. The uptake of zinc by plants in the form of divalent cations requires zinc solubilizing microbes to release from unavailable forms. They decrease the pH of the soil and secrete IAA, siderophores, and ammonia, thus increasing the organic content of the soil and increasing the rate of enzymatic activities [106]. These bacteria have been reported to increase the yield of wheat and other cereals [107], soybean [108], chickpea [109], tomato [110], and several other crops [111]. Iron is present in extremely high amounts in the soil, but due to its poor solubility with water, it remains in low amounts. Mycorrhiza and other PGPR are known to secrete siderophores which leads to enhanced iron uptake and transport by various crop plants [100].

Due to the increase in demand, a need to boost agricultural yields has become essential in today's scenario. To achieve this, farmers use chemical fertilizers which cause depletion of minerals and biodiversity of the soil and degrade fertility. Microbial inoculants, thus are considered a major alternative for nutrient cycling for maintaining healthy agricultural soils and, in turn, protect plants from various deficiencies and stress and are promising in terms of a sustainable farming system [70].

Role of Biofertilizers in Soil Reclamation of Stressed/ Degraded/ Deserted Land

Saline Soil

Salinity is major abiotic stress limiting plant growth and development by reducing the osmotic potential of the soil solution, deteriorating the physical structure of the soil, and increasing the concentration of certain ions, having an inhibitory effect on the plant by causing direct toxicity. Three types of soil affected by salinity are saline soil, sodic soil, and saline-sodic soil [112, 113]. The application of biofertilizer, as reported by Wafaa et al. [114], showed improved soil properties of saline soil. They studied cyanobacteria applied with organic acids, compogypsum and gypsum, increased nutrient availability in saline soil, and increased N, P, K uptake in wheat and rice-grown fields. Badar et al. [115] have shown biofertilizer interaction with salinity, wherein increasing salt concentration has reduced the root length. On contrarily, saline-affected plants treated with biofertilizer showed a significant increase in root and shoot length, suggesting improved properties of plants when salt-effected soil was applied with biofertilizer. Li et al. [116] reviewed cyanobacteria application and challenges of soil remediation in salt-affected areas. They argued that cyanobacterial strain selection, co-culture establishment, large-scale cultivation, and contaminant control are major issues inhibiting wide-scale application. Cyanobacteria improve soil structure and make more channels in the soil that makes salts move deeper through these channels. Also, Na^+ is absorbed by cyanobacteria, thus decreasing the soluble Na^+ in soil. Biofertilizer application improved wheat yield by 15-25% over control and inferred that carrier-assisted PGPR (*Pseudomonas moraviensis* and *Bacillus cereus*) have increased the shelf life of PGPR and promoted crop productivity under salt stress [117]. Nisha et al. [118] studied two indigenous cyanobacterial strains, *Nostoc ellipsosporum* HH-205 and *Nostoc punctiforme* HH-206, isolated from the salt-affected area and utilized it as biofertilizer in saline-affected soil having high pH and electrical conductivity. Experiments conducted concluded increased carbon, phosphate, nitrogen, potassium, magnesium, cation exchange capacity, hydraulic conductivity, and a simultaneous decrease in sodium ion and electrical conductivity. There was an increase in microbial activity, leading to improved soil aggregation and stability. Biofertilizer treatment increased carbon mineralization as compared to the control. Autotrophic cyanobacteria produce biomass and add organic matter to the soil and are considered to be an efficient sink for P. They are considered to increase phosphate availability by mobilizing different insoluble forms of inorganic phosphate in rice plants [119].

Mine Degraded Soil

Mine land is deficient in nutrient, is eroded, have a low moisture retention tendency, and lack biological activities. Several reports have claimed that biofertilizers use help in the reclamation of barren lands, thus improving water holding capacity and creating topsoil for sustaining vegetation. This leads to the development of low-cost technology in rehabilitating mining areas and increasing their productivity. Dubey et al. [120] have explored the application of microbes in rehabilitating, restoring, and increasing the productivity of silica mine sites through revegetation. They recorded increased germination frequency of selected plant species under varied biofertilizer combinations such as *Rhizobium, Azotobacter*, PSM, VAM combinations, Blue-green algae, *etc.* compared to the control. They concluded that biofertilizer inoculation led to the establishment of plant seedlings under stress by ensuring a continuous nutrient supply. Another report by Mukhopadhyay and Maiti [121], reviewed in 2009, showed/reported/concluded that mycorrhizal inoculation has increased the productivity of degraded land, ensuring sustained phosphorus availability and enhanced drought tolerance of plants. Also, mycorrhizal inoculation increases nutrient uptake, improves water holding capacity, increases salt and heavy metal tolerance, protects from the pathogen, *etc.*

Heavy Metal affected soil

Human activities like mining, mineral, and smelting contaminate the soil with Cd, Cr, Zn, Pb, and Cu [122]. Heavy metal pollution becomes serious as it does not affect the productivity of crops but influences the quality of the atmosphere [123]. A group conducted pot experiments to evaluate the effect of different concentrations of heavy metal on the growth of *Jatropha curcas* when the soil was amended with biosludge and biofertilizer. These findings suggest that amendments in soil improved the survival rate along with enhanced growth of the plant. They concluded that biosludge in combination with biofertilizer made the plant survive in highly zinc-contaminated soil as compared to the control [124]. Another report by Neerja et al. [125] utilized *Azobacter* as a biofertilizer to evaluate its effect on the planting of *Jatropha curcas* in the reclamation of metal-contaminated soil. They inferred that the survival rate of plants in heavy metal-contaminated soil increased with biofertilizer addition.

Calcareous soil

Higher $CaCO_3$ soil has low organic matter and lacks nitrogen. Calcareous soil has greater than 15% $CaCO_3$ in powdery, nodule, and crust form. This type of soil covers more than 30% of the earth [126]. Egyptian soil constitutes 25-30% of the calcareous soil of the total area [127]. A study conducted to improve the soil character and help plant growth utilized a number of biofertilizers like *Nostoc*, diatomaceous earth, sewage effluent, and yeast, in addition to Hoagland medium. These supplements to calcareous soil showed enhanced growth and biomass of *Nigella sativa,* a medicinal plant. They showed a significant reduction in stress markers like malondialdehyde, proline content, and antioxidant activity compared to control grown on calcareous soil treated with biofertilizer. Thus, they inferred biofertilizer application, extinguishing the oxidative stress triggered by calcareous soil [128]. Another study aimed to assess the utility of phosphate solubilizing bacteria (PSB) and cyanobacteria (Cyan) in combination with mineral phosphate fertilizer in calcareous soil for maize production. They showed a significant increase in all growth and yield parameters. N, P, K uptake was increased by grain and straw than the control treatment. Carbohydrate content under PSB was higher in grain than under Cyanobacteria, whereas protein content showed the opposite result [129].

PLANT GROWTH PROMOTING RHIZOBACTERIA IN ENHANCING SOIL FERTILITY AND PLANT GROWTH

Numerous soil bacteria that flourish in the rhizosphere of plants and stimulate plant growth are termed PGPR (plant growth-promoting rhizobacteria). PGPR is more efficient and is replacing chemical fertilizers, pesticide use continuously in agriculture [130, 131], thus improving plant and soil health. PGPRs have been studied to play a major role in fixing N_2, increasing nutrient availability, influencing root growth and morphology, and promoting beneficial plant-microbe symbiosis. This group of microbial bacteria enhances plant growth and development by production of a variety of agroactive substances such as siderophores, phytohormones (indoleacetic acid, gibberellic acid), 1-aminocyclopropane-1-carboxylate deaminase (ACC), and enzymes to degrade fungal cell wall [132, 133]. These growth-promoting substances are produced in large quantities and have numerous beneficial effects on host plant morphology [134]. Recent reports suggest that sustainable agriculture depends on the use and diversity of PGPR. Their colonizing capacity and mechanism of action are used in the application of PGPR in the management of sustainable agriculture [131, 135].

Environmental stress (biotic and abiotic) hijacks plant growth and lead to poor germination and yield characters. *Azotobacter* spp., a PGPR is considered to promote and improve plant health *via* the production of phytohormone, pesticide and oil globules degradation, ob*via*tion of stressor, heavy metals metabolization, *etc.* Additionally, bacterial application has been shown to be helpful in the reclamation of soil [136]. Another PGPR, actinomycetes, has been seen to display biocontrol traits against root pathogenic fungi [137]. Some PGPR exhibit bifunctional properties, *i.e.*, biofertilizers or biopesticides such as strains of *Burkholderia cepacia* have been shown to control *Fusarium sp.* and are capable of producing siderophores which boost maize growth in iron-deficient land [29].

Phytoremediation by PGPR to remove organic pollutants like petroleum hydrocarbons from contaminated soil is implemented widely with great success [138]. A study aimed to investigate petroleum-contaminated soil phytoremediation enhancement by two PGPR strains *Klebsiella* sp. and *Pseudomonas* sp. concluded that PGPR inoculation has increased tall fescue biomass along with the removal of petroleum hydrocarbon [139]. Another study by Li et al. [140] showed that multifunctional PGPR microbial inoculants which contained four compatible strains of *Providencia rettgeri* P2, *Advenella incenata* P4, *Acinetobacter calcoaceticus* P19, and *Serratia plymuthica* P35 having the ability to solubilize phosphate and fix nitrogen. These PGPR inoculants improved the growth, physiology, and soil properties of oat, alfalfa, and cucumber seedlings and also increased the activity of peroxidase, catalase, and superoxide dismutase in the seedlings. They concluded that PGPR inoculants can be used as an alternative approach to biocontrol plant disease and improve plant health. Another interesting utilization of PGPR in PGPR-Enhanced Phytoremediation Systems (PEPS) has been observed on plant health in salt-stressed soil. It has been shown that PGPR protects against plant membrane damage and photosynthesis inhibition, suggesting conferring tolerance to plants under salt stress. They showed significant improvement in sodium chloride uptake and plant growth and provided evidence that phytoremediation of saline soil is feasible using PEPS [141]. Xun and the group investigated PGPR and arbuscular mycorrhizal fungi (AMF) on phytoremediation in saline-alkali petroleum-contaminated soil. PGPRs application on oat plants was measured for plant physiological parameters and petroleum degradation rate. It was found that PGPR (*Acinetobacter* sp.) in combination with AMF (*Glomus intraradices*) improved the growth of the plant and suppressed the increase in antioxidant enzyme activities caused by petroleum stress as compared to control, hence making it tolerant to hydrocarbon contaminants [142]. Several halo-tolerant PGPR has the potential to assist plants to survive saline conditions by the production of organic and inorganic compounds like IAA, ACC, volatile organic compounds, antioxidants, *etc* [143]. Co-application of titanium dioxide nanoparticles (TiO$_2$ NPs) and PGPR in promoting phytoremediation of Cd-contaminated soil. They reported that TiO$_2$ NPs and the PGPR combination enhanced plant growth and chlorophyll content of *Trifolium repens*. This addition also increased the accumulation capacity of *T. repens*. Thus, they argued intelligent association of plants, PGPR, and nanomaterials have great prospects in soil remediation. Another study investigated the effect of *Phyllobacterium myrsinacearum* strain RC6b on the growth and phytoextraction efficiency of heavy metals by a Zn/Cd hyperaccumulator (*Sedum alfredii*) and alfalfa (*Medicago sativa* L.). They showed RC6b inoculation has increased shoot biomass for alfalfa and *S. alfredii,* and also, there was increased phytoextraction of Pb, Cd, and Zn by shoots after six consecutive harvests in a co-cropping system as compared to control. Further, they suggested that RC6b inoculation increased soil microbial activity and the carbon utilization ability of soil [144]. Kumar and colleagues argued that surfactants produced by PGPR can allev*ia*te the problem of petroleum contamination and promote plant growth. They studied three *Pseudomonas* strains, characterized for biosurfactants, and essayed for rhamnolipids. The inoculation of petrol-contaminated soil with these strains helped in promoting the growth of *Withania somnifera,* and the presence of rhamnolipids helped in lowering petrol oil toxicity. This study concludes that rhamnolipids producing PGPR could be a better alternative for the reclamation of petrol-contaminated soil and growing medicinal plants [145]. A study conducted to reveal a possible mechanism of plant growth-promotion strain SNB6 on enhancing Cd phytoextraction of vetiver grass showed SNB6 encodes numerous genes for Cd tolerance, mobilization, and plant growth promotion. They enhance antioxidant enzymatic activity and improve the biochemical properties of rhizospheric soil. Thus, increasing Cd uptake and biomass of accumulator further enhances Cd Phytoextraction [146]. The usefulness of *L. sativus* to boost an efficient reclamation of Pb and Cd contaminated soil by 15 inoculums formed by mixing heavy metal resistant PGPR that contributed to restoring the fertility and quality of polluted soil [147].

The biofertilizer properties of PGPR are attributed to their role in nitrogen fixation, ability to increase the bioavailability of inorganic and organic phosphorus, and formation of mycorrhizal associations. Thus, PGPR can be used as an alternative to chemicals for plant growth in many contaminated soils. They fulfill diverse plant functions and help in plant growth hence are a promising solution to numerous emerging modern problems such as climate change and food productivity.

Role of Biotechnology

Advances in biotechnology and genome studies have led to the development of microbes with improved functions, referred to as microbial biotechnology [148]. DNA sequencing of various biofertilizers, along with meta-omics and computational tools, has helped in understanding the genes involved in various critical functions and also the microbial community and structure prevailing at a specific location. Though PGPRs are a promising approach for sustainable agriculture, limitations in varying environmental conditions, poor colonization, and persistence need to be overcome by genetic engineering and synthetic biology. Several methods, like horizontal gene transfer, mobile genetic elements, and phages, have been utilized to introduce traits of interest in microbes [149]. Over the years, the engineering of microbes has been done to enhance the levels of macronutrients NPK, proper growth of crops, suppressing biotic and abiotic stresses, and improve the quality of the soil. These modified microbes are then inoculated at a specific location for a particular crop to deliver the advantages of the desired PGP traits [150].

At the Biological Level

Recently, the Artificial Microbial Consortium (AMC), consisting of several important microorganisms artificially added, has been utilized for the reconstitution of the plant microbiome. SynComs (microbial synthetic communities) derived from compost enhanced Arabidopsis and tomato growth and suppressed the symptoms of *Fusarium* wilt pathogen [151]. Since colonization is still a major limitation, a genetically engineered strain of *B. velenzeus* expressing deqQ gene under a xylose-inducible promoter was developed. Enhancing xylose levels led to effective colonization due to biofilm production and antimicrobial properties [152]. Quorum sensing property of bacteria to respond co-ordinately as a single unit when a certain threshold of population density is acquired has been applied in genetic engineering. It is a communication process of several bacteria in a population *via* signalling molecules; the most common of all is Acyl Homoserine lactones (AHLs). This property has been exploited for bioremediation through biofilm production, minimizing pathogens and virus populations in the soil involved in biotic stress, thereby restructuring the microbial community, designing synthetic consortiums and others. However, several chemicals inhibiting quorum sensing behaviour are limitations that need to be further explored for its successful application [153 - 155].

Heterologous Expression of Gene Clusters

Sequencing of various microbial genomes has helped in identifying various genes or gene clusters responsible for the production of natural products or metabolites beneficial for soil sustainability. Expression in heterologous hosts has been found to increase the expression of gene clusters like PKS gene clusters were found to have an increasing production under strong inducible promoters, and thaxtomins had 18-fold improvement in their activity when expressed in another host *S. albus,* as compared to its original host *S. scabei* without any manipulation [156, 157]. Recently, a peptide Thirucin, 17 produced by *B. thuringenesis* has been identified to promote the growth of plants in biotic and abiotic stresses and stimulate the niche size in plant roots. This peptide can be used as a potential candidate to be introduced into other microbes [158]. Heterologous expression is useful for gaining information on various genes and pathways, like the involvement of nitrate assimilatory pathway in salt-stressed colonized microbe *B. megaterium* [159]. Owing to the limitations in identifying a heterologous host and laborious cloning strategies, tools like CRISPR-Cas are gaining importance for studying functional genomics in microbes. This genetic tool has been utilized in modifying pathways involved in ethanol, carotene, and plant growth-promoting siderophore and for deleting the susceptibility factors in pathogenic bacteria, which otherwise can prove beneficial to the rhizosphere [160].

Increasing the Production of PGRs

Since IAA (auxin) is considered the major regulator of plant development, engineering the IAA pathway in microbes has led to a beneficial effect on plants. Due to the limitation of tryptophan in nature, an artificial de novo biosynthesis pathway from glucose to tryptophan to IAA was established in *E. coli,* which can be further applied to other microbes for the production of excess IAA [161]. Using synthetic biology, a non-PGPR *Cupriavidus pinatubonesis* engineered to produce quorum-sensing dependent IAA hormone showed better acquisition of nutrients and enhanced root growth in *A. thaliana* [161]. Excessive production of auxin by expression of *nifH* gene in model bacteria increased the efficiency of nitrogen fixation. ACC deaminase is a promising candidate gene that decreases the levels of ethylene-induced understress, thereby elevating plant growth. Since this gene is widespread in various bacteria, its overexpression has been useful for several stresses like salinity and water deficit stress [162].

Improving Nutrient Availability

Phosphorus

A glucose dehydrogenase (*gcd*) gene involved in the solubilization of phosphate was discovered to produce gluconic acid in *Pseudomonas,* which can be further applied in phosphate-deficient soils. Several root-associated strains engineered with phytase solubilizing genes were effective in improving phosphorus availability in the rhizosphere plane [163]. Transcriptome profiling of a PSB, *Burkholderia* sp, provided insights into the molecular mechanisms involved in growth and phosphate solubilization at different concentrations [164]. Recently, a *Pseudomnonas putida* strain was shown to improve the growth of *Arabidopsis* in saline soils under phosphorus deficiency [165]. Various genes like *pqq, gabY, suc2, ppc, gdh, vgb and mps* have been characterized to be involved in phosphate solubilization or mineralization through the secretion of acids or quinone [166].

Nitrogen

Over the years, several genes involved in nitrogen fixation were discovered through forward and reverse genetics approaches; among them, the Nitrogenase complex plays the most crucial role in fixing nitrogen [80]. Mutagenesis and the creation of deletion mutants have enabled microbes to enhance their capability of fixing nitrogen. Mutations in the glutamine synthetase in *Azospirillium*, cyanide resistant pathway in *Rhizobium*, SA1 mutant in *Anabena,*and zinc finger regulators are some of the genes where the mutagenesis approach increased the amount of nitrogen fixation [167]. Similarly, disrupting the glycogen metabolism in *Cyanothece* (delta glgx strain) reported a high rate of fixation along with faster growth of the microbe [168]. Not only mutagenesis but the transfer of an entire set of nitrogenase genes to a biocontrol non-nitrogen fixing *P. protogens* ret S strain rendered it with both antifungal and biological fixing properties. Similarly, thirty *nif* genes were transferred to a non-diazotrophic cyanobacterium [169, 170]. Deletion of a negative regulator *nifL* gene and expression of a positive regulator constitutively in *Azotobacter* increased the yield of wheat crop by 60% without any urea application [171].

Bioremediation

Over the years, the isolation of stress-tolerant microbes and studying in-depth mechanisms at the cellular level have made it feasible to improve microbe robustness under stress conditions. Trans kingdom heterologous expression of *Hsp* gene from nematode to *E. coli* made the engineered strain grow at 50 degrees Celsius as compared to its optimum temperature 37°C, thus adapting to high-temperature stress [156]. Overexpression of *rpoS* and *htg* genes in a biocontrol agent *Pseudomonas fluorescens* with preadapted conditions resulted in higher survival rates in salt-stressed conditions [172]. Heavy metal stress is a major contamination in various soils, and several PGP microbes have been isolated to accumulate, transform or detoxify various heavy metals. An engineered *P. aeruginosa* strain was found to immobilize cadmium and, at the same time, promote plant growth [173]. Introducing a gene involved in metal capturing which assembled de novo with a type-IV secretion system in *E. coli* showed high-efficiency removal (> 80%) of cadmium and led heavy metals.

The complete genome sequences: metabolomics, systems biology, refactoring several genes for rearranging the regulatory units, engineering the biosphere, media engineering, and enhancement of the function of biostimulants along with statistical tools with multivariate analysis have enhanced our understanding of the intricacies of genetic engineering beyond classical approaches. Though GEM has higher potential, tight regulations, and unawareness need to be overcome to get them into commercial use.

CHALLENGES AND FUTURE SOLUTIONS

India relies on extensive agricultural practices, mainly in rural areas, and efforts need to be applied for safe and environment-friendly methods in farming. Though fertilizers have been extremely useful for farmers, there is an urgent need to awaken farmers and develop eco-friendly approaches as an alternative to chemical fertilizers. Biofertilizers enhance soil health as well as crop productivity, but still, they have not been widely accepted due to the difficulty in reproducing their beneficial effects on plants in a natural environment. Due to the wide array of environmental conditions, it is not possible to replicate their beneficial effects always. The major challenges include the lack of awareness of the eco-friendly importance of microbial biofertilizers, storage and transportation of inoculants, and selective action of biofertilizers [174]. Biofertilizers have immense potential for their use in sustainable agriculture. However, its widespread availability and use are challenged by a number of factors like formulation of inoculants, providing a safe environment during storage, and ensuring survival and establishment after introduction in soil. The key issue during the formulation and development of inoculants is the quality control of products at every stage. Other technical difficulties are related to the large-scale production of inoculants. The mode of cultivation of bacteria, the nature of growing media, conditions of incubation (temperature, pH, time), and the required space for large-scale usage are major concerns and drawbacks which limit their usability [175]. The practical application of biofertilizers is challenged by a lack of standardized testing and evaluation guidelines in addition to disregard to protect intellectual property and technology transfer.

Hence, the efficient training of operators is crucial to ensuring the right methods are implemented. The cost of production should be minimized, and the purity of culture has to be maintained [176]. The formulation is an important area of future research. Incubation of carriers is a crucial and critical step and new carriers are needed to overcome the limitation of peat availability and toxicity, and to provide more optimum conditions for microorganisms to grow and enhance their *via*bility and fitness during storage [177]. Successful commercialization relies on the linkages between researchers to formulate the best inoculants, the private sector to scale up productivity and establish a sustainable chain, and finally, building farmer's confidence in the usage of biofertilizers for sustainable agriculture. To achieve that quality, biofertilizers need to be improved so that farmers are convinced of their efficacy and are more willing to buy instead of expensive chemical fertilizers. Demonstration trails and regular training of operators and farmers on fields will help in a significant increase in the use of biofertilizers [29, 175]. Also, public and private sectors, encouraged by proper regulatory bodies and policymakers, will ensure the utility of biofertilizers in soil sustainability. A commercially available list of biofertilizers is mentioned in Table 1.

Table 1. List of important biofertilizers commercially available in India and the world over the last decades [modified from 201].

Name	Species	Organism Type	Target	Reference
BactoFil	*Azospirillum brasilense, Azotobacter vinelandii, Bacillus megaterium, B. polymyxa, Pseudomonas fluorescens*	Mix	Enhances nutrient acquisition	[178]
Bio-P	*Bacillus megaterium*	PGPR	Enhances phosphorus acquisition	[179]
Bio-K	*Frateuria aurantia*	PGPR	Enhances potassium acquisition	[179]
Bio-N	*Azotobacter sp.*	PGPR	Enhances nutrient acquisition	[179]

(Table 1) cont.....

Name	Species	Organism Type	Target	Reference
Bioagro	*Pseudomonas fluorescens, Candida tropicalis, Bacillus amyloliquefaciens, Bacillus subtilis*	Mix	Nitrogen fixation, phosphorous solubilization, organic matter mineralization	[180]
Botrycid	*Burkholderia vietnamiensis*	PGPR	Controls soil fungal pathogens, plant pathogenic bacteria, and nematodes	[181]
CataPult SuperFine	*Glomus intraradices, Bacillus spp.*	Mix	Phosphorous solubilization and mineralization, drought resistance	[182]
Jumpstart	*Penicillium bilaie*	Soil fungi	Helps the plant in phosphate acquisition	[183]
Micofert	*Glomus claroideum, Acaulospora kentinensis, Diversi sporaspurca, Glomus etunicatum*	Mycorrhiza	Enhances nutrient availability	[184]
Myc 800	*Rhizophagus intraradices*	Mycorrhiza	Enhances nutrient acquisition	[185]
MycoApply Endo, VAM80	*Glomus intraradices*	Mycorrhiza	Enhances phosphate availability	[186]
MycoUp, Resid	*Glomus iranicum var. tenuihypharum*	Mycorrhiza	Enhances phosphate availability	[187]
Mykoflor	*Rhizophagus irregularis, Funneliformis mosseae, Claroideo glomusetunicatum*	Mycorrhiza	Enhances nutrient acquisition	[185]
Nitragin Gold	*Rhizobium meliloti*	Rhizobia	Provides nitrogen fixation	[188]
Nitroguard	*Azospirillum brasilense NAB317, Azorhizobium caulinodens NAB38, Azoarcusindigens NAB04, Bacillus sp.*	Mix	Nitrogen fixation, phosphate solubilization	[189]
Nitrofix	*Azospirillum str. Az39*	PGPR	Improves nitrogen acquisition	[190]
Phylazonit-M	*Pseudomonas putida, Azotobacter chroococcum, Bacillus circulans, Bacillus megaterium*	Mix	Enhances phosphate availability	[191]
Proradix	*Pseudomonas sp. DSMZ 13134*	PGPR	Produces organic and siderophore acids, highly effective in chelating metal cations (zinc, copper and iron)	[192]

(Table 1) cont.....

Name	Species	Organism Type	Target	Reference
Rhizocell	*Bacillus amyloliquefaciens IT45*	PGPR	Increases phosphorous solubilization and mineralization	[193]
RhizoMyco, RhizoMyx, RhizoPlex	*A mixture of 18 species of endoandectomycorrhizal fungi*	mycorrhiza	Enhances nutrient and water availability	[194]
Rhizosum N	*Azotobacter vinelandi, Rhizophagus irregularis*	Mix	Enhances nitrogen and phosphate	[195]
Rhizosum PK	*Bacillus megaterium, Rhizophagus irregularis, Frateuria aurantia*	Mix	Improve phosphate and potassiumavailability	[195]
RizofosLiqMaiz	*Pseudomonas fluorescens*	PGPR	Phosphoroussolubilization and mineralization, production of phytohormones,siderophore and antibiotics	[196]
Suma Grow	*14 bacterial (Rhizobium spp., Pseudomonas spp., Bacillus spp.) 7 fungal species (Trichoderma spp.)*	Mix	Enhances nutrient acquisition in several crops	[197]
Symbion-N	*Azospirillum, Rhizobium, Acetobacter, Azotobacter*	Mix	Nitrogen fixation on field crops	[198]
Symbion-P	*Bacillus megaterium var. phosphaticum*	PGPR	Phosphorous solubilization and mineralization	[198]
Symbion-K	*Frateuria aurantia*	PGPR	Potassium solubilization	[198]
Tagteam	*Penicillium bilaii, Rhizobium leguminosarum*	Mix	Enhances nitrogen availability	[199]
Trichoderma X	*Trichoderma asperellum*	Soil fungi	Controls soil fungal pathogens on seeds, roots and flowers	[200]
Twin N	*Azospirillum brasilense NAB317, Azorhizobium caulinodens NAB38, Azoarcus indigens NAB0*	Mix	Nitrogen fixation	[189]

Over the years, biofertilizers have emerged as a key player in enhancing soil and crop sustainability through various ways. There are certain challenges like short shelf life, colonization of microorganisms, adaptation to new stress environment, and influence of the soil microbes, which inhibit the growth of beneficial microorganisms and decrease the efficiency of biofertilizers.

Liquid biofertilizers have gained importance over the years as a suitable substitute for solid biofertilizers due to properties like the increased shelf life of about 2 years, greater potential to fight the native microbes, better survival, high microbial density, and easy to use by farmers. They contain a dormant form of microorganisms which, on reaching the soil, become active and grow rapidly to release bioactive molecules involved in soil and plant health [201, 202]. Nanofertilizers or nano-biofertilizers are the most technically advanced method developed recently to deliver nutrients to plants in nano-scale amounts with the help of molecular aggregates called nanoparticles [24]. Biofertilizers are encapsulated in nanoscale polymers, chitosan, or zeolite, which improve the shelf life, allow a controlled and steady release of nutrients, and are easy to formulate. Nanobiofertilizers have been reported to promote the growth of various crops, enhance nutrient absorption, minimize leaching or runoff and resistance to several pathogens. Indian scientists developed the world's first nano fertilizer, which was projected to bring down the application of fertilization upto 80-100 times, and over some years, IFFCO has released several nanotechnology-based nutrient products [203]. In 2021, the world's first liquid biofertilizer using nanotechnology, Nano Urea, was introduced which can reduce the requirement of urea of upto 50%, generally used as a substitute for nitrogen. Certain limitations like low cost and development of novel techniques for increasing their production need to be overcome. In the near future, addressing these constraints will lead to nano biofertilizers as a revolutionary approach in modern agriculture [204].

CONCLUSION

Biofertilizers are a promising and potent method to achieve sustainable agriculture and prevent deterioration of environmental sectors, such as soil and land, and create a balance in the ecosystem. They show the potential to prevent the agro-ecosystem from chemical pollution, sequester toxic chemicals, mitigate the high demand for food from the growing world population, and to convert nutritionally essential elements into highly assimilated forms from non-usable forms without any harmful effect on the environment. They are a crucial part of the Integrated Plant Nutrient System (IPNS) [205, 206].The possible risk arising from conventional farming practices can be nullified by using biological inoculants in forthcoming years. Micro-organisms that belong to the rhizosphere genus have been known to accelerate a plethora of plant growth activities, but the number is still very low in order to use them in the form of biofertilizers. Therefore, advanced biotechnological techniques are required to manifest microorganisms as biofertilizers and to attain the objective of sustainable agriculture in the coming years.

REFERENCES

[1]   Food and Agriculture Organization of the United Nations. FAOSTAT statistical database. FAO 2021. Rome

[2]   Savci S. An agricultural pollutant: chemical fertilizer. Int J Environ Sci Dev 2012; 3: 73-80.
[http://dx.doi.org/10.7763/IJESD.2012.V3.191]

[3]   Goenadi DH, Mustafa AB, Santi LP. Bio-organo-chemical fertilizers: a new prospecting technology for improving fertilizer use efficiency (FUE). IOP Conf S Earth Environ Sci Jakarta, Indonesia. 2018; 183:1-11.

[4]   Liu CW, Sung Y, Chen BC, Lai HY. Effects of nitrogen fertilizers on the growth and nitrate content of lettuce (*Lactuca sativa* L.). Int J Environ Res Public Health 2014; 11(4): 4427-40.
[http://dx.doi.org/10.3390/ijerph110404427] [PMID: 24758896]

[5]   Zaidi A, Khan MS, Rizvi A, Saif S, Ahmad B, Shahid M. Role of phosphate-solubilizing bacteria in legume improvement.Microbes for legume improvement. 2nd ed. Springer International Publishing 2017; pp. 175-97.
[http://dx.doi.org/10.1007/978-3-319-59174-2_8]

[6]   Atafar Z, Mesdaghinia A, Nouri J, *et al.* Effect of fertilizer application on soil heavy metal concentration. Environ Monit Assess 2010; 160(1-4): 83-9.
[http://dx.doi.org/10.1007/s10661-008-0659-x] [PMID: 19058018]

[7]   Gizaki LJ, Alege AA, Iwuchukwu JC. Farmer's perception of sustainable alternatives to the use of chemical fertilizers to enhance crop yield in Bauchi state Nigeria. Int J Sci Res Sci Technol 2015; 1: 242-50.

[8]   Mahanty T, Bhattacharjee S, Goswami M, *et al.* Biofertilizers: a potential approach for sustainable agriculture development. Environ Sci Pollut Res Int 2017; 24(4): 3315-35.
[http://dx.doi.org/10.1007/s11356-016-8104-0] [PMID: 27888482]

[9]   Santos VB, Araújo ASF, Leite LFC, Nunes LAPL, Melo WJ. Soil microbial biomass and organic matter fractions during transition from conventional to organic farming systems. Geoderma 2012; 170: 227-31.
[http://dx.doi.org/10.1016/j.geoderma.2011.11.007]

[10]   Vasile AJ, Popescu C, Ion RA, Dobre I. From conventional to organic in Romanian agriculture – Impact assessment of a land use changing paradigm. Land Use Policy 2015; 46: 258-66.
[http://dx.doi.org/10.1016/j.landusepol.2015.02.012]

[11]   Rahman K, Zhang D. Effects of fertilizer broadcasting on the excessive use ofinorganic fertilizers and environmental sustainability. Sustainability (Basel) 2018; 10(3): 759.
[http://dx.doi.org/10.3390/su10030759]

[12]   Niggli U. Sustainability of organic food production: challenges and innovations. Proc Nutr Soc 2015; 74(1): 83-8.
[http://dx.doi.org/10.1017/S0029665114001438] [PMID: 25221987]

[13]   Brenes-Muñoz T,. Lakner S Br€ummer B, Ger J Agric Econ. What influences the growth of organic farms? Evidence from a panel of organic farms in Germany. 2016; 65:1-15.

[14]   Xiang W, Zhao L, Xu X, Qin Y, Yu G. Mutual information flow between beneficial microorganisms and the roots of host plants determined the bio-functions of biofertilizers. Am J Plant Sci 2012; 3(8): 1115-20.
[http://dx.doi.org/10.4236/ajps.2012.38134]

[15]   Mishra P, Dash D. Rejuvenation of biofertilizer for sustainable agriculture and economic development.

Consilience 2014; 11: 41-61.

[16]   Toyota K, Watanabe T. Recent trends in microbial inoculants in agriculture. Microbes Environ 2013; 28(4): 403-4.
[http://dx.doi.org/10.1264/jsme2.ME2804rh] [PMID: 24366038]

[17]   Raja N. Biopesticides and biofertilizers: ecofriendly sources for sustainable agriculture. Journal of Fertilizers & Pesticides 2013; 4(1): 1-2.
[http://dx.doi.org/10.4172/2155-6202.1000e112]

[18]   El-Lattief EAA. Use of Azospirillum and Azobacter bacteria as biofertilizers in cerealcrops: a review. Int J Res Eng Appl Sci 2016; 6: 36-44.

[19]   Siddiquee S, Aishah SN, Azad SA, Shafawati SN, Naher L. Tolerance andbiosorption capacity of Zn^{2-}, Pb^{2+}, Ni^{3+} and Cu^{2+} by filamentous fungi (*Trichoderma harzianum, T. aureoviride* and *T. virens*). Adv Biosci Biotechnol 2013; 4: 570-83.
[http://dx.doi.org/10.4236/abb.2013.44075]

[20]   Pelcastre MI, Ibarra JRV, Navarrete AM, Rosas JC, Ramirez CAG, Sandoval OAA. Bioremediation perspectives using autochthonous species of *Trichoderma sp.* for degradation of atrazine in agricultural soil from the Tulancingo valley, Hidalgo, Mexico. Trop Subtrop Agroecosystems 2013; 16: 265-76.

[21]   Fragoeiro S, Magan N. Impact of *Trametes versicolor* and *Phanerochaete chrysosporium* on differential breakdown of pesticide mixtures in soil microcosms at two water potentials and associated respiration and enzyme activity. Int Biodeterior Biodegradation 2008; 62(4): 376-83.
[http://dx.doi.org/10.1016/j.ibiod.2008.03.003]

[22]   Swapna G, Divya M, Brahmaprakash GP. Survival of microbial consortium ingranular formulations, degradation and release of microorganisms in soil. Ann Plant Sci 2016; 5: 1348-52.
[http://dx.doi.org/10.21746/aps.2016.05.004]

[23]   Vessey JK. Plant growth promoting rhizobacteria as biofertilizers. Plant Soil 2003; 255(2): 571-86.
[http://dx.doi.org/10.1023/A:1026037216893]

[24]   El-Ghamry AM, Mosa AA, Alshaal TA, El-Ramady HR. Nanofertilizers vs biofertilizers: new insights. Environ Biodiv Soil Sec 2018; 2: 1-22.

[25]   Macik M, Gryta A, Frac M. Biofertilizers in agriculture: an overview on concepts, strategies and effects on soil microorganisms.Adv Agron Academic Press. 2020; pp. 31-87.

[26]   Okon Y, Labandera-Gonzalez CA. Agronomic applications of *azospirillum*: An evaluation of 20 years worldwide field inoculation. Soil Biol Biochem 1994; 26(12): 1591-601.
[http://dx.doi.org/10.1016/0038-0717(94)90311-5]

[27]   Fuentes-Ramirez LE, Caballero-Mellado J. Bacterial biofertilizers.Siddiqui ZA) PGPR. Dordrecht, The Netherlands: Springer 2005; pp. 143-72.

[28]   Somers E, Vanderleyden J, Srinivasan M. Rhizosphere bacterial signalling: a love parade beneath our feet. Crit Rev Microbiol 2004; 30(4): 205-40.
[http://dx.doi.org/10.1080/10408410490468786] [PMID: 15646398]

[29]   Bhattacharyya PN, Jha DK. Plant growth-promoting rhizobacteria (PGPR): emergence in agriculture. World J Microbiol Biotechnol 2012; 28(4): 1327-50.
[http://dx.doi.org/10.1007/s11274-011-0979-9] [PMID: 22805914]

[30]   Mishra DJ, Singh R, Mishra UK, Kumar SS. Role of bio-fertilizer in organic agriculture: a review. Res J Recent Sci 2013; 2: 39-41.

[31]   Schütz L, Gattinger A, Meier M, *et al.* Improving Crop Yield and Nutrient Use Efficiency *via* Biofertilization—A Global Meta-analysis. Front Plant Sci 2018; 8: 2204.
[http://dx.doi.org/10.3389/fpls.2017.02204] [PMID: 29375594]

[32]   Rashid MI, Mujawar LH, Shahzad T, Almeelbi T, Ismail IMI, Oves M. Bacteria and fungi can

contribute to nutrients bioavailability and aggregate formation in degraded soils. Microbiol Res 2016; 183: 26-41.
[http://dx.doi.org/10.1016/j.micres.2015.11.007] [PMID: 26805616]

[33]   Xun F, Xie B, Liu S, Guo C. Effect of plant growth-promoting bacteria (PGPR) and arbuscular mycorrhizal fungi (AMF) inoculation on oats in saline-alkali soil contaminated by petroleum to enhance phytoremediation. Environ Sci Pollut Res Int 2015; 22(1): 598-608.
[http://dx.doi.org/10.1007/s11356-014-3396-4] [PMID: 25091168]

[34]   Simarmata T. Hersanti, Turmuktini T, Fitriatin BN, Setiawati MR, Purwanto. Application of bioameliorant and biofertilizers to increase the soil health and rice productivity. Biosci 2016; 23(4): 181-4.

[35]   Dey R, Pal KK, Tilak KVBR. Plant growth promoting rhizobacteria in crop protectionand challenges.Future challenges in Crop protection against fungal pathogens. NewYork: Springer 2014; pp. 31-59.
[http://dx.doi.org/10.1007/978-1-4939-1188-2_2]

[36]   Patil H, Solanki MK. Microbial inoculant: modern era of fertilizers and pesticides. 2016.
[http://dx.doi.org/10.1007/978-81-322-2647-5_19]

[37]   García-Fraile P, Menéndez E, Rivas R. Role of bacterial biofertilizers in agriculture and forestry. AIMS Bioeng 2015; 2(3): 183-205.
[http://dx.doi.org/10.3934/bioeng.2015.3.183]

[38]   Tellen VA, Yerima BPK. Correction to: Effects of land use change on soil physicochemical properties in selected areas in the North West region of Cameroon. Environ Syst Res 2018; 7(1): 12.
[http://dx.doi.org/10.1186/s40068-018-0116-y]

[39]   Al-Suhaibani N, Selim M, Alderfasi A, El-Hendawy S, Selim M. Comparative performance of integrated nutrient management between composted agricultural wastes, chemical fertilizers, and biofertilizers in improving soil quantitative and qualitative properties and crop yields under arid conditions. Agronomy (Basel) 2020; 10(10): 1503.
[http://dx.doi.org/10.3390/agronomy10101503]

[40]   Massah J, Azadegan B. Effect of Chemical Fertilizers on Soil Compaction and Degradation Development of portable enzyme biosensors View project Design and Construction of a Prototype Harvesting Robot for Greenhouses Products View project Effect of Chemical Fertilizers on Soil Compaction and Degradation. 2016.

[41]   Palakit K, Duangsathaporn K, Lumyai P, Sangram N, Sikareepaisarn P, Khantawan C. Efficiency of biochar and bio-fertilizers derived from maize debris as soil amendments. Environ Nat Resour J 2018; 16: 79-90.

[42]   Saxena J, Rana G, Pandey M. Impact of addition of biochar along with Bacillus sp. on growth and yield of French beans. Sci Hortic (Amsterdam) 2013; 162: 351-6.
[http://dx.doi.org/10.1016/j.scienta.2013.08.002]

[43]   Lim SL, Wu TY, Lim PN, Shak KPY. The use of vermicompost in organic farming: overview, effects on soil and economics. J Sci Food Agric 2015; 95(6): 1143-56.
[http://dx.doi.org/10.1002/jsfa.6849] [PMID: 25130895]

[44]   Youssef MA, Eissa MA. Comparison between organic and inorganic nutrition for tomato. J Plant Nutr 2017; 40(13): 1900-7.
[http://dx.doi.org/10.1080/01904167.2016.1270309]

[45]   Li R, Tao R, Ling N, Chu G. Chemical, organic and bio-fertilizer management practices effect on soil physicochemical property and antagonistic bacteria abundance of a cotton field: Implications for soil biological quality. Soil Tillage Res 2017; 167: 30-8.
[http://dx.doi.org/10.1016/j.still.2016.11.001]

[46]   Meddich A, Oufdou K, Boutasknit A, Raklami A, Tahiri A, Ben-Laouane R, *et al.* Use of organic and

biological fertilizers as strategies to improve crop biomass, yields and physicochemical parameters of soil.Nutrient dynamics for sustainable crop production. Singapore: Springer 2019; pp. 247-88.

[47]   Costa OYA, Raaijmakers JM, Kuramae EE. Microbial extracellular polymeric substances: ecological function and impact on soil aggregation. Front Microbiol 2018; 9: 1636.
[http://dx.doi.org/10.3389/fmicb.2018.01636] [PMID: 30083145]

[48]   Bharadwaj A. Role of microbial extracellular polymeric substances in soil fertility.Microbial polymers. Singapore: Springer 2021; pp. 341-54.
[http://dx.doi.org/10.1007/978-981-16-0045-6_15]

[49]   Sandhya A, Vijaya T, Narasimha G. Effect of microbial inoculants (VAM and PSB) on soil physico-chemical properties. Indian J Biotechnol 2013; 7(8): 320-4.

[50]   Crouzet O, Consentino L, Pétraud JP, *et al.* Soil photosynthetic microbial communities mediate aggregate stability: influence of cropping systems and herbicide use in an agricultural soil. Front Microbiol 2019; 10: 1319.
[http://dx.doi.org/10.3389/fmicb.2019.01319] [PMID: 31258520]

[51]   Ibraheem IBM. Cyanobacteria as alternative biological conditioners for bioremediation of barren soil; 2007.
[http://dx.doi.org/10.21608/egyjs.2007.114548]

[52]   Kumar J, Singh D, Tyagi MB, Kumar A. Cyanobacteria: applications in Biotechnology in Cyanobacteria: From Basic Science to Applications. Elsevier 2018; pp. 327-46.

[53]   Wang Y, Zhang Z, Liang Y, Han Y, Han Y, Tan J. High potassium application rate increased grain yield of shading-stressed winter wheat by improving photosynthesis and photosynthate translocation. Front Plant Sci 2020; 11: 134.
[http://dx.doi.org/10.3389/fpls.2020.00134] [PMID: 32184793]

[54]   Wang Z, Chai L, Yang Z, Wang Y, Wang H. Identifying sources and assessing potential risk of heavy metals in soils from direct exposure to children in a mine-impacted city, Changsha, China. J Environ Qual 2010; 39(5): 1616-23.
[http://dx.doi.org/10.2134/jeq2010.0007] [PMID: 21043267]

[55]   Andrade BO, Bonilha CL, Ferreira PMA, Boldrini II, Overbeck GE. Overbeck Highland grasslands at the southern tip of the Atlantic Forest biome: management options and conservation challenges. Oecol Aust 2016; 20(2): 175-99.
[http://dx.doi.org/10.4257/oeco.2016.2002.04]

[56]   Doran JW, Zeiss MR. Soil health and sustainability: managing the biotic component of soil quality. Appl Soil Ecol 2000; 15(1): 3-11.
[http://dx.doi.org/10.1016/S0929-1393(00)00067-6]

[57]   Yilmaz E, Sönmez M. The role of organic/bio–fertilizer amendment on aggregate stability and organic carbon content in different aggregate scales. Soil Tillage Res 2017; 168: 118-24.
[http://dx.doi.org/10.1016/j.still.2017.01.003]

[58]   Kalayu G. Phosphate solubilizing microorganisms: promising approach as biofertilizers. Int J Agron 2019; 2019: 1-7.
[http://dx.doi.org/10.1155/2019/4917256]

[59]   Khan MS, Zaidi A, Ahmad E. Mechanism of phosphate solubilization and physiological functions of phosphate-solubilizing microorganisms Phosphate Solubilizing Microorganisms Princ Appl Microphos Technol. Springer Int Publ 2014; pp. 31-62.
[http://dx.doi.org/10.1007/978-3-319-08216-5_2]

[60]   Altomare C, Norvell WA, Björkman T, Harman GE. Solubilization of phosphates and micronutrients by the plant-growth-promoting and biocontrol fungus *trichoderma harzianum* rifai 1295-22. Appl Environ Microbiol 1999; 65(7): 2926-33.
[http://dx.doi.org/10.1128/AEM.65.7.2926-2933.1999] [PMID: 10388685]

[61] Zeng Q, Wu X, Wang J, Ding X. Phosphate solubilization and gene expression of phosphate-solubilizing bacterium *Burkholderia multivorans* WS-FJ9 under different levels of soluble phosphate. J Microbiol Biotechnol 2017; 27(4): 844-55.
[http://dx.doi.org/10.4014/jmb.1611.11057] [PMID: 28138122]

[62] Ditta A, Imtiaz M, Mehmood S, *et al.* Rock phosphate-enriched organic fertilizer with phosphate-solubilizing microorganisms improves nodulation, growth, and yield of legumes. Commun Soil Sci Plant Anal 2018; 49(21): 2715-25.
[http://dx.doi.org/10.1080/00103624.2018.1538374]

[63] Valetti L, Iriarte L, Fabra A. Growth promotion of rapeseed (*Brassica napus*) associated with the inoculation of phosphate solubilizing bacteria. Appl Soil Ecol 2018; 132: 1-10.
[http://dx.doi.org/10.1016/j.apsoil.2018.08.017]

[64] Parastesh F, Alikhani HA, Etesami H. Vermicompost enriched with phosphate–solubilizing bacteria provides plant with enough phosphorus in a sequential cropping under calcareous soil conditions. J Clean Prod 2019; 221: 27-37.
[http://dx.doi.org/10.1016/j.jclepro.2019.02.234]

[65] Boubekri K, Soumare A, Mardad I, *et al.* The screening of potassium- and phosphate-solubilizing *Actinobacteria* and the assessment of their ability to promote wheat growth parameters. Microorganisms 2021; 9(3): 470.
[http://dx.doi.org/10.3390/microorganisms9030470] [PMID: 33668691]

[66] Pande A, Pandey P, Mehra S, Singh M, Kaushik S. Phenotypic and genotypic characterization of phosphate solubilizing bacteria and their efficiency on the growth of maize. J Genet Eng Biotechnol 2017; 15(2): 379-91.
[http://dx.doi.org/10.1016/j.jgeb.2017.06.005] [PMID: 30647676]

[67] Dash N, Pahari A, Dangar TK. Functionalities of phosphate- solubilizing bacteria of rice rhizosphere: techniques and perspectives.Recent advances in applied microbiology. Singapore: Springer 2017; pp. 151-63.
[http://dx.doi.org/10.1007/978-981-10-5275-0_7]

[68] Soumare A, Diedhiou AG, Thuita M, *et al.* Exploiting biological nitrogen fixation: A route towards a sustainable agriculture. Plants 2020; 9(8): 1011.
[http://dx.doi.org/10.3390/plants9081011] [PMID: 32796519]

[69] Bueno Batista M, Dixon R. Manipulating nitrogen regulation in diazotrophic bacteria for agronomic benefit. Biochem Soc Trans 2019; 47(2): 603-14.
[http://dx.doi.org/10.1042/BST20180342] [PMID: 30936245]

[70] Asoegwu CR, Awuchi CG, Nelson KCT, *et al.* A review on the role of biofertilizers in reducing soil pollution and increasing soil nutrients. Hmlyn J Agr 2020; 1: 34-8.

[71] Mahmud K, Makaju S, Ibrahim R, Missaoui A. Current progress in nitrogen fixing plants and microbiome research. Plants 2020; 9(1): 97.
[http://dx.doi.org/10.3390/plants9010097] [PMID: 31940996]

[72] Mylona P, Pawlowski K, Bisseling T. Symbiotic nitrogen fixation. Plant Cell 1995; 7(7): 869-85.
[http://dx.doi.org/10.2307/3870043] [PMID: 12242391]

[73] Knops JMH, Bradley KL, Wedin DA. Mechanisms of plant species impacts on ecosystem nitrogen cycling. Ecol Lett 2002; 5(3): 454-66.
[http://dx.doi.org/10.1046/j.1461-0248.2002.00332.x]

[74] James EK. Nitrogen fixation in endophytic and associative symbiosis. Field Crops Res 2000; 65(2-3): 197-209.
[http://dx.doi.org/10.1016/S0378-4290(99)00087-8]

[75] Hayatsu M, Tago K, Saito M. Various players in the nitrogen cycle: Diversity and functions of the microorganisms involved in nitrification and denitrification. Soil Sci Plant Nutr 2008; 54(1): 33-45.

[http://dx.doi.org/10.1111/j.1747-0765.2007.00195.x]

[76]   Püschel D, Janoušková M, Voříšková A, Gryndlerová H, Vosátka M, Jansa J. Arbuscular mycorrhiza stimulates biological nitrogen fixation in two Medicago spp. through improved phosphorus acquisition. Front Plant Sci 2017; 8: 390.
[http://dx.doi.org/10.3389/fpls.2017.00390] [PMID: 28396674]

[77]   Verzeaux J, Hirel B, Dubois F, Lea PJ, Tétu T. Agricultural practices to improve nitrogen use efficiency through the use of arbuscular mycorrhizae: Basic and agronomic aspects. Plant Sci 2017; 264: 48-56.
[http://dx.doi.org/10.1016/j.plantsci.2017.08.004] [PMID: 28969802]

[78]   Cassán F, Coniglio A, López G, *et al.* Everything you must know about *Azospirillum* and its impact on agriculture and beyond. Biol Fertil Soils 2020; 56(4): 461-79.
[http://dx.doi.org/10.1007/s00374-020-01463-y]

[79]   Dixon R, Kahn D. Genetic regulation of biological nitrogen fixation. Nat Rev Microbiol 2004; 2(8): 621-31.
[http://dx.doi.org/10.1038/nrmicro954] [PMID: 15263897]

[80]   Roy S, Liu W, Nandety RS, *et al.* Celebrating 20 years of genetic discoveries in legume nodulation and symbiotic nitrogen fixation. Plant Cell 2020; 32(1): 15-41.
[http://dx.doi.org/10.1105/tpc.19.00279] [PMID: 31649123]

[81]   Zúñiga A, Fuente F, Federici F, *et al.* An engineered device for indoleacetic acid production under quorum sensing signals enables *Cupriavidus pinatubonensis* JMP134 to stimulate plant Growth. ACS Synth Biol 2018; 7(6): 1519-27.
[http://dx.doi.org/10.1021/acssynbio.8b00002] [PMID: 29746094]

[82]   Choudhury ATMA, Kennedy IR. Nitrogen fertilizer losses from rice soils and control of environmental pollution problems. Commun Soil Sci Plant Anal 2005; 36(11-12): 1625-39.
[http://dx.doi.org/10.1081/CSS-200059104]

[83]   Sun B, Bai Z, Bao L, *et al. Bacillus subtilis* biofertilizer mitigating agricultural ammonia emission and shifting soil nitrogen cycling microbiomes. Environ Int 2020; 144105989
[http://dx.doi.org/10.1016/j.envint.2020.105989] [PMID: 32739514]

[84]   Kakraliya SK, Singh U, Bohra A, Choudhary KK, Kumar S, Meena RS, *et al.* Nitrogen and legumes: A meta-analysis.Legumes for soil health and sustainable management. Singapore: Springer 2018; pp. 277-314.
[http://dx.doi.org/10.1007/978-981-13-0253-4_9]

[85]   Liu J, Yu X, Qin Q, Dinkins RD, Zhu H. The impacts of domestication and breeding on nitrogen fixation symbiosis in legumes. Front Genet 2020; 11: 00973.
[http://dx.doi.org/10.3389/fgene.2020.00973] [PMID: 33014021]

[86]   Luna Castellanos L. L, lopez BS, D, Alfonso García Peña J, AroldoEspitia Montes A 2020. Effect of the inoculation of nitrogen-fixing rhizobacteria in the sweet potato crop (*Ipomoea batatas* Lam.). Horticult Int J 2020; 4(1): 35-40.

[87]   Pereira W, Oliveira RP, Pereira A, *et al.* Nitrogen acquisition and 15N-fertiliser recovery efficiency of sugarcane cultivar RB92579 inoculated with five diazotrophs. Nutr Cycl Agroecosyst 2021; 119(1): 37-50.
[http://dx.doi.org/10.1007/s10705-020-10100-x]

[88]   Sarbadhikary SB, Mandal NC. Elevation of plant growth parameters in two solanaceous crops with the application of endophytic fungus. Indian J Agric Res 2018; 52(Of): 424-8.
[http://dx.doi.org/10.18805/IJARe.A-4784]

[89]   Singh P, Sindhu SS, Parmar P, Sindhu SS. Potassium solubilization by rhizosphere bacteria: influence of nutritional and environmental conditions biological control of plant pathogens and insects View project Potassium Solubilization by rhizosphere Bacteria: influence of Nutritional and Environmental

Conditions. J Microbiol Res (Rosemead Calif) 2013; 2013: 25-31.

[90]　Meena VS, Zaid A, Maurya BR, *et al.* Evaluation of potassium solubilizing rhizobacteria (KSR): enhancing K-bioavailability and optimizing K-fertilization of maize plants under Indo-Gangetic Plains of India. Environ Sci Pollut Res Int 2018; 25(36): 36412-24.
[http://dx.doi.org/10.1007/s11356-018-3571-0] [PMID: 30368711]

[91]　Raghavendra MP, Nayaka SC, Nuthan BR. Role of rhizosphere microflora in potassium solubilization.Potassium solubilizing microorganisms for sustainable agriculture. India: Springer 2016; pp. 43-59.
[http://dx.doi.org/10.1007/978-81-322-2776-2_4]

[92]　Saha M, Maurya BR, Meena VS, Bahadur I, Kumar A. Identification and characterization of potassium solubilizing bacteria (KSB) from Indo-Gangetic Plains of India. Biocatal Agric Biotechnol 2016; 7: 202-9.
[http://dx.doi.org/10.1016/j.bcab.2016.06.007]

[93]　Masood S, Bano A. Mechanism of potassium solubilization in the agricultural soils by the help of soil microorganisms.Potassium solubilizing microorganisms for sustainable agriculture. India: Springer 2016; pp. 137-47.
[http://dx.doi.org/10.1007/978-81-322-2776-2_10]

[94]　Sattar A, Naveed M, Ali M, *et al.* Perspectives of potassium solubilizing microbes in sustainable food production system: A review. Appl Soil Ecol 2019; 133: 146-59.
[http://dx.doi.org/10.1016/j.apsoil.2018.09.012]

[95]　Nath D, Maurya BR, Meena VS. Documentation of five potassium- and phosphorus-solubilizing bacteria for their K and P-solubilization ability from various minerals. Biocatal Agric Biotechnol 2017; 10: 174-81.
[http://dx.doi.org/10.1016/j.bcab.2017.03.007]

[96]　Suyal DC, Singh M, Singh D, *et al.* Phosphate-solubilizing fungi: current perspective and future need for agricultural sustainability. Fungal Biol 2021; 109-33.

[97]　Ashfaq M, Hassan HM, Ghazali AHA, Ahmad M. Halotolerant potassium solubilizing plant growth promoting rhizobacteria may improve potassium availability under saline conditions. Environ Monit Assess 2020; 192(11): 697.
[http://dx.doi.org/10.1007/s10661-020-08655-x] [PMID: 33043403]

[98]　Chavoshi S, Nourmohamadi G, Madani H, Heidari H, Abad S, Fazel MA. Role of potassium solublizing bacteria on nutrients uptake in red bean (*Phaseolus vulgaris* L. cv. Goli) under water deficit condition. Legume Res 2018; 41: 416-21.

[99]　Dotaniya ML, Panwar NR. V. D. Meena D, CK, Regar KL, Lata M, et al. Bioremediation of metal contaminated soil for sustainable crop production.Role of rhizospheric microbes in soil: stress management and agricultural sustainability. Singapore: Springer 2018; pp. 143-73.
[http://dx.doi.org/10.1007/978-981-10-8402-7_6]

[100]　Liu C, Ye Y, Liu J, Pu Y, Wu C. Iron biofortification of crop food by symbiosis with beneficial microorganisms. J Plant Nutr 2021; 44(18): 2793-810.
[http://dx.doi.org/10.1080/01904167.2021.1927089]

[101]　Feng K, Cai Z, Ding T, Yan H, Liu X, Zhang Z. Effects of potassium⬚solubilizing and photosynthetic bacteria on tolerance to salt stress in maize. J Appl Microbiol 2019; 126(5): 1530-40.
[http://dx.doi.org/10.1111/jam.14220] [PMID: 30758905]

[102]　Youseff MMA, Eissa MFM. Biofertilizers and their role in management of plant parasitic nematodes. A review.E3 J Biotechnol. Pharm Res 2014; 5: 1-6.

[103]　Wu P, Wang Z, Zhu Q, *et al.* Stress preadaptation and overexpression of rpoS and hfq genes increase stress resistance of *Pseudomonas fluorescens* ATCC13525. Microbiol Res 2021; 250126804
[http://dx.doi.org/10.1016/j.micres.2021.126804] [PMID: 34144508]

[104] Pramanik P, Kalita C, Borah K, Kalita P. Combined application of mica waste and *Bacillus pseudomycoides* as a potassium solubilizing bio-fertilizer reduced the dose of potassium fertilizer in tea-growing soil. Agroecol Sustain Food Syst 2021; 45(5): 732-44.
[http://dx.doi.org/10.1080/21683565.2020.1847232]

[105] Dong X, Lv L, Wang W, *et al.* Differences in distribution of potassium-solubilizing bacteria in forest and plantation soils in Myanmar. Int J Environ Res Public Health 2019; 16(5): 700.
[http://dx.doi.org/10.3390/ijerph16050700] [PMID: 30818756]

[106] Kushwaha P, Kashyap PL, Pandiyan K, Bhardwaj AK. Zinc-solubilizing microbes for sustainable crop production: current understanding, opportunities, and challenges.Phytobiomes: current insights and future vistas. Singapore: Springer 2020; pp. 281-98.
[http://dx.doi.org/10.1007/978-981-15-3151-4_11]

[107] Kamran S, Shahid I, Baig DN, Rizwan M, Malik KA, Mehnaz S. Contribution of zinc solubilizing bacteria in growth promotion and zinc content of wheat. Front Microbiol 2017; 8: 2593.
[http://dx.doi.org/10.3389/fmicb.2017.02593] [PMID: 29312265]

[108] Ramesh A, Sharma SK, Sharma MP, Yadav N, Joshi OP. Inoculation of zinc solubilizing Bacillus aryabhattai strains for improved growth, mobilization and biofortification of zinc in soybean and wheat cultivated in Vertisols of central India. Appl Soil Ecol 2014; 73: 87-96.
[http://dx.doi.org/10.1016/j.apsoil.2013.08.009]

[109] Ullah A, Farooq M, Nadeem F, *et al.* Zinc application in combination with zinc solubilizing Enterobacter sp. MN17 improved productivity, profitability, zinc efficiency, and quality of Desi chickpea. J Soil Sci Plant Nutr 2020; 20(4): 2133-44.
[http://dx.doi.org/10.1007/s42729-020-00281-3]

[110] Karnwal A. *Pseudomonas* spp., a zinc-solubilizing vermicompost bacteria with plant growth-promoting activity moderates zinc biofortification in tomato. Int J Veg Sci 2021; 27(4): 398-412.
[http://dx.doi.org/10.1080/19315260.2020.1812143]

[111] Kumar A, Dewangan S, Lawate P, Bahadur I, Prajapati S. Zinc-solubilizing bacteria: A boon for sustainable agriculture. Microorganisms Sustainability 2019; pp. 139-55.

[112] Amini S, Ghadiri H, Chen C, Marschner P. Salt-affected soils, reclamation, carbon dynamics, and biochar: a review. J Soils Sediments 2016; 16(3): 939-53.
[http://dx.doi.org/10.1007/s11368-015-1293-1]

[113] Nouri H, Chavoshi Borujeni S, Nirola R, *et al.* Application of green remediation on soil salinity treatment: A review on halophytoremediation. Process Saf Environ Prot 2017; 107: 94-107.
[http://dx.doi.org/10.1016/j.psep.2017.01.021]

[114] Wafaa MT, Eletr FM, Ghazal AA, Mahmoud G, Yossef H. Responses of wheat–rice cropping system to cyanobacteria inoculation and different soil conditioners sources under saline soil. Nat Sci 2013; 11: 118-29.

[115] Badar R, Batool B, Ansari A, Mustafa S, Ajmal A, Perveen S. Amelioration of salt affected soils for cowpea growth by application of organic amendments. J Pharmacogn Phytochem 2015; 3(6)

[116] Li H, Zhao Q, Huang H. Current states and challenges of salt-affected soil remediation by cyanobacteria. Sci Total Environ 2019; 669: 258-72.
[http://dx.doi.org/10.1016/j.scitotenv.2019.03.104] [PMID: 30878933]

[117] Hassan TU, Bano A. Role of carrier-based biofertilizer in reclamation of saline soil and wheat growth Arch Agron 2015; 61(12):1719-31.

[118] Nisha R, Kiran B, Kaushik A, Kaushik CP. Bioremediation of salt affected soils using cyanobacteria in terms of physical structure, nutrient status and microbial activity. Int J Environ Sci Technol 2018; 15(3): 571-80.
[http://dx.doi.org/10.1007/s13762-017-1419-7]

[119] Mandal B, Vlek PLG, Mandal LN. Beneficial effects of blue-green algae and *Azolla*, excluding supplying nitrogen, on wetland rice fields: a review. Biol Fertil Soils 1999; 28(4): 329-42.
[http://dx.doi.org/10.1007/s003740050501]

[120] Dubey K, Singh VK, Mishra CM, Kumar A. Use of biofertilizer for reclamation of silica mining area. Makalahdisampaikan pada Billings Land Reclamation Symposium. 4-8.

[121] Mukhopadhyay S, Maiti SK. Biofertilizer: VAM fungi–A future prospect for biological reclamation of mine degraded lands. Indian J Environ Prot 2009; 29(9): 801-8.

[122] Akoto O, Ephraim JH, Darko G. Heavy metals pollution in surface soils in the vicinity of abundant railway service workshop in Kumasi, Ghana. Int J Environ Res 2008; 2: 359-64.

[123] Woo SL, Ruocco M, Vinale F, *et al. Trichoderma* based products and their widespread use in agriculture. Open Mycol J 2014; 8(1): 71-126.
[http://dx.doi.org/10.2174/1874437001408010071]

[124] Juwarkar AA, Yadav SK, Kumar P, Singh SK. Effect of biosludge and biofertilizer amendment on growth of *Jatropha curcas* in heavy metal contaminated soils. Environ Monit Assess 2008; 145(1-3): 7-15.
[http://dx.doi.org/10.1007/s10661-007-0012-9] [PMID: 17973198]

[125] Neerja S, Sharma S, Rani A, Sharma M. Phytoremediation enhancement of Jatropha curcas for decontamination of heavy metals by use of biofertilizer. J Microb World 2014.

[126] Marschner H. Mineral nutrition of higher plants. 2nd ed., London: Academic Press 1995.

[127] The dynamic changes in chemical and mineralogical characteristics of calcids soils as affected by natural soil amendments 2002.

[128] El-Zohri M, Medhat N, Saleh FE, El-Maraghy S. Some biofertilizers relieved the stressful drawbacks of calcareous soil upon black seed (*Nigella sativa L.*) through inhibiting stress markers and antioxidant enzymes with enhancing plant growth. Egypt J Bot 2017; 57(1): 75-92.
[http://dx.doi.org/10.21608/ejbo.2017.262.1006]

[129] Ganzour S, Ghabour T, Hemeid N, Khatab K. Impact of Biofertilizers on Maize (*Zea mays* L.) Growth and yield under calcareous soil conditions. Egyptian Journal of Soil Science 2020; 0(0): 0.
[http://dx.doi.org/10.21608/ejss.2020.45922.1392]

[130] Ansari RA, Rizvi R, Sumbul A, Mahmood I. PGPR: current vogue in sustainable crop production.Probiotics and plant health. Singapore: Springer 2017; pp. 455-72.
[http://dx.doi.org/10.1007/978-981-10-3473-2_21]

[131] Ansari RA, Mahmood I, Eds. 2019.

[132] Lugtenberg B, Kamilova F. Plant-Growth-Promoting Rhizobacteria. Annu Rev Microbiol 2009; 63(1): 541-56.
[http://dx.doi.org/10.1146/annurev.micro.62.081307.162918] [PMID: 19575558]

[133] Majeed A, Abbasi MK, Hameed S, Imran A, Rahim N. Isolation and characterization of plant growth-promoting rhizobacteria from wheat rhizosphere and their effect on plant growth promotion. Front Microbiol 2015; 6: 198.
[http://dx.doi.org/10.3389/fmicb.2015.00198] [PMID: 25852661]

[134] Gouda S, Kerry RG, Das G, Paramithiotis S, Shin HS, Patra JK. Revitalization of plant growth promoting rhizobacteria for sustainable development in agriculture. Microbiol Res 2018; 206: 131-40.
[http://dx.doi.org/10.1016/j.micres.2017.08.016] [PMID: 29146250]

[135] Antonella Di Benedetto N, Rosaria Corbo M, Campaniello D, *et al.* The role of Plant Growth Promoting Bacteria in improving nitrogen use efficiency for sustainable crop production: a focus on wheat. AIMS Microbiol 2017; 3(3): 413-34.
[http://dx.doi.org/10.3934/microbiol.2017.3.413] [PMID: 31294169]

[136] Sumbul A, Ansari RA, Rizvi R, Mahmood I. *Azotobacter:* A potential bio-fertilizer for soil and plant health management. Saudi J Biol Sci 2020; 27(12): 3634-40.
[http://dx.doi.org/10.1016/j.sjbs.2020.08.004] [PMID: 33304174]

[137] Franco-Correa M, Quintana A, Duque C, Suarez C, Rodríguez MX, Barea JM. Evaluation of actinomycete strains for key traits related with plant growth promotion and mycorrhiza helping activities. Appl Soil Ecol 2010; 45(3): 209-17.
[http://dx.doi.org/10.1016/j.apsoil.2010.04.007]

[138] Fester T, Giebler J, Wick LY, Schlosser D, Kästner M. Plant–microbe interactions as drivers of ecosystem functions relevant for the biodegradation of organic contaminants. Curr Opin Biotechnol 2014; 27: 168-75.
[http://dx.doi.org/10.1016/j.copbio.2014.01.017] [PMID: 24583828]

[139] Hou J, Liu W, Wang B, Wang Q, Luo Y, Franks AE. PGPR enhanced phytoremediation of petroleum contaminated soil and rhizosphere microbial community response. Chemosphere 2015; 138: 592-8.
[http://dx.doi.org/10.1016/j.chemosphere.2015.07.025] [PMID: 26210024]

[140] Li H, Qiu Y, Yao T, Ma Y, Zhang H, Yang X. Effects of PGPR microbial inoculants on the growth and soil properties of *Avena sativa, Medicago sativa,* and *Cucumis sativus* seedlings. Soil Tillage Res 2020; 199104577
[http://dx.doi.org/10.1016/j.still.2020.104577]

[141] Gerhardt KE, MacNeill GJ, Gerwing PD, Greenberg BM. Phytoremediation of salt-impacted soils and use of plant growth-promoting rhizobacteria (PGPR) to enhance phytoremediation.Phytoremediation. Cham: Springer 2017; pp. 19-51.
[http://dx.doi.org/10.1007/978-3-319-52381-1_2]

[142] Yaghoubi Khanghahi M, Pirdashti H, Rahimian H, Nematzadeh G, Ghajar Sepanlou M. The role of potassium solubilizing bacteria (KSB) inoculations on grain yield, dry matter remobilization and translocation in rice (*Oryza sativa* L.). J Plant Nutr 2019; 42(10): 1165-79.
[http://dx.doi.org/10.1080/01904167.2019.1609511]

[143] Abbas R, Rasul S, Aslam K, *et al.* Halotolerant PGPR: A hope for cultivation of saline soils. J King Saud Univ Sci 2019; 31(4): 1195-201.
[http://dx.doi.org/10.1016/j.jksus.2019.02.019]

[144] Liu Z, Ge H, Li C, Zhao Z, Song F, Hu S. Enhanced phytoextraction of heavy metals from contaminated soil by plant co-cropping associated with PGPR. Water Air Soil Pollut 2015; 226(3): 29.
[http://dx.doi.org/10.1007/s11270-015-2304-y]

[145] Kumar R, Das AJ, Juwarkar AA. Reclamation of petrol oil contaminated soil by rhamnolipids producing PGPR strains for growing *Withania somnifera* a medicinal shrub. World J Microbiol Biotechnol 2015; 31(2): 307-13.
[http://dx.doi.org/10.1007/s11274-014-1782-1] [PMID: 25480735]

[146] Xie L, Lehvävirta S, Timonen S, Kasurinen J, Niemikapee J, Valkonen JPT. Species-specific synergistic effects of two plant growth—promoting microbes on green roof plant biomass and photosynthetic efficiency. PLoS One 2018; 13(12)e0209432
[http://dx.doi.org/10.1371/journal.pone.0209432] [PMID: 30596699]

[147] Abdelkrim S, Jebara SH, Saadani O, *et al.* In situ effects of *Lathyrus sativus*- PGPR to remediate and restore quality and fertility of Pb and Cd polluted soils. Ecotoxicol Environ Saf 2020; 192110260
[http://dx.doi.org/10.1016/j.ecoenv.2020.110260] [PMID: 32050135]

[148] Mitter EK, Tosi M, Obregón D, Dunfield KE, Germida JJ. Rethinking crop nutrition in times of modern microbiology: innovative biofertilizer technologies. Front Sustain Food Syst 2021; 5606815
[http://dx.doi.org/10.3389/fsufs.2021.606815]

[149] Ke J, Wang B, Yoshikuni Y. Microbiome engineering: synthetic biology of plant-associated microbiomes in sustainable agriculture. Trends Biotechnol 2021; 39(3): 244-61.

[http://dx.doi.org/10.1016/j.tibtech.2020.07.008] [PMID: 32800605]

[150]   Quiza L, St-Arnaud M, Yergeau E. Harnessing phytomicrobiome signaling for rhizosphere microbiome engineering. Front Plant Sci 2015; 6: 507.
[http://dx.doi.org/10.3389/fpls.2015.00507] [PMID: 26236319]

[151]   Tsolakidou MD, Stringlis IA, Fanega-Sleziak N, Papageorgiou S, Tsalakou A, Pantelides IS. Rhizosphere-enriched microbes as a pool to design synthetic communities for reproducible beneficial outputs. FEMS Microbiol Ecol 2019; 95(10)fiz138
[http://dx.doi.org/10.1093/femsec/fiz138] [PMID: 31504462]

[152]   Yadav AN, Kaur T, Kour D, *et al.* Functional annotation and biotechnological applications of soil microbiomes: current research and future challenges. Sustainable Dev Biodiversity 2021; pp. 605-34.

[153]   Liang X, Wagner RE, Li B, Zhang N, Radosevich M. Quorum sensing signals alter in vitro soil virus abundance and bacterial community composition. Front Microbiol 2020; 11: 1287.
[http://dx.doi.org/10.3389/fmicb.2020.01287] [PMID: 32587586]

[154]   Mangwani N, Kumari S, Das S. Bacterial biofilms and quorum sensing: fidelity in bioremediation technology. Biotechnol Genet Eng Rev 2016; 32(1-2): 43-73.
[http://dx.doi.org/10.1080/02648725.2016.1196554] [PMID: 27320224]

[155]   Stephens K, Bentley WE. Synthetic biology for manipulating quorum sensing in microbial consortia. Trends Microbiol 2020; 28(8): 633-43.
[http://dx.doi.org/10.1016/j.tim.2020.03.009] [PMID: 32340782]

[156]   Jia H, Fan Y, Feng X, Li C. Enhancing stress-resistance for efficient microbial biotransformations by synthetic biology. Front Bioeng Biotechnol 2014; 2: 44.
[http://dx.doi.org/10.3389/fbioe.2014.00044] [PMID: 25368869]

[157]   Kang HS, Kim ES. Recent advances in heterologous expression of natural product biosynthetic gene clusters in *Streptomyces* hosts. Curr Opin Biotechnol 2021; 69: 118-27.
[http://dx.doi.org/10.1016/j.copbio.2020.12.016] [PMID: 33445072]

[158]   Nazari M, Smith DL. A PGPR-produced bacteriocin for sustainable agriculture: a review of thuricin 17 characteristics and applications. Front Plant Sci 2020; 11: 916.
[http://dx.doi.org/10.3389/fpls.2020.00916] [PMID: 32733506]

[159]   Chu S, Zhang D, Wang D, Zhi Y, Zhou P. Heterologous expression and biochemical characterization of assimilatory nitrate and nitrite reductase reveals adaption and potential of *Bacillus megaterium* NCT-2 in secondary salinization soil. Int J Biol Macromol 2017; 101: 1019-28.
[http://dx.doi.org/10.1016/j.ijbiomac.2017.04.009] [PMID: 28389402]

[160]   Alok A, Tiwari S, Kaur J. CRISPR/Cas9 mediated genome engineering in microbes and its application in plant beneficial effects.Molecular aspects of plant beneficial microbes in agriculture. Elsevier 2020; pp. 351-9.
[http://dx.doi.org/10.1016/B978-0-12-818469-1.00028-6]

[161]   Guo D, Kong S, Chu X, Li X, Pan H. De novo biosynthesis of indole-3-acetic acid in engineered Escherichia coli. J Agric Food Chem 2019; 67(29): 8186-90.
[http://dx.doi.org/10.1021/acs.jafc.9b02048] [PMID: 31272146]

[162]   Orozco-Mosqueda MC, Glick BR, Santoyo G. ACC deaminase in plant growth-promoting bacteria (PGPB): An efficient mechanism to counter salt stress in crops. Microbiol Res 2020; 235126439
[http://dx.doi.org/10.1016/j.micres.2020.126439] [PMID: 32097862]

[163]   Shulse CN, Chovatia M, Agosto C, *et al.* Engineered root bacteria release plant-available phosphate from phytate. Appl Environ Microbiol 2019; 85(18)e01210-19
[http://dx.doi.org/10.1128/AEM.01210-19] [PMID: 31285192]

[164]   Zhu N, Zhang B, Yu Q. Genetic engineering-facilitated coassembly of synthetic bacterial cells and magnetic nanoparticles for efficient heavy metal removal. ACS Appl Mater Interfaces 2020; 12(20): 22948-57.

[http://dx.doi.org/10.1021/acsami.0c04512] [PMID: 32338492]

[165] Srivastava S, Srivastava S. Prescience of endogenous regulation in Arabidopsis thaliana by *Pseudomonas putida* MTCC 5279 under phosphate starved salinity stress condition. Sci Rep 2020; 10(1): 5855.
[http://dx.doi.org/10.1038/s41598-020-62725-1] [PMID: 32246044]

[166] Kour D, Rana KL, Kaur T, Yadav N, Halder SK, Yadav AN, *et al.* Potassium solubilizing and mobilizing microbes: biodiversity, mechanisms of solubilization, and biotechnological implication for alleviations of abiotic stress.New and future developments in microbial biotechnology and bioengineering. Elsevier 2020; pp. 177-202.
[http://dx.doi.org/10.1016/B978-0-12-820526-6.00012-9]

[167] Vaishnav A, Kumari S, Awasthi S, Singh S, Varma A, Choudhary DK. Bacterial mutants for enhanced nitrogen fixation. Soil Biol 2021; pp. 349-58.

[168] [95] Liberton M, Bandyopadhyay A, Pakrasi HB. Enhanced nitrogen fixation in a glgX-deficient strain of Cyanothece sp. strain ATCC 51142, a unicellular nitrogen-fixing cyanobacterium. Appl Environ Microbiol 2019; 85(7)

[169] Jing X, Cui Q, Li X, *et al.* Engineering *Pseudomonas protegens* Pf☐5 to improve its antifungal activity and nitrogen fixation. Microb Biotechnol 2020; 13(1): 118-33.
[http://dx.doi.org/10.1111/1751-7915.13335] [PMID: 30461205]

[170] Liu D, Liberton M, Yu J, Pakrasi HB, Bhattacharyya-Pakrasi M. Engineering nitrogen fixation activity in an oxygenic phototroph. Bio 2018; 9(3).
[http://dx.doi.org/10.1128/mBio.01029-18]

[171] Bageshwar UK, Srivastava M, Pardha-Saradhi P, *et al.* An environmentally friendly engineered *Azotobacter* strain that replaces a substantial amount of urea fertilizer while sustaining the same wheat yield. Appl Environ Microbiol 2017; 83(15)e00590-17
[http://dx.doi.org/10.1128/AEM.00590-17] [PMID: 28550063]

[172] Wu B, He T, Wang Z, *et al.* Insight into the mechanisms of plant growth promoting strain SNB6 on enhancing the phytoextraction in cadmium contaminated soil. J Hazard Mater 2020; 385121587
[http://dx.doi.org/10.1016/j.jhazmat.2019.121587] [PMID: 31744727]

[173] Huang J, Liu Z, Li S, *et al.* Isolation and engineering of plant growth promoting rhizobacteria Pseudomonas aeruginosa for enhanced cadmium bioremediation. J Gen Appl Microbiol 2016; 62(5): 258-65.
[http://dx.doi.org/10.2323/jgam.2016.04.007] [PMID: 27725404]

[174] Fasusi OA, Cruz C, Babalola OO. Agricultural sustainability: microbial biofertilizers in rhizosphere management. Agriculture 2021; 11(2): 163.
[http://dx.doi.org/10.3390/agriculture11020163]

[175] Herrmann L, Lesueur D. Challenges of formulation and quality of biofertilizers for successful inoculation. Appl Microbiol Biotechnol 2013; 97(20): 8859-73.
[http://dx.doi.org/10.1007/s00253-013-5228-8] [PMID: 24037408]

[176] Malusá E, Sas-Paszt L, Ciesielska J. Technologies for beneficial microorganisms inocula used as biofertilizers. ScientificWorldJournal 2012; 2012: 1-12.
[http://dx.doi.org/10.1100/2012/491206] [PMID: 22547984]

[177] Deaker R, Roughley RJ, Kennedy IR. Legume seed inoculation technology?a review. Soil Biol Biochem 2004; 36(8): 1275-88.
[http://dx.doi.org/10.1016/j.soilbio.2004.04.009]

[178] Katai J, Sandor Z, Tallai M. The effect of an artificial and a bacterium fertilizer on some soil characteristics and on the biomass of the rye-grass (*Lolium perenne* L.). Cereal Res Commun 2008; 36: 1171-4.

[179] Zand AD, Mikaeili Tabrizi A, Vaezi Heir A. Application of titanium dioxide nanoparticles to promote

phytoremediation of Cd-polluted soil: contribution of PGPR inoculation. Bioremediat J 2020; 24(2-3): 171-89.
[http://dx.doi.org/10.1080/10889868.2020.1799929]

[180] Hien NT, Toan PV, Choudhury ATMA, Rose MT, Roughley RJ, Kennedy IR. Field application strategies for the inoculant biofertilizer BioGro supplementing fertilizer nitrogen application in rice production. J Plant Nutr 2014; 37(11): 1837-58.
[http://dx.doi.org/10.1080/01904167.2014.911320]

[181] Harman GE, Obregón MA, Samuels GJ, Lorito M. Changing models for commercialization and implementation of biocontrol in the developing and the developed world. Plant Dis 2010; 94(8): 928-39.
[http://dx.doi.org/10.1094/PDIS-94-8-0928] [PMID: 30743493]

[182] Mishra J, Arora NK. Bioformulations for plant growth promotion and combating phytopathogens: a sustainable approach.Bioformulations: for sustainable agriculture. New Delhi: Springer 2016; pp. 3-33.
[http://dx.doi.org/10.1007/978-81-322-2779-3_1]

[183] Leggett M, Cross J, Hnatowich G, Holloway G. Challenges in commercializing a phosphate-solubilizing microorganism: Penicillium bilaiae, a case history in first International Meeting on Microbial Phosphate Solubilization Velázquez E, Rodríguez-Barrueco C, editors. Vol. 102. The Netherlands: Springer. 2007; 215-22

[184] Ley-Rivas JF, Aliagar L, Moron C. Furrazola-G mez E. Efecto del biofertilizante MICOFERT en la producci n de dos variedades de lechuga en Per. Acta Botanica Cubana 2011; 213: 36-9.

[185] Mikiciuk G, Sas-Paszt L, Mikiciuk M, *et al.* Mycorrhizal frequency, physiological parameters, and yield of strawberry plants inoculated with endomycorrhizal fungi and rhizosphere bacteria. Mycorrhiza 2019; 29(5): 489-501.
[http://dx.doi.org/10.1007/s00572-019-00905-2] [PMID: 31264099]

[186] Corkidi L, Allen EB, Merhaut D, *et al.* Assessing the infectivity of commercial mycorrhizal inoculants in plant nursery conditions. J Environ Hortic 2004; 22(3): 149-54.
[http://dx.doi.org/10.24266/0738-2898-22.3.149]

[187] Martín FF, Molina JJ, Nicolás EN, *et al.* Application of arbuscular Mycorrhizae *Glomus iranicum* var. tenuihypharum var. nova in Intensive Agriculture: a study case. J Agric Sci Technol B 2017; 7: 221-47.

[188] Smith R. Inoculant formulations and applications to meet changing needs.Nitrogen fixation: fundamentals and applications. Dordrecht, Boston: Springer 1995; pp. 653-7.
[http://dx.doi.org/10.1007/978-94-011-0379-4_76]

[189] Le Mire G, Nguyen M, Fassotte B, *et al.* Implementing biostimulants and biocontrol strategies in the agroecological management of cultivated ecosystems. Biotechnol Agron Soc Environ 2016; 20: 1-15.

[190] Okon Y, Labandera-Gonzales C, Lage M, Lage P. Agronomic applications of Azospirillum and other PGPR biological nitrogen fixation 2015; 921-33.

[191] Ingle KP, Padole DA. Phosphate solubilizing microbes: an overview. Int J Curr Microbiol Appl Sci 2017; 6(1): 844-52.
[http://dx.doi.org/10.20546/ijcmas.2017.601.099]

[192] Fröhlich A, Buddrus-Schiemann K, Durner J, Hartmann A, von Rad U. Response of barley to root colonization by *Pseudomonas* sp. DSMZ 13134 under laboratory, greenhouse, and field conditions. J Plant Interact 2012; 7(1): 1-9.
[http://dx.doi.org/10.1080/17429145.2011.597002]

[193] Xu Z, Xie J, Zhang H, Wang D, Shen Q, Zhang R. Enhanced control of plant wilt disease by a xylose-inducible degq gene engineered into bacillus velezensis strain SQR9XYQ. Phytopathology 2019; 109(1): 36-43.

[http://dx.doi.org/10.1094/PHYTO-02-18-0048-R] [PMID: 29927357]

[194] Poddar R, Sen A, Kundu R, Das H, Bandopadhyay P. Response of various mycorrhizal inoculants on rice growth, productivity and nutrient uptake. Int J Stress Manag 2020; 11(2): 171-7.

[195] Dal Cortivo C, Ferrari M, Visioli G, *et al.* Effects of seed-applied biofertilizers on rhizosphere biodiversity and growth of common wheat (*Triticum aestivum* L.) in the field. Front Plant Sci 2020; 11: 72.
[http://dx.doi.org/10.3389/fpls.2020.00072] [PMID: 32174929]

[196] Deambrosi E, Mendez R, Avila S. Evaluaci n de efectos del uso de rizofosen el cultivo de arroz. INIA Serie Actividades de Difusion 2014; pp. 39-42.

[197] Janarthanam L. Bioprotectant with multifunctional microorganisms: A new dimension in plant protection. J Biopesticides 2013; 6(2): 219.

[198] Abd El Ghafour A, Darwish MA, Azoz SN, Abd-Alla AM, Elsayed SI. Effect of mineral, bio and organic fertilizers on productivity, essential oil composition and fruit anatomy of two dill cultivars (*Anethum graveolens* L.). Sci 2017; 7(03): 532-50.

[199] Vaishnavi SJ, Jeyakumar P. Bioinoculant on microbial population, biochemical characters and yield of cowpea. Environ Ecol 2016; 34(1): 129-31.

[200] Willer H, Lernoud J. The world of organic agriculture statistics and EmergingTrends Research Institute of Organic Agriculture (FiBL). Bonn: Frick, and IFOAM-Organics International 2018.

[201] Pirttilä AM, Mohammad Parast Tabas H, Baruah N, Koskimäki JJ. Mohammad ParastTabas H, Baruah N, Koskimäki JJ. Biofertilizers and biocontrol agents for agriculture: how to identify and develop new potent microbial strains and traits. Microorganisms 2021; 9(4): 817.
[http://dx.doi.org/10.3390/microorganisms9040817] [PMID: 33924411]

[202] Ansari MF, Tipre DR, Dave SR. Efficiency evaluation of commercial liquid biofertilizers for growth of *Cicer aeritinum* (chickpea) in pot and field study. Biocatal Agric Biotechnol 2015; 4(1): 17-24.
[http://dx.doi.org/10.1016/j.bcab.2014.09.010]

[203] Kumari R, Singh DP. Nano-biofertilizer: an emerging eco-friendly approach for sustainable agriculture. Proc Natl Acad Sci, India, Sect B Biol Sci 2020; 90(4): 733-41.
[http://dx.doi.org/10.1007/s40011-019-01133-6]

[204] Thirugnanasambandan T. Advances and trends in nano-biofertilizers. SSRN Electron J 2019.

[205] Alley MM, Vanlauwe B. The role of fertilizers in integrated plant nutrient management. Paris, France: International Fertilizer Industry Association 2009.

[206] Motsara MR, Bhattacharyya P, Srivastava B. Biofertiliser technology, marketing and usage: a sourcebook-cum-glossary. Fertiliser Development and consultation Org. 1995.

SUBJECT INDEX

A

B

www.ingramcontent.com/pod-product-compliance
Lightning Source LLC
Chambersburg PA
CBHW050805220326
41598CB00006B/125